MEASURE AND INTEGRAL

PURE AND APPLIED MATHEMATICS

A Program of Monographs, Textbooks, and Lecture Notes

MONOGRAPHS AND TEXTBOOKS IN PURE AND APPLIED MATHEMATICS

1. *K. Yano,* Integral Formulas in Riemannian Geometry (1970) *(out of print)*
2. *S. Kobayashi,* Hyperbolic Manifolds and Holomorphic Mappings (1970) *(out of print)*
3. *V. S. Vladimirov,* Equations of Mathematical Physics (A. Jeffrey, editor; A. Littlewood, translator) (1970) *(out of print)*
4. *B. N. Pshenichnyi,* Necessary Conditions for an Extremum (L. Neustadt, translation editor; K. Makowski, translator) (1971)
5. *L. Narici, E. Beckenstein, and G. Bachman,* Functional Analysis and Valuation Theory (1971)
6. *D. S. Passman,* Infinite Group Rings (1971)
7. *L. Dornhoff,* Group Representation Theory (in two parts). Part A: Ordinary Representation Theory. Part B: Modular Representation Theory (1971, 1972)
8. *W. Boothby and G. L. Weiss (eds.),* Symmetric Spaces: Short Courses Presented at Washington University (1972) *(out of print)*
9. *Y. Matsushima,* Differentiable Manifolds (E. T. Kobayashi, translator) (1972)
10. *L. E. Ward, Jr.,* Topology: An Outline for a First Course (1972) *(out of print)*
11. *A. Babakhanian,* Cohomological Methods in Group Theory (1972)
12. *R. Gilmer,* Multiplicative Ideal Theory (1972) *(out of print)*
13. *J. Yeh,* Stochastic Processes and the Wiener Integral (1973) *(out of print)*
14. *J. Barros-Neto,* Introduction to the Theory of Distributions (1973) *(out of print)*
15. *R. Larsen,* Functional Analysis: An Introduction (1973) *(out of print)*
16. *K. Yano and S. Ishihara,* Tangent and Cotangent Bundles: Differential Geometry (1973) *(out of print)*
17. *C. Procesi,* Rings with Polynomial Identities (1973)
18. *R. Hermann,* Geometry, Physics, and Systems (1973) *(out of print)*
19. *N. R. Wallach,* Harmonic Analysis on Homogeneous Spaces (1973) *(out of print)*
20. *J. Dieudonné,* Introduction to the Theory of Formal Groups (1973)

21. *I. Vaisman,* Cohomology and Differential Forms (1973)
22. *B.-Y. Chen,* Geometry of Submanifolds (1973) *(out of print)*
23. *M. Marcus,* Finite Dimensional Multilinear Algebra (in two parts) (1973, 1975)
24. *R. Larsen,* Banach Algebras: An Introduction (1973)
25. *R. O. Kujala and A. L. Vitter (eds.),* Value Distribution Theory: Part A; Part B. Deficit and Bezout Estimates by Wilhelm Stoll (1973)
26. *K. B. Stolarsky,* Algebraic Numbers and Diophantine Approximation (1974)
27. *A. R. Magid,* The Separable Galois Theory of Commutative Rings (1974)
28. *B. R. McDonald,* Finite Rings with Identity (1974)
29. *J. Satake,* Linear Algebra (S. Koh, T. Akiba, and S. Ihara, translators) (1975)
30. *J. S. Golan,* Localization of Noncommutative Rings (1975)
31. *G. Klambauer,* Mathematical Analysis (1975)
32. *M. K. Agoston,* Algebraic Topology: A First Course (1976)
33. *K. R. Goodearl,* Ring Theory: Nonsingular Rings and Modules (1976)
34. *L. E. Mansfield,* Linear Algebra with Geometric Applications (1976)
35. *N. J. Pullman,* Matrix Theory and its Applications: Selected Topics (1976)
36. *B. R. McDonald,* Geometric Algebra Over Local Rings (1976)
37. *C. W. Groetsch,* Generalized Inverses of Linear Operators: Representation and Approximation (1977)
38. *J. E. Kuczkowski and J. L. Gersting,* Abstract Algebra: A First Look (1977)
39. *C. O. Christenson and W. L. Voxman,* Aspects of Topology (1977)
40. *M. Nagata,* Field Theory (1977)
41. *R. L. Long,* Algebraic Number Theory (1977)
42. *W. F. Pfeffer,* Integrals and Measures (1977)
43. *R. L. Wheeden and A. Zygmund,* Measure and Integral: An Introduction to Real Analysis (1977)

MEASURE AND INTEGRAL

An Introduction to Real Analysis

Richard L. Wheeden

Department of Mathematics
Rutgers, the State University
of New Jersey
New Brunswick, New Jersey

Antoni Zygmund

Department of Mathematics
University of Chicago
Chicago, Illinois

CRC Press
Taylor & Francis Group
Boca Raton London New York

CRC Press is an imprint of the
Taylor & Francis Group, an **informa** business

Published in 1977 by
CRC Press
Taylor & Francis Group
6000 Broken Sound Parkway NW, Suite 300
Boca Raton, FL 33487-2742

Printed in the United States of America on acid-free paper
30 29 28 27 26 25

International Standard Book Number-10: 0-8247-6499-4 (Hardcover)
International Standard Book Number-13: 978-0-8247-6499-9 (Hardcover)
Library of Congress Card Number 77-14167

Library of Congress Cataloging-in-Publication Data

Wheeden, Richard L.
Measure and integral.
(Monographs and textbooks in pure and applied mathematics ; 43)
Includes index.
ISBN 0-8247-6499-4
1. Measure theory. 2. Integrals, Generalized. I. Zygmund, Antoni, joint author. II. Title.
QA312.W43
515'.42 77-14167

To our families

Introduction

The modern theory of measure and integration was created, primarily through the work of Lebesgue, at the turn of this century. Although the basic ideas are by now well established, there are ever widening applications which have made the theory one of the central parts of mathematical analysis. However, different applications require different emphasis on various aspects of the theory. For example, certain facts are of primary interest for real and complex analysis, others for functional analysis, and still others for probability and statistics. This text is written from the point of view of real variables, and treats the theory primarily as a modern calculus.

The book presupposes that the reader has a feeling for rigor and some knowledge of elementary facts from calculus. Some material which is no doubt familiar to many readers has been included; its inclusion seemed desirable in order to make the presentation clear and self-contained.

The approach of the book is to develop the theory of measure and integration first in the simple setting of Euclidean space. In this case, there is a rich theory having a close relation to familiar facts from calculus and generalizing those facts. Later on, we introduce a more general treatment based on abstract notions characterized by axioms and with less geometric content. We have chosen this approach purposely, even though it leads to some repetition, since considering a special case first usually helps in developing a better understanding of the general situation. Anyway, we all "learn by repetition."

The outline of the book is as follows. Chapter 1 is primarily a collection of various background information, including elementary definitions and results that will be taken for granted later in the book; the reader should already be familiar with most of this material. Very few proofs are given in Chapter 1. Actual presentation of the theory begins in Chapter 2, which treats notions associated with functions of bounded variation, such as the Riemann-Stieltjes integral. Strictly speaking, a reading of Chapter 2 could be postponed until Chapter 5, where we use the Riemann-Stieltjes integral as a way of representing the Lebesgue integral.

Chapter 3 deals with Lebesgue measure in Euclidean space, via the notion of outer measure. Chapter 4 gives the theory of measurable functions, and Chapter 5 considers the Lebesgue integral, again in Euclidean space. In Chapter 6, we study repeated integration, the central result being Fubini's theorem. Chapter 7 treats the process which is the inverse of integration, namely, differentiation. Here, we consider the differentiation of integrals treated as set functions, as well as the differentiation of real-valued functions of a single variable, such as the differentiability of monotone functions. Chapters 3–7 complete the treatment of the general theory of integration in Euclidean spaces.

In Chapters 8 and 9 we consider special classes of functions, like L^2 and L^p, and special results for these classes, such as the behavior of convolution operators, the Hardy-Littlewood maximal function, and the integral of Marcinkiewicz.

In Chapters 10 and 11 we give an abstract treatment of Lebesgue measure and integration. Here, there are several possible approaches. We have chosen to start with an abstract definition of measure and develop the theory of integration following the pattern of earlier chapters. This is done in Chapter 10. It is natural to ask how such abstract measures actually arise. This question is answered to some extent in Chapter 11, where we use the notion of abstract outer measure to construct some specific examples of measures.

Chapter 12 plays a special role and can be read immediately after Chapter 9. It deals with an application of the Lebesgue integral to a specific branch of analysis—harmonic analysis. This is a very broad field, and we consider only a few problems indicative of the role that Lebesgue integration plays in applications. Harmonic analysis also happens to be a field whose development had a great impact on the theory of integration.

At the end of each chapter, we list a number of problems as exercises, sometimes with parenthetical hints at solutions. Some relatively important results are given in the exercises, but as a rule, the text does not require facts which have appeared earlier only as exercises.

We would like to express our thanks to the Departments of Mathematics of Rutgers University and the University of Chicago, and in particular to Professor William H. Meyer for the friendly help he gave us during the preparation of the manuscript of the book. Special thanks also go to Joanne Darken and Dr. Edward Lotkowski, both of whom proofread almost the entire manuscript and offered many helpful comments, and to Michele Ginouves for her help with the cover design. Finally, thanks to Mrs. Annette Roselli, our typist, for a really excellent job.

Richard L. Wheeden and Antoni Zygmund

Contents

Chapter 1

Preliminaries

This book is devoted to Lebesgue integration and related topics, a basic part of modern analysis. There are classical and abstract approaches to the integral, and we have chosen the classical one, postponing a more abstract treatment until later in the book. The classical approach is based on the theory of measure (while in some modern treatments the integral is introduced as a linear functional). Measure can be defined and studied in various spaces, but we will primarily consider n-dimensional Euclidean space, $\mathbf{R}^n$. A prerequisite, undertaken in this chapter, is a review of elementary notions about $\mathbf{R}^n$. We have not attempted to present these in a thorough manner, but only to list some of the definitions and notation that will be used throughout the book, and state some background facts that a reader should know. We assume a knowledge of various properties of the real line $\mathbf{R}^1$ and of functions defined on $\mathbf{R}^1$, and leave as exercises the proofs of many facts which are either similar to or derivable from their one-dimensional analogues.

1. Points and Sets in $\mathbf{R}^n$

Let n be a positive integer. By *n-dimensional Euclidean space* $\mathbf{R}^n$, we mean the collection of all n-tuples $\mathbf{x} = (x_1, \ldots, x_n)$ of real numbers x_k, $-\infty < x_k < +\infty$, $k = 1, \ldots, n$. If $\mathbf{x} = (x_1, \ldots, x_n)$ and $\mathbf{y} = (y_1, \ldots, y_n)$ are points of $\mathbf{R}^n$, we say that $\mathbf{x} = \mathbf{y}$ if $x_k = y_k$ for $1 \le k \le n$. $\mathbf{R}^n$ is a vector space over the reals if for $\mathbf{x} = (x_1, \ldots, x_n)$, $\mathbf{y} = (y_1, \ldots, y_n)$, and $\alpha \in \mathbf{R}^1$ we define $\mathbf{x} + \mathbf{y} = (x_1 + y_1, \ldots, x_n + y_n)$ and $\alpha\mathbf{x} = (\alpha x_1, \ldots, \alpha x_n)$. The point each of whose coordinates is zero is called the *origin* and denoted $0 = (0, \ldots, 0)$. By the *vector* emanating from $\mathbf{x}$ and terminating at $\mathbf{y}$, we mean the line segment connecting $\mathbf{x}$ and $\mathbf{y}$, directed from $\mathbf{x}$ to $\mathbf{y}$. The points of this segment are of the form $(1 - t)\mathbf{x} + t\mathbf{y}$, $0 \le t \le 1$. We will identify vectors that have equal length and direction. We will also identify $\mathbf{x}$ with the vector emanating from 0 and terminating at $\mathbf{x}$.

If E is a set of points of $\mathbf{R}^n$, we use the notation $CE = \mathbf{R}^n - E$ for the *complement of E*. The complement of $\mathbf{R}^n$ is the *empty set* $\varnothing$. If $\mathscr{F} = \{E\}$ is a family of subsets of $\mathbf{R}^n$, the *union* and *intersection* of the sets E in $\mathscr{F}$ are defined respectively by

$$\bigcup_{E\in\mathscr{F}} E = \{\mathbf{x} : \mathbf{x} \in E \text{ for some } E \in \mathscr{F}\},$$

$$\bigcap_{E\in\mathscr{F}} E = \{\mathbf{x} : \mathbf{x} \in E \text{ for all } E \in \mathscr{F}\}.$$

(Here, and systematically below, we use the notation $\{\mathbf{x} : \ldots\}$ to denote the set of points $\mathbf{x}$ which satisfy. . . .)

If $\mathscr{F}$ is countable (i.e., finite or countably infinite), it will be called a *sequence of sets*, and denoted $\mathscr{F} = \{E_k : k = 1, 2, \ldots\}$. The corresponding union and intersection will be written $\bigcup_k E_k$ and $\bigcap_k E_k$. A sequence $\{E_k\}$ of sets is said to *increase* to $\bigcup_k E_k$ if $E_k \subset E_{k+1}$ for all k, and to *decrease* to $\bigcap_k E_k$ if $E_k \supset E_{k+1}$ for all k; we use the notations $E_k \nearrow \bigcup_k E_k$ and $E_k \searrow \bigcap_k E_k$ to denote these two possibilities. If $\{E_k\}_{k=1}^{\infty}$ is a sequence of sets, we define

$$\text{(1.1)} \qquad \limsup E_k = \bigcap_{j=1}^{\infty}\left(\bigcup_{k=j}^{\infty} E_k\right), \quad \liminf E_k = \bigcup_{j=1}^{\infty}\left(\bigcap_{k=j}^{\infty} E_k\right),$$

noting that the sets $U_j = \bigcup_{k=j}^{\infty} E_k$ and $V_j = \bigcap_{k=j}^{\infty} E_k$ satisfy $U_j \searrow \limsup E_k$ and $V_j \nearrow \liminf E_k$. We leave it as a simple exercise to verify that $\limsup E_k$ consists of those points of $\mathbf{R}^n$ which belong to infinitely many E_k, and $\liminf E_k$ of those which belong to all E_k for $k \geq k_0$ (where k_0 may vary from point to point). Thus, $\liminf E_k \subset \limsup E_k$.

If E_1 and E_2 are two sets, we define $E_1 - E_2$ by $E_1 - E_2 = E_1 \cap CE_2$, and call it the *difference* of E_1 and E_2, or the *relative complement* of E_2 in E_1. We will often have occasion to use the *De Morgan laws*, which govern relations between complements, unions, and intersections; these state that

$$C\Big(\bigcup_{E\in\mathscr{F}} E\Big) = \bigcap_{E\in\mathscr{F}} CE, \qquad C\Big(\bigcap_{E\in\mathscr{F}} E\Big) = \bigcup_{E\in\mathscr{F}} CE,$$

and are easily verified. The set-theoretic notions discussed above are not confined to $\mathbf{R}^n$ and hold for subsets of an arbitrary set S.

2. $\mathbf{R}^n$ as a Metric Space

$\mathbf{R}^n$ also has, of course, a metric space structure. If $\mathbf{x} = (x_1, \ldots, x_n)$ and $\mathbf{y} = (y_1, \ldots, y_n)$, we define their *inner (dot) product* by

$$\mathbf{x}\cdot\mathbf{y} = \sum_{k=1}^{n} x_k y_k.$$

We have $\mathbf{x}\cdot\mathbf{y} = \mathbf{y}\cdot\mathbf{x}$, $\alpha\mathbf{x}\cdot\mathbf{y} = \alpha(\mathbf{x}\cdot\mathbf{y})$ for real α, and $\mathbf{x}\cdot(\mathbf{y} + \mathbf{z}) = \mathbf{x}\cdot\mathbf{y} +$

$\mathbf{x}\cdot\mathbf{z}$. Noting that $\mathbf{x}\cdot\mathbf{x} \geq 0$, we define the *absolute value* of $\mathbf{x}$, or the *length* of $\mathbf{x}$, by

$$|\mathbf{x}| = (\mathbf{x}\cdot\mathbf{x})^{1/2} = \left(\sum_{k=1}^{n} x_k^2\right)^{1/2}.$$

We will use this notation regardless of the dimension n. Thus, if $x \in \mathbf{R}^1$, $|x|$ means the usual one-dimensional absolute value of x. Then in any dimension $|\mathbf{x}|$ has the following properties:

(i) $|\mathbf{x}| \geq 0$ and $|\mathbf{x}| = 0$ if and only if $\mathbf{x} = 0$
(ii) $|\alpha\mathbf{x}| = |\alpha||\mathbf{x}|$ for $\alpha \in \mathbf{R}^1$
(iii) $|\mathbf{x} + \mathbf{y}| \leq |\mathbf{x}| + |\mathbf{y}|$ (the *triangle inequality*)

To verify (iii), observe that if we square both sides of the inequality, (iii) is equivalent to showing that $(\mathbf{x} + \mathbf{y})\cdot(\mathbf{x} + \mathbf{y}) \leq |\mathbf{x}|^2 + 2|\mathbf{x}||\mathbf{y}| + |\mathbf{y}|^2$. Since $(\mathbf{x} + \mathbf{y})\cdot(\mathbf{x} + \mathbf{y}) = |\mathbf{x}|^2 + 2(\mathbf{x}\cdot\mathbf{y}) + |\mathbf{y}|^2$, the problem reduces to showing that $\mathbf{x}\cdot\mathbf{y} \leq |\mathbf{x}||\mathbf{y}|$; i.e., that

$$\text{(1.2)} \qquad \sum_{k=1}^{n} x_k y_k \leq \left(\sum_{k=1}^{n} x_k^2\right)^{1/2}\left(\sum_{k=1}^{n} y_k^2\right)^{1/2}.$$

This important inequality is called the *Schwarz* (or *Cauchy-Schwarz*) *inequality* and can be proved as follows. For $\alpha,\beta \in \mathbf{R}^1$, the fact that $(\alpha - \beta)^2 \geq 0$ gives $\alpha\beta \leq \frac{1}{2}\alpha^2 + \frac{1}{2}\beta^2$. Therefore, $\sum_{k=1}^{n} x_k y_k \leq \sum_{k=1}^{n} (\frac{1}{2}x_k^2 + \frac{1}{2}y_k^2) = \frac{1}{2}(|\mathbf{x}|^2 + |\mathbf{y}|^2)$. Inequality (1.2) follows immediately if $|\mathbf{x}| = |\mathbf{y}| = 1$, since then $\sum_{k=1}^{n} x_k y_k \leq \frac{1}{2}(1 + 1) = 1 = |\mathbf{x}||\mathbf{y}|$. Moreover, (1.2) is obvious if either $|\mathbf{x}| = 0$ or $|\mathbf{y}| = 0$ since then both sides must be zero. Finally, if $|\mathbf{x}| > 0$ and $|\mathbf{y}| > 0$, let $x_k' = x_k/|\mathbf{x}|$, $y_k' = y_k/|\mathbf{y}|$, $\mathbf{x}' = (x_1', \ldots, x_n') = \mathbf{x}/|\mathbf{x}|$ and $\mathbf{y}' = (y_1', \ldots, y_n') = \mathbf{y}/|\mathbf{y}|$. Then $|\mathbf{x}'| = |\mathbf{y}'| = 1$, so that by the case already proved, $\sum_{k=1}^{n} x_k' y_k' \leq 1$; that is, $\sum_{k=1}^{n} x_k y_k \leq |\mathbf{x}||\mathbf{y}|$, as claimed.

If we now define the *distance between two points* $\mathbf{x}$ and $\mathbf{y}$ by $d(\mathbf{x},\mathbf{y}) = |\mathbf{x} - \mathbf{y}|$, we immediately obtain the characteristic metric space properties:

(i) $d(\mathbf{x},\mathbf{y}) = d(\mathbf{y},\mathbf{x})$
(ii) $d(\mathbf{x},\mathbf{y}) \geq 0$, and $d(\mathbf{x},\mathbf{y}) = 0$ if and only if $\mathbf{x} = \mathbf{y}$
(iii) $d(\mathbf{x},\mathbf{y}) \leq d(\mathbf{x},\mathbf{z}) + d(\mathbf{z},\mathbf{y})$

We have used the symbol x_k to denote the kth coordinate of $\mathbf{x}$. When no confusion should arise, we will also use $\{\mathbf{x}_k\}$ to denote a *sequence of points* of $\mathbf{R}^n$. If $\mathbf{x} \in \mathbf{R}^n$, we say that a sequence $\{\mathbf{x}_k\}$ *converges* to $\mathbf{x}$, or that $\mathbf{x}$ is the *limit point* of $\{\mathbf{x}_k\}$, if $|\mathbf{x} - \mathbf{x}_k| \to 0$ as $k \to \infty$. We denote this by writing either $\mathbf{x} = \lim_{k\to\infty} \mathbf{x}_k$ or $\mathbf{x}_k \to \mathbf{x}$ as $k \to \infty$. A point $\mathbf{x} \in \mathbf{R}^n$ is called a *limit point of a set E* if it is the limit point of a sequence of distinct points of E. A point $\mathbf{x} \in E$ is called an *isolated point* of E if it is not the limit of any sequence in E (excluding the trivial sequence $\{\mathbf{x}_k\}$ where $\mathbf{x}_k = \mathbf{x}$ for all k). It follows that $\mathbf{x}$

is isolated if and only if there is a $\delta > 0$ such that $|\mathbf{x} - \mathbf{y}| > \delta$ for every $\mathbf{y} \in E$, $\mathbf{y} \neq \mathbf{x}$.

For sequences $\{x_k\}$ in $\mathbf{R}^1$, we will write $\lim_{k\to\infty} x_k = +\infty$, or $x_k \to +\infty$ as $k \to \infty$, if given $M > 0$, there is an integer K such that $x_k \geq M$ whenever $k \geq K$. A similar definition holds for $\lim_{k\to\infty} x_k = -\infty$.

A sequence $\{\mathbf{x}_k\}$ in $\mathbf{R}^n$ is called a *Cauchy sequence* if given $\varepsilon > 0$, there is an integer K such that $|\mathbf{x}_k - \mathbf{x}_j| < \varepsilon$ for all $k,j \geq K$. We leave it as an exercise to prove that $\mathbf{R}^n$ is a *complete metric space*, that is, that every Cauchy sequence in $\mathbf{R}^n$ converges to a point of $\mathbf{R}^n$.

A set $E \subset E_1$ is said to be *dense* in E_1 if for every $\mathbf{x}_1 \in E_1$ and $\varepsilon > 0$, there is a point $\mathbf{x} \in E$ such that $0 < |\mathbf{x} - \mathbf{x}_1| < \varepsilon$. Thus, E is dense in E_1 if every point of E_1 is a limit point of E. If $E = E_1$, we say E is *dense in itself.* As an example, the set of points of $\mathbf{R}^n$ each of whose coordinates is a rational number is dense in $\mathbf{R}^n$. Since this set is also countable, it follows that $\mathbf{R}^n$ is *separable*, by which we mean that $\mathbf{R}^n$ has a countable dense subset.

For subsets E of $\mathbf{R}^1$, we use the standard notations $\sup E$ and $\inf E$ for the *supremum* (*least upper bound*) and *infimum* (*greatest lower bound*) of E. In case $\sup E$ belongs to E, it will be called *max* E; similarly, $\inf E$ will be called *min* E if it belongs to E.

If $\{a_k\}_{k=1}^{\infty}$ is a sequence of points in $\mathbf{R}^1$, let $b_j = \sup_{k\geq j} a_k$ and $c_j = \inf_{k\geq j} a_k$, $j = 1, 2, \ldots$. Then $-\infty \leq c_j \leq b_j \leq +\infty$, and $\{b_j\}$ and $\{c_j\}$ are monotone decreasing and increasing, respectively; that is, $b_j \geq b_{j+1}$ and $c_j \leq c_{j+1}$. Define $\limsup_{k\to\infty} a_k$ and $\liminf_{k\to\infty} a_k$ by

$$\limsup_{k\to\infty} a_k = \lim_{j\to\infty} b_j = \lim_{j\to\infty} \{\sup_{k\geq j} a_k\},$$

(1.3)

$$\liminf_{k\to\infty} a_k = \lim_{j\to\infty} c_j = \lim_{j\to\infty} \{\inf_{k\geq j} a_k\}.$$

We leave it as an exercise to show that $-\infty \leq \liminf_{k\to\infty} a_k \leq \limsup_{k\to\infty} a_k \leq +\infty$, and that the following characterizations hold.

(1.4) Theorem

(a) $L = \limsup_{k\to\infty} a_k$ *if and only if* (i) *there is a subsequence* $\{a_{k_j}\}$ *of* $\{a_k\}$ *which converges to* L, *and* (ii) *if* $L' > L$, *there is an integer* K *such that* $a_k < L'$ *for* $k \geq K$.

(b) $l = \liminf_{k\to\infty} a_k$ *if and only if* (i) *there is a subsequence* $\{a_{k_j}\}$ *of* $\{a_k\}$ *which converges to* l, *and* (ii) *if* $l' < l$, *there is an integer* K *such that* $a_k > l'$ *for* $k \geq K$.

Thus, when they are finite, $\limsup_{k\to\infty} a_k$ and $\liminf_{k\to\infty} a_k$ are the largest and smallest limit points of $\{a_k\}$, respectively. We leave it as a problem to show that $\{a_k\}$ converges to a, $-\infty \leq a \leq +\infty$, if and only if $\limsup_{k\to\infty} a_k = \liminf_{k\to\infty} a_k = a$.

We can also use the metric on $\mathbf{R}^n$ to define the *diameter of a set E* by letting

$$\delta(E) = \text{diam } E = \sup \{|\mathbf{x} - \mathbf{y}| : \mathbf{x},\mathbf{y} \in E\}.$$

If the diameter of E is finite, E is said to be *bounded.* Equivalently, E is bounded if there is a finite constant M such that $|\mathbf{x}| \leq M$ for all $\mathbf{x} \in E$. If E_1 and E_2 are two sets, the *distance between* E_1 *and* E_2 is defined by

$$d(E_1,E_2) = \inf \{|\mathbf{x} - \mathbf{y}| : \mathbf{x} \in E_1, \mathbf{y} \in E_2\}.$$

3. Open and Closed Sets in $\mathbf{R}^n$; Special Sets

For $\mathbf{x} \in \mathbf{R}^n$ and $\delta > 0$, the set

$$B(\mathbf{x};\delta) = \{\mathbf{y} : |\mathbf{x} - \mathbf{y}| < \delta\}$$

is called the *open ball with center* $\mathbf{x}$ *and radius* δ. A point $\mathbf{x}$ of a set E is called an *interior point of E* if there exists $\delta > 0$ such that $B(\mathbf{x};\delta) \subset E$. The collection of all interior points of E is called the *interior of E*, and denoted $\mathring{E}$. A set E is said to be *open* if $E = \mathring{E}$; that is, E is open if for each $\mathbf{x} \in E$ there exists $\delta > 0$ such that $B(\mathbf{x};\delta) \subset E$. The empty set $\emptyset$ is open by convention. The whole space $\mathbf{R}^n$ is clearly open, and we leave it as an exercise to prove that $B(\mathbf{x};\delta)$ is open. We will generally denote open sets by the letter G.

A set E is called *closed* if CE is open. Note that $\emptyset$ and $\mathbf{R}^n$ are closed. Closed sets will generally be denoted by the letter F. The union of a set E and all its limit points is called the *closure of E*, and written $\bar{E}$. We leave it to the reader to prove the following facts.

(1.5) Theorem

(i) $\overline{B(\mathbf{x},\delta)} = \{\mathbf{y} : |\mathbf{x} - \mathbf{y}| \leq \delta\}$.
(ii) *E is closed if and only if $E = \bar{E}$; that is, E is closed if and only if it contains all its limit points.*
(iii) *$\bar{E}$ is closed, and $\bar{E}$ is the smallest closed set containing E; that is, if F is closed and $E \subset F$, then $\bar{E} \subset F$.*

By the *boundary* of E, we mean the set $\bar{E} - \mathring{E}$.

The open subsets of $\mathbf{R}^n$ satisfy the conditions listed in the next theorem.

(1.6) Theorem

(i) *The union of any number of open sets is open.*
(ii) *The intersection of a finite number of open sets is open.*

Verification is left to the reader. Using De Morgan's laws, we obtain the following equivalent statements.

(1.7) Theorem

(i) *The intersection of any number of closed sets is closed.*
(ii) *The union of a finite number of closed sets is closed.*

A subset E_1 of E is said to be *relatively open with respect to E* if it can be written $E_1 = E \cap G$ for some open set G. Similarly, E_1 *is relatively closed with respect to E* if $E_1 = E \cap F$ for some closed F. Note that the relative complement of a relatively open set is relatively closed. A useful alternate characterization of relatively closed is as follows.

(1.8) Theorem *A set $E_1 \subset E$ is relatively closed with respect to E if and only if $E_1 = E \cap \bar{E}_1$, that is, if and only if every limit point of E_1 which lies in E is in E_1.*

The proof is left as an exercise.

Consider a collection $\{A\}$ of sets A. Then a set is said to be of *type A_δ* if it can be written as a countable intersection of sets A, and to be of *type A_σ* if it can be written as a countable union of sets A. Thus, "δ" stands for intersection and "σ" for union. The most common uses of this notation are G_δ and F_σ, where $\{G\}$ denotes the open sets in $\mathbf{R}^n$ and $\{F\}$ the closed sets. Hence, H is of *type G_δ* if

$$H = \bigcap_k G_k, \ G_k \text{ open},$$

and H is of *type F_σ* if

$$H = \bigcup_k F_k, \ F_k \text{ closed}.$$

The complement of a G_δ set is an F_σ set, and vice versa. A G_δ (F_σ) set is of course not generally open (closed); in fact, any closed (open) set in $\mathbf{R}^n$ is of type G_δ (F_σ): see exercise 1(j). These two special types of sets will be very useful later in the measure approximation of general sets.

Another special type of set which we will have occasion to use is a *perfect set*, by which we mean a closed set C each of whose points is a limit point of C. Thus, a perfect set is a closed set which is dense in itself. One particular property of perfect sets we will use is stated in the following theorem. The proof is postponed until Section 4 of this chapter.

(1.9) Theorem *A perfect set is uncountable.*

Other special sets which will be important are n-dimensional intervals. When $n = 1$ and $a < b$, we will use the usual notations $[a,b] = \{x : a \le x \le b\}$, $(a,b) = \{x : a < x < b\}$, $[a,b) = \{x : a \le x < b\}$ and $(a,b] = \{x : a < x \le b\}$ for *closed, open*, and *partly open intervals*. Whenever we use just the word interval, we generally mean closed interval. An *n-dimensional interval* I is a subset of $\mathbf{R}^n$ of the form $I = \{\mathbf{x} = (x_1, \ldots, x_n): a_k \le x_k \le b_k, k = 1, \ldots, n\}$, where $a_k < b_k$, $k = 1, \ldots, n$. An interval is thus closed, and we say it has edges parallel to the coordinate axes. If the edge lengths $b_k - a_k$ are all equal, I will be called an *n-dimensional cube* with edges parallel to the coordinate axes. Cubes will usually be denoted by the letter Q. Two intervals I_1 and I_2 are said to be *nonoverlapping* if their interiors are disjoint, i.e., if the most they have in common is some part of their boundaries. A set equal to an interval minus some part of its boundary will be called a *partly open interval*. By definition, the *volume* $v(I)$ of the interval $I = \{(x_1, \ldots, x_n): a_k \le x_k \le b_k, k = 1, \ldots, n\}$ is

$$v(I) = \prod_{k=1}^{n} (b_k - a_k).$$

Somewhat more generally, if $\{\mathbf{e}_k\}_{k=1}^n$ is any given set of n vectors emanating from a point in $\mathbf{R}^n$, we will consider the closed *parallelepiped*

$$P = \{\mathbf{x} : \mathbf{x} = \sum_{k=1}^{n} t_k\mathbf{e}_k, \ 0 \le t_k \le 1\}.$$

Note that the edges of P are parallel translates of the $\mathbf{e}_k$. Thus, P is an interval if the $\mathbf{e}_k$ are parallel to the coordinate axes. The *volume* $v(P)$ of P is *by definition* the absolute value of the $n \times n$ determinant having $\mathbf{e}_1, \ldots, \mathbf{e}_n$ as rows.* In case P is an interval, this definition agrees with the one given above. A linear transformation T of $\mathbf{R}^n$ transforms a parallelpiped P into a parallelpiped P' with volume $v(P') = |\det T| v(P)$.† In particular, a rotation of axes in $\mathbf{R}^n$ (which is an orthogonal linear transformation) does not change the volume of a parallelepiped. We will assume simple facts about volume: e.g., if $I \subset \bigcup_1^N I_k$, N finite, then $v(I) \leqslant \sum_1^N v(I_k)$, and if $\{I_k\}_1^N$ are nonoverlapping intervals contained in a parallelepiped P, then $\sum_1^N v(I_k) \leqslant v(P)$.

We shall use the notion of interval to obtain a basic decomposition of open sets in $\mathbf{R}^n$. We consider first the case $n = 1$, which is somewhat simpler than $n > 1$.

(1.10) Theorem *Every open set in* $\mathbf{R}^1$ *can be written as a countable union of disjoint open intervals.*

Proof. Let G be an open set in $\mathbf{R}^1$. For $x \in G$, let I_x denote the maximal open interval containing x which is in G; that is, I_x is the union of all open intervals which contain x and which lie in G. If $x, x' \in G$ and $x \neq x'$, then I_x and $I_{x'}$ must either be disjoint or identical, since if they intersect, their union is an open interval containing x and x'. Clearly, $G = \bigcup_{x \in G} I_x$. Since each I_x contains a rational number, the number of distinct I_x must be countable, and the theorem follows.

*See, e.g., G. Birkhoff and S. Mac Lane, *A Survey of Modern Algebra*, 3d ed., Macmillan, New York, 1965, Theorem 8, p. 290.

†*Ibid.*, Theorem 9, p. 290.

The construction used in this proof fails in $\mathbf{R}^n$ if $n > 1$, since the union of (overlapping) intervals is not generally an interval. The theorem itself fails when $n > 1$, as is easily seen by considering any open ball. As a substitute, we have the following useful result.

(1.11) Theorem *Every open set in* $\mathbf{R}^n$, $n \geq 1$, *can be written as a countable union of nonoverlapping (closed) cubes.*

Proof. Consider the lattice of points of $\mathbf{R}^n$ with integral coordinates and the corresponding net K_0 of cubes with edge length 1 and vertices at these lattice points. Bisecting each edge of a cube in K_0, we obtain from it 2^n subcubes of edge length $\frac{1}{2}$. The total collection of these subcubes for every cube in K_0 forms a net K_1 of cubes. If we continue bisecting, we obtain finer and finer nets K_j of cubes such that each cube in K_j has edge length 2^{-j} and is the union of 2^n nonoverlapping cubes in K_{j+1}.

Now let G be any open set in $\mathbf{R}^n$. Let S_0 be the collection of all cubes in K_0 which lie entirely in G. Let S_1 be those cubes in K_1 which lie in G but which are not subcubes of any cube in S_0. More generally, for $j \geq 1$, let S_j be the cubes in K_j which lie in G but which are not subcubes of any cube in $S_0, \ldots, S_{j-1}$. If S denotes the total collection of cubes from all the S_j, then S is countable since each K_j is countable, and the cubes in S are nonoverlapping by construction. Moreover, since G is open and the cubes in K_j become arbitrarily small as $j \to \infty$, each point of G will eventually be caught in a cube in some S_j. Hence, $G = \bigcup_{Q \in S} Q$, and the proof is complete.

The collection $\{Q : Q \in K_j, j = 1, 2, \ldots\}$ constructed above is called a family of dyadic cubes. In general, by *dyadic cubes* we mean the family of cubes obtained from repeated bisection of any initial net of cubes in $\mathbf{R}^n$.

It follows from (1.10) that any closed set in $\mathbf{R}^1$ can be constructed by deleting a countable number of open disjoint intervals from $\mathbf{R}^1$. A perfect set results by removing the intervals in such a way as to create no isolated points; thus, we would not remove any two open intervals with a common endpoint.

4. Compact Sets; the Heine-Borel Theorem

By a *cover* of a set E, we mean a family $\mathscr{F}$ of sets A such that $E \subset \bigcup_{A \in \mathscr{F}} A$. A *subcover* $\mathscr{F}_1$ of a cover $\mathscr{F}$ is a cover with the property that $A_1 \in \mathscr{F}$ whenever $A_1 \in \mathscr{F}_1$. A cover $\mathscr{F}$ is called an *open cover* if each set in $\mathscr{F}$ is open. We say E is *compact* if every open cover of E has a finite subcover. Two equivalent statements, whose proofs are left as exercises, are as follows.

(1.12) Theorem

(i) (*The Heine-Borel theorem*) *A set* $E \subset \mathbf{R}^n$ *is compact if and only if it is closed and bounded.*

(ii) *A set $E \subset \mathbf{R}^n$ is compact if and only if every sequence of points of E has a subsequence which converges to a point of E.*

We leave it as an exercise to show that the distance between nonempty, compact, disjoint sets is positive, and that the intersection of a countable sequence of decreasing, nonempty, compact sets is nonempty. Thus, a nested sequence of closed intervals has a nonempty intersection.

With these facts, we can now prove (1.9).

Proof of theorem (1.9). Let C be a perfect set in $\mathbf{R}^n$, and suppose that C is countable: $C = \{\mathbf{c}_k\}_{k=1}^{\infty}$. Let $C_k = C - \{\mathbf{c}_k\}$, $k \geq 1$. Given $\mathbf{x}_1 \in C_1$, let Q_1 be a (closed) cube with center $\mathbf{x}_1$ such that $\mathbf{c}_1 \notin Q_1$. Then $Q_1 \cap C$ is compact (closed and bounded) and not empty. Since $\mathbf{x}_1 \in C$ and C is perfect, $\mathbf{x}_1$ is a limit point of C, and so also of C_2. It follows that $C_2 \cap \mathring{Q}_1$ is not empty. Let $\mathbf{x}_2 \in C_2 \cap \mathring{Q}_1$ and choose a cube Q_2 with center $\mathbf{x}_2$ such that $Q_2 \subset Q_1$ and $\mathbf{c}_2 \notin Q_2$. Then $Q_2 \cap C$ is a compact, nonempty subset of $Q_1 \cap C$. Continuing in this way, we obtain a decreasing sequence $Q_k \cap C$ of compact, nonempty sets such that $\mathbf{c}_k \notin Q_k$. It follows that $\bigcap_k (Q_k \cap C)$ is a nonempty subset of C which contains no $\mathbf{c}_k$. This contradiction proves that C must be countable and establishes the theorem.

5. Functions

By a function $f = f(\mathbf{x})$ defined for $\mathbf{x}$ in a set $E \subset \mathbf{R}^n$, we will always mean a *real-valued* function, unless explicitly stated otherwise. We allow f to take the values $\pm\infty$; if $|f(\mathbf{x})| < +\infty$ for all $\mathbf{x} \in E$, we say f is *finite* (or *finite-valued*) on E. A finite function f is said to be *bounded* on E if there is a finite number M such that $|f(\mathbf{x})| \leq M$ for $\mathbf{x} \in E$; that is, f is bounded on E if $\sup_{\mathbf{x}\in E} |f(\mathbf{x})|$ is finite. A sequence $\{f_k\}$ of functions is said to be *uniformly bounded* on E if there is a finite M such that $|f_k(\mathbf{x})| \leq M$ for $\mathbf{x} \in E$ and all k.

By the *support* of f, we mean the closure of the set where f is not zero. Thus, the support of a function is always closed. It follows that a function defined in $\mathbf{R}^n$ has *compact support* if and only if it vanishes outside some bounded set.

A function f defined on an interval I in $\mathbf{R}^1$ is called *monotone increasing* (*decreasing*) if $f(x) \leq f(y)$ $[f(x) \geq f(y)]$ whenever $x < y$ and $x,y \in I$. By *strictly* monotone increasing (decreasing), we mean that $f(x) < f(y)$ $[f(x) > f(y)]$ if $x < y$ and $x,y \in I$.

Let f be defined on $E \subset \mathbf{R}^n$ and let $\mathbf{x}_0$ be a limit point of E. Let $B'(\mathbf{x}_0, \delta) = B(\mathbf{x}_0;\delta) - \{\mathbf{x}_0\}$ denote the punctured ball with center $\mathbf{x}_0$ and radius δ, and let

$$M(\mathbf{x}_0;\delta) = \sup_{\mathbf{x}\in B'(\mathbf{x}_0;\delta)\cap E} f(\mathbf{x}), \qquad m(\mathbf{x}_0;\delta) = \inf_{\mathbf{x}\in B'(\mathbf{x}_0;\delta)\cap E} f(\mathbf{x}).$$

As $\delta \to 0$, $M(\mathbf{x}_0;\delta)$ decreases and $m(\mathbf{x}_0;\delta)$ increases, and we define

$$\text{(1.13)}\qquad \begin{aligned} \operatorname*{limsup}_{\mathbf{x}\to\mathbf{x}_0;\mathbf{x}\in E} f(\mathbf{x}) &= \lim_{\delta\to 0} M(\mathbf{x}_0;\delta), \\ \operatorname*{liminf}_{\mathbf{x}\to\mathbf{x}_0;\mathbf{x}\in E} f(\mathbf{x}) &= \lim_{\delta\to 0} m(\mathbf{x}_0;\delta). \end{aligned}$$

We leave it as an exercise to show that the following characterizations are valid.

(1.14) Theorem

(a) $M = \text{limsup}_{\mathbf{x}\to\mathbf{x}_0;\mathbf{x}\in E} f(\mathbf{x})$ *if and only if* (i) *there exists* $\{\mathbf{x}_k\}$ *in* E *such that* $\mathbf{x}_k \to \mathbf{x}_0$ *and* $f(\mathbf{x}_k) \to M$, *and* (ii) *if* $M' > M$, *there exists* $\delta > 0$ *such that* $f(\mathbf{x}) < M'$ *for* $\mathbf{x} \in B'(\mathbf{x}_0;\delta) \cap E$.

(b) $m = \text{liminf}_{\mathbf{x}\to\mathbf{x}_0;\mathbf{x}\in E} f(\mathbf{x})$ *if and only if* (i) *there exists* $\{\mathbf{x}_k\}$ *in* E *such that* $\mathbf{x}_k \to \mathbf{x}_0$ *and* $f(\mathbf{x}_k) \to m$, *and* (ii) *if* $m' < m$, *there exists* $\delta > 0$ *such that* $f(\mathbf{x}) > m'$ *for* $\mathbf{x} \in B'(\mathbf{x}_0;\delta) \cap E$.

We also define $\text{limsup}_{|\mathbf{x}|\to\infty;\mathbf{x}\in E} f(\mathbf{x})$ and $\text{liminf}_{|\mathbf{x}|\to\infty;\mathbf{x}\in E} f(\mathbf{x})$. For example, $M = \text{limsup}_{|\mathbf{x}|\to\infty;\mathbf{x}\in E} f(\mathbf{x})$ means (i) there exist $\{\mathbf{x}_k\}$ in E such that $|\mathbf{x}_k| \to \infty$ and $f(\mathbf{x}_k) \to M$, and (ii) if $M' > M$, there exists N such that $f(\mathbf{x}) < M'$ if $|\mathbf{x}| > N$ and $\mathbf{x} \in E$. These notions should not be confused with $\limsup_{k\to\infty} f_k(\mathbf{x})$ and $\liminf_{k\to\infty} f_k(\mathbf{x})$, which denote the limsup and liminf of the sequence $\{f_k(\mathbf{x})\}$.

6. Continuous Functions and Transformations

A function f defined in a neighborhood of $\mathbf{x}_0$ is said to be *continuous* at $\mathbf{x}_0$ if $f(\mathbf{x}_0)$ is finite and $\lim_{\mathbf{x}\to\mathbf{x}_0} f(\mathbf{x}) = f(\mathbf{x}_0)$. If f is not continuous at $\mathbf{x}_0$, it follows that unless $f(\mathbf{x}_0)$ is infinite, either $\lim_{\mathbf{x}\to\mathbf{x}_0} f(\mathbf{x})$ does not exist or is different from $f(\mathbf{x}_0)$.

For functions on $\mathbf{R}^1$, we will use the notation

$$f(x_0+) = \lim_{x\to x_0;x>x_0} f(x) \quad\text{and}\quad f(x_0-) = \lim_{x\to x_0;x<x_0} f(x)$$

for the *right* and *left-hand limits* of f at x_0, when they exist. If $f(x_0+)$, $f(x_0-)$, and $f(x_0)$ exist and are finite, but f is not continuous at x_0, then either $f(x_0+) \neq f(x_0-)$ or $f(x_0+) = f(x_0-) \neq f(x_0)$. In the first case, x_0 is called a *jump discontinuity* of f, and in the second, a *removable discontinuity* of f (since by changing the value of f at x_0, we can make it continuous there). Such discontinuities are said to be of the *first kind*, as distinguished from those of the *second kind*, for which either $f(x_0+)$ or $f(x_0-)$ does not exist or for which $f(x_0+)$, $f(x_0-)$, or $f(x_0)$ is infinite.

If f is defined only in a set E containing $\mathbf{x}_0$, $E \subset \mathbf{R}^n$, then f is said to be *continuous at* $\mathbf{x}_0$ *relative to* E if $f(\mathbf{x}_0)$ is finite and either $\mathbf{x}_0$ is an isolated point of E or $\mathbf{x}_0$ is a limit point of E and $\lim_{\mathbf{x}\to\mathbf{x}_0;\mathbf{x}\in E} f(\mathbf{x}) = f(\mathbf{x}_0)$. If $E_1 \subset E$, a function is said to be *continuous in* E_1 *relative to* E if it is continuous relative to E at every point of E_1. The proofs of the following basic facts are left as exercises.

(1.15) Theorem *Let E be a compact set in $\mathbf{R}^n$ and f be continuous in E relative to E. Then*

(i) *f is bounded on E; that is,* $\sup_{\mathbf{x}\in E} |f(\mathbf{x})| < \infty$.
(ii) *f attains its supremum and infimum on E; i.e., there exist* $\mathbf{x}_1,\mathbf{x}_2 \in E$ *such that* $f(\mathbf{x}_1) = \sup_{\mathbf{x}\in E} f(\mathbf{x})$, $f(\mathbf{x}_2) = \inf_{\mathbf{x}\in E} f(\mathbf{x})$.
(iii) *f is uniformly continuous on E relative to E; i.e., given $\varepsilon > 0$, there exists $\delta > 0$ such that $|f(\mathbf{x}) - f(\mathbf{y})| < \varepsilon$ if $|\mathbf{x} - \mathbf{y}| < \delta$ and $\mathbf{x},\mathbf{y} \in E$.*

A sequence of functions $\{f_k\}$ defined on E is said to *converge uniformly* on E to a finite f if given $\varepsilon > 0$, there exists K such that $|f_k(\mathbf{x}) - f(\mathbf{x})| < \varepsilon$ for $k \geq K$ and $\mathbf{x} \in E$. We will use the following fact, whose proof is again left to the reader.

(1.16) Theorem *Let $\{f_k\}$ be a sequence of functions defined on E which are continuous in E relative to E and which converge uniformly on E to a finite f. Then f is continuous in E relative to E.*

A *transformation T of a set $E \subset \mathbf{R}^n$ into* $\mathbf{R}^n$ is a mapping $\mathbf{y} = T\mathbf{x}$ which carries points $\mathbf{x} \in E$ into points $\mathbf{y} \in \mathbf{R}^n$. If $\mathbf{y} = (y_1, \ldots, y_n)$, then T can be identified with the collection of coordinate functions $y_k = f_k(\mathbf{x})$, $k = 1, \ldots, n$, which are induced by T. The *image* of E under T is the set $\{\mathbf{y} : \mathbf{y} = T\mathbf{x}$ for some $\mathbf{x} \in E\}$. T is continuous at $\mathbf{x}_0 \in E$ relative to E (by which we mean $\lim_{\mathbf{x}\to\mathbf{x}_0;\mathbf{x}\in E} T\mathbf{x} = T\mathbf{x}_0$) if and only if each f_k is continuous at $\mathbf{x}_0$ relative to E. We will use the following result in Chapter 3.

(1.17) Theorem *Let $\mathbf{y} = T\mathbf{x}$ be a transformation of $\mathbf{R}^n$ which is continuous in E relative to E. If E is compact, then so is its image TE.*

7. The Riemann Integral

We shall see that the Lebesgue integral is more general than the Riemann integral, in the sense that whenever the Riemann integral of a function exists then so does its Lebesgue integral, and the two are equal [theorem (5.52)]. The Riemann integral is nonetheless useful, its significance being simplicity and computability.

If f is defined and bounded on an interval $I = \{\mathbf{x} : \mathbf{x} = (x_1, \ldots, x_n), a_k \leq x_k \leq b_k, k = 1, \ldots, n\}$ in $\mathbf{R}^n$, its Riemann integral will be denoted by

$$(1.18)\quad (R)\int_{a_1}^{b_1} \cdots \int_{a_n}^{b_n} f(x_1, \ldots, x_n)\, dx_1 \cdots dx_n \quad \text{or} \quad (R)\int_I f(\mathbf{x})\, d\mathbf{x},$$

and is defined as follows. Partition I into a finite collection Γ of nonoverlapping intervals, $\Gamma = \{I_k\}_{k=1}^N$, and define the *norm* $|\Gamma|$ of Γ by $|\Gamma| = \max_k (\operatorname{diam} I_k)$. Select a point ξ_k in I_k for $k \geq 1$, and let

(1.19)
$$R_\Gamma = R_\Gamma(\xi_1, \ldots, \xi_N) = \sum_{k=1}^{N} f(\xi_k)v(I_k),$$
$$U_\Gamma = \sum_{k=1}^{N} [\sup_{\mathbf{x}\in I_k} f(\mathbf{x})]v(I_k), \qquad L_\Gamma = \sum_{k=1}^{N} [\inf_{\mathbf{x}\in I_k} f(\mathbf{x})]v(I_k).$$

We then define the Riemann integral by saying that $A = (R)\int_I f(\mathbf{x})\,d\mathbf{x}$ if and only if $\lim_{|\Gamma|\to 0} R_\Gamma$ exists and equals A; i.e., if and only if given $\varepsilon > 0$, there exists $\delta > 0$ such that $|A - R_\Gamma| < \varepsilon$ for any Γ and any chosen $\{\xi_k\}$, provided only that $|\Gamma| < \delta$. This definition is actually equivalent to the statement that

(1.20)
$$\inf_\Gamma U_\Gamma = \sup_\Gamma L_\Gamma = A.$$

The integral of course exists if f is continuous on I. Proofs of these facts are left as exercises; the treatment given in Chapter 2 for Riemann-Stieltjes integrals should serve as a review for many facts about Riemann integrals. See also Theorem 5.54.

Exercises

1. Prove the following facts, which were left as exercises above.
   (a) For a sequence of sets $\{E_k\}$, $\limsup E_k$ consists of those points which belong to infinitely many E_k, and $\liminf E_k$ consists of those points which belong to all E_k from some k on.
   (b) The De Morgan laws.
   (c) Every Cauchy sequence in $\mathbf{R}^n$ converges to a point of $\mathbf{R}^n$. (This can be deduced from its analogue in $\mathbf{R}^1$ by noting that the entries in a given coordinate position of the points in a Cauchy sequence in $\mathbf{R}^n$ form a Cauchy sequence in $\mathbf{R}^1$.)
   (d) Theorem 1.4.
   (e) A sequence $\{a_k\}$ in $\mathbf{R}^1$ converges to a, $-\infty \leq a \leq +\infty$, if and only if $\limsup_{k\to\infty} a_k = \liminf_{k\to\infty} a_k = a$.
   (f) $B(\mathbf{x}; \delta)$ is open.
   (g) Theorem 1.5.
   (h) Theorems 1.6 and 1.7.
   (i) Theorem 1.8.
   (j) Any closed (open) set in $\mathbf{R}^n$ is of type $G_\delta(F_\sigma)$. [If F is closed, consider the sets $\{\mathbf{x}; \operatorname{dist}(\mathbf{x},F) < (1/k)\}$, $k = 1, 2, \ldots$]
   (k) Theorem 1.12.

(l) The distance between two nonempty, compact, disjoint sets in $\mathbf{R}^n$ is positive.
(m) The intersection of a countable sequence of decreasing, nonempty, compact sets is nonempty.
(n) Theorem 1.14.
(o) Theorem 1.15.
(p) Theorem 1.16.
(q) Theorem 1.17.
(r) The Riemann integral $A = (R) \int_I f(\mathbf{x})\, d\mathbf{x}$ of a bounded f over an interval I exists if and only if $\inf_\Gamma U_\Gamma = \sup_\Gamma L_\Gamma = A$.
(s) If f is continuous on an interval I, then $(R)\int_I f(\mathbf{x})\, d\mathbf{x}$ exists.

2. Find limsup E_k and liminf E_k if $E_k = [-(1/k),1]$ for k odd and $E_k = [-1,(1/k)]$ for k even.
3. (a) Show that $C(\limsup E_k) = \liminf CE_k$.
   (b) Show that if $E_k \nearrow E$ or $E_k \searrow E$, then $\limsup E_k = \liminf E_k = E$.
4. (a) Show that $\limsup_{k\to\infty} (-a_k) = -\liminf_{k\to\infty} a_k$.
   (b) Show that $\limsup_{k\to\infty} (a_k + b_k) \le \limsup_{k\to\infty} a_k + \limsup_{k\to\infty} b_k$, provided that the expression on the right does not have the form $\infty + (-\infty)$ or $-\infty + \infty$.
   (c) If $\{a_k\}$ and $\{b_k\}$ are nonnegative, bounded sequences, show that $\limsup_{k\to\infty} (a_k b_k) \le (\limsup_{k\to\infty} a_k)(\limsup_{k\to\infty} b_k)$.
   (d) Give examples for which the inequalities in parts (b) and (c) are not equalities. Show that if either $\{a_k\}$ or $\{b_k\}$ converges, equality holds in (b) and (c).
5. Find analogues of the statements in exercise 4 for $\limsup_{\mathbf{x}\to\mathbf{x}_0;\mathbf{x}\in E} f(\mathbf{x})$.
6. Compare $\limsup_{k\to\infty} a_k$ and $\limsup\, (-\infty,a_k)$.
7. Show that $\mathring{E}_1 \cap \mathring{E}_2 = (E_1 \cap E_2)^\circ$, and $\mathring{E}_1 \cup \mathring{E}_2 \subset (E_1 \cup E_2)^\circ$. Give an example when $\mathring{E}_1 \cup \mathring{E}_2 \ne (E_1 \cup E_2)^\circ$.
8. Let E be relatively open with respect to an interval I. Show that E can be written as a countable union of nonoverlapping intervals.
9. Prove that any closed subset of a compact set is compact.
10. Let $\{\mathbf{x}_k\}$ be a bounded infinite sequence in $\mathbf{R}^n$. Show that $\{\mathbf{x}_k\}$ has a limit point. (This is the Bolzano-Weierstrass theorem in $\mathbf{R}^n$.)
11. Give an example of a decreasing sequence of nonempty closed sets in $\mathbf{R}^n$ whose intersection is empty.
12. Give an example of two disjoint closed sets F_1 and F_2 in $\mathbf{R}^n$ for which $\operatorname{dist}(F_1,F_2) = 0$.
13. If f is defined and uniformly continuous on E, show there is a function $\bar{f}$ defined and continuous on $\bar{E}$ such that $\bar{f} = f$ on E.
14. If f is defined and uniformly continuous on a bounded set E, show that f is bounded on E.
15. Show that a bounded f is Riemann integrable on I if and only if given $\varepsilon > 0$, there is a partition Γ of I such that $0 \le U_\Gamma - L_\Gamma < \varepsilon$.
16. If $\{f_k\}$ is a sequence of bounded, Riemann integrable functions on an interval I

which converges uniformly on I to f, show that f is Riemann integrable on I and that

$$(R)\int_I f_k(\mathbf{x})\,d\mathbf{x} \to (R)\int_I f(\mathbf{x})\,d\mathbf{x}.$$

17. Let f be a finite function on $\mathbf{R}^n$ and define

$$\omega(\delta) = \sup\{|f(\mathbf{x}) - f(\mathbf{y})|: |\mathbf{x} - \mathbf{y}| < \delta\},$$

$\delta > 0$, to be the *modulus of continuity* of f. Show that $\omega(\delta)$ decreases as δ decreases to 0 and that f is uniformly continuous if and only if $\omega(\delta) \to 0$ as $\delta \to 0$.

18. Let F be a closed subset of $(-\infty, +\infty)$, and let f be continuous relative to F. Show that there is a continuous function g on $(-\infty, +\infty)$ which equals f in F. If $|f(x)| \leq M$ for $x \in F$, show that g can be chosen so that $|g(x)| \leq M$ for $-\infty < x < +\infty$. (This is the Tietze extension theorem for the real line.)

Chapter 2

Functions of Bounded Variation; the Riemann-Stieltjes Integral

In the chapters ahead, we will study the Lebesgue integral. In this chapter, we introduce the Riemann-Stieltjes integral and, as a natural preliminary step, study functions of bounded variation. The justification for doing so is that Lebesgue integration is intimately connected with Riemann-Stieltjes integration, although this is not apparent from the definitions. We shall see in (5.43) that Lebesgue integrals can be represented as Riemann-Stieltjes integrals.

1. Functions of Bounded Variation

Let $f(x)$ be a real-valued function which is defined and finite for all x in a closed bounded interval $a \le x \le b$. Let

$$\Gamma = \{x_0, x_1, \ldots, x_m\}$$

be a *partition* of $[a,b]$; that is, Γ is a collection of points x_i, $i = 0, 1, \ldots, m$, satisfying $x_0 = a$, $x_m = b$, and $x_{i-1} < x_i$ for $i = 1, \ldots, m$. With each partition Γ, we associate the sum

$$S_\Gamma = S_\Gamma[f; a,b] = \sum_{i=1}^{m} |f(x_i) - f(x_{i-1})|.$$

The *variation of f over* $[a,b]$ is defined as

$$V = V[f; a,b] = \sup_\Gamma S_\Gamma,$$

where the supremum is taken over all partitions Γ of $[a,b]$. Since $0 \le S_\Gamma < +\infty$, we have $0 \le V \le +\infty$. If $V < +\infty$, f is said to be of *bounded variation on* $[a,b]$; if $V = +\infty$, f is of *unbounded variation on* $[a,b]$.

We list several simple examples.

Example 1. Suppose f is monotone in $[a,b]$. Then, clearly, each S_Γ equals $|f(b) - f(a)|$, and therefore $V = |f(b) - f(a)|$.

Example 2. Suppose the graph of f can be split into a finite number of monotone arcs; that is, suppose $[a,b] = \bigcup_{i=1}^{k} [a_i,a_{i+1}]$ and f is monotone in each $[a_i,a_{i+1}]$. Then $V = \sum_{i=1}^{k} |f(a_{i+1}) - f(a_i)|$. To see this, we use the result of Example 1 and the fact, to be proved in Theorem 2.2, that $V = V[a,b] = \sum_{i=1}^{k} V[a_i,a_{i+1}]$.

Example 3. Let f be defined by $f(x) = 0$ when $x \neq 0$ and $f(0) = 1$, and let $[a,b]$ be any interval containing 0 in its interior. Then S_Γ is either 2 or 0, depending on whether or not $x = 0$ is a partitioning point of Γ. Thus, $V[a,b] = 2$.

If $\Gamma = \{x_0, x_1, \ldots, x_m\}$ is a partition of $[a,b]$, let $|\Gamma|$, the *norm of* Γ, be defined as the length of a longest subinterval of Γ:

$$|\Gamma| = \max_i (x_i - x_{i-1}).$$

If f is continuous on $[a,b]$ and $\{\Gamma_j\}$ is a sequence of partitions of $[a,b]$ with $|\Gamma_j| \to 0$, we shall see in (2.9) that $V = \lim_{j\to\infty} S_{\Gamma_j}$. Example 3 shows that this equality may fail for functions which are discontinuous even at a single point: if we take f and $[a,b]$ as in Example 3 and choose the Γ_j such that $x = 0$ is never a partitioning point, then $\lim S_{\Gamma_j} = 0$, while if we choose the Γ_j such that $x = 0$ alternately is and is not a partitioning point, then $\lim S_{\Gamma_j}$ does not exist.

Example 4. Let f be the *Dirichlet function*, defined by $f(x) = 1$ for rational x and $f(x) = 0$ for irrational x. Then, clearly, $V[a,b] = +\infty$ for any interval $[a,b]$.

Example 5. A function which is continuous on an interval is not necessarily of bounded variation on the interval. To see this, let $\{a_j\}$ and $\{d_j\}$, $j = 1, 2, \ldots$, be two monotone decreasing sequences in $(0,1]$ with $a_1 = 1$, $\lim_{j\to\infty} a_j = \lim_{j\to\infty} d_j = 0$ and $\sum d_j = +\infty$. Construct a continuous f as follows. On each subinterval $[a_{j+1},a_j]$, the graph of f consists of the sides of the isosceles triangle with base $[a_{j+1},a_j]$ and height d_j. Thus $f(a_j) = 0$, and if m_j denotes the midpoint of $[a_{j+1},a_j]$, then $f(m_j) = d_j$. If we further define $f(0) = 0$, then f is continuous on $[0,1]$. Taking Γ_k to be the partition defined by the points 0, $\{a_j\}_{j=1}^{k+1}$, and $\{m_j\}_{j=1}^{k}$, we see that $S_{\Gamma_k} = 2\sum_{j=1}^{k} d_j$. Hence, $V[f; 0,1] = +\infty$.

We mention here that there exist functions which are continuous on an interval but which are not of bounded variation on any subinterval. See Exercise 26 of Chapter 3.

Example 6. A function f defined on $[a,b]$ is said to satisfy a *Lipschitz condition* on $[a,b]$, or to be a *Lipschitz function* on $[a,b]$, if there is a constant C such that

$$|f(x) - f(y)| \leq C|x - y| \text{ for all } x,y \in [a,b].$$

Such a function is clearly of bounded variation, with $V[f;a,b] \leq C(b - a)$. For example, if f has a continuous derivative on $[a,b]$, then (by the mean-value theorem) f satisfies a Lipschitz condition on $[a,b]$.

For more examples of functions of bounded variation, see the exercises at the end of the chapter.

In the next two theorems, we summarize some of the simplest properties of functions of bounded variation. The proof of the first theorem is left as an exercise.

(2.1) Theorem

(i) *If f is of bounded variation on $[a,b]$, then f is bounded on $[a,b]$.*

(ii) *Let f and g be of bounded variation on $[a,b]$. Then cf (for any real constant c), $f + g$ and fg are of bounded variation on $[a,b]$. Moreover, f/g is of bounded variation on $[a,b]$ if there exists an $\varepsilon > 0$ such that $|g(x)| \geq \varepsilon$ for $x \in [a,b]$.*

Before stating the second result, we note that if $\bar{\Gamma}$ is a *refinement* of Γ, that is, if $\bar{\Gamma}$ contains all the partitioning points of Γ plus some additional points, then $S_\Gamma \leq S_{\bar{\Gamma}}$. This follows from the triangle inequality, and is most easily seen in the case when $\bar{\Gamma}$ consists of all the points of Γ plus one additional point. The case of general $\bar{\Gamma}$ can be reduced to this simple case by adding one point at a time to Γ.

(2.2) Theorem

(i) *If $[a',b']$ is a subinterval of $[a,b]$, then $V[a',b'] \leq V[a,b]$; that is, variation increases with interval.*

(ii) *If $a < c < b$ then $V[a,b] = V[a,c] + V[c,b]$; that is, variation is additive on adjacent intervals.*

Proof. (i) Let $I = [a,b]$, $I' = [a',b']$, $V = V[a,b]$, and $V' = V[a',b']$. If Γ' is any partition of I', let Γ be Γ' with the points a and b adjoined. Then Γ is a partition of I and $S_{\Gamma'}[I'] \leq S_\Gamma[I]$. Thus $S_{\Gamma'}[I'] \leq V$, and therefore $V' \leq V$.

(ii) Let $I = [a,b]$, $I_1 = [a,c]$, $I_2 = [c,b]$, $V = V[a,b]$, $V_1 = V[a,c]$, and $V_2 = V[c,b]$. If Γ_1 and Γ_2 are any partitions of I_1 and I_2 respectively, then $\Gamma = \Gamma_1 \cup \Gamma_2$ is one of I, and $S_\Gamma[I] = S_{\Gamma_1}[I_1] + S_{\Gamma_2}[I_2]$. Thus, $S_{\Gamma_1}[I_1] + S_{\Gamma_2}[I_2] \leq V$. Therefore, taking the supremum over Γ_1 and Γ_2 separately, we obtain $V_1 + V_2 \leq V$.

Conversely, if Γ is any partition of I, let $\bar{\Gamma}$ be Γ with c adjoined. Then

$S_\Gamma[I] \le S_{\bar\Gamma}[I]$, and $\bar\Gamma$ splits into partitions Γ_1 of I_1 and Γ_2 of I_2. Thus, we have

$$S_\Gamma[I] \le S_{\bar\Gamma}[I] = S_{\Gamma_1}[I_1] + S_{\Gamma_2}[I_2] \le V_1 + V_2.$$

Therefore, $V \le V_1 + V_2$, which completes the proof of (ii).

For any real number x, define

$$x^+ = \begin{cases} x \text{ if } x > 0, \\ 0 \text{ if } x \le 0, \end{cases} \qquad x^- = \begin{cases} 0 \text{ if } x > 0. \\ -x \text{ if } x \le 0. \end{cases}$$

These are called the *positive* and *negative parts of* x, respectively, and satisfy the relations

(2.3) $$x^+, x^- \ge 0; \qquad |x| = x^+ + x^-; \qquad x = x^+ - x^-.$$

Given a function f and a partition $\Gamma = \{x_i\}_{i=0}^m$ of $[a,b]$, define

$$P_\Gamma = P_\Gamma[f; a,b] = \sum_{i=1}^m [f(x_i) - f(x_{i-1})]^+,$$
$$N_\Gamma = N_\Gamma[f; a,b] = \sum_{i=1}^m [f(x_i) - f(x_{i-1})]^-.$$

Thus, P_Γ is the sum of the positive terms of S_Γ, and $-N_\Gamma$ is the sum of the negative terms of S_Γ. In particular, by (2.3), $P_\Gamma \ge 0$, $N_\Gamma \ge 0$,

(2.4) $$P_\Gamma + N_\Gamma = S_\Gamma,$$

(2.5) $$P_\Gamma - N_\Gamma = f(b) - f(a).$$

The *positive variation* P and the *negative variation* N of f are defined by

$$P = P[f; a,b] = \sup_\Gamma P_\Gamma,$$
$$N = N[f; a,b] = \sup_\Gamma N_\Gamma.$$

Thus, $0 \le P,N \le +\infty$.

(2.6) Theorem *If any one of P, N, or V is finite, then all three are finite. Moreover, we then have*

$$P + N = V, \qquad P - N = f(b) - f(a),$$

or equivalently,

$$P = \tfrac{1}{2}[V + f(b) - f(a)], \qquad N = \tfrac{1}{2}[V - f(b) + f(a)].$$

Proof. By (2.4), $P_\Gamma + N_\Gamma \le V$, and therefore, since P_Γ and N_Γ are non-negative, $P \le V$ and $N \le V$. In particular, P and N are finite if V is. By (2.4) again, $S_\Gamma \le P + N$ and, therefore, $V \le P + N$. However, if either P or N is finite, so is the other by (2.5), and therefore, so is V. This gives the first part of the theorem.

Now choose a sequence of partitions Γ_k so that $P_{\Gamma_k} \to P$. Then $N_{\Gamma_k} \to N$ since P_Γ and N_Γ differ by a constant [see (2.5)]. Letting $k \to \infty$ in the relations $P_{\Gamma_k} - N_{\Gamma_k} = f(b) - f(a)$ and $P_{\Gamma_k} + N_{\Gamma_k} \le V$, we obtain $P - N = f(b) - f(a)$ and $P + N \le V$. Since $V \le P + N$ was shown above, we have $V = P + N$, and the theorem follows.

(2.7) Corollary (*Jordan's Theorem*) *A function f is of bounded variation on $[a,b]$ if and only if it can be written as the difference of two bounded increasing functions on $[a,b]$.*

Proof. Suppose $f = f_1 - f_2$, where f_1 and f_2 are bounded and increasing on $[a,b]$. Then f_1 and f_2 are of bounded variation on $[a,b]$, and therefore, by (2.1)(ii), so is f.

Conversely, suppose f is of bounded variation on $[a,b]$. By (2.2)(i), f is of bounded variation on every interval $[a,x]$, $a \le x \le b$. Let $P(x)$ and $N(x)$ denote the positive and negative variations of f on $[a,x]$, respectively. By the analogue of (2.2)(i) for P and N (see Exercise 3), it follows that $P(x)$ and $N(x)$ are bounded and increasing on $[a,b]$. Moreover, by (2.6) applied to $[a,x]$, $f(x) = [P(x) + f(a)] - N(x)$ when $a \le x \le b$. Since $P(x)$ is bounded and increasing, so is $P(x) + f(a)$, and the corollary follows.

Note that since the negative of an increasing function is decreasing, corollary (2.7) may be rephrased to say that f is of bounded variation if and only if it is the sum of a bounded increasing function and a bounded decreasing function.

We remark here that there exist continuous functions of bounded variation which are not monotone in any subinterval. See Exercise 27 of Chapter 3.

In the next theorem, we consider a continuity property of functions of bounded variation. We recall from Chapter 1 that a discontinuity is said to be of the first kind if it is either a jump or a removable discontinuity.

(2.8) Theorem *Every function of bounded variation has at most a countable number of discontinuities, and they are all of the first kind.*

Proof. Let f be of bounded variation on $[a,b]$. By (2.7), we may assume f is bounded and increasing on $[a,b]$. Then the only discontinuities of f are of the first kind; in fact, they are all jump discontinuities. If D denotes the set of all discontinuities of f, then $D = \bigcup_{k=1}^{\infty} \{x : f(x+) - f(x-) \ge 1/k\}$. Since f is bounded, each set on the right is finite (or empty); therefore D is countable.

We now discuss a property of the variation of a continuous function.

(2.9) Theorem *If f is continuous on $[a,b]$, then $V = \lim_{|\Gamma| \to 0} S_\Gamma$; that is, given M satisfying $M < V$, there exists $\delta > 0$ such that $S_\Gamma > M$ for any partition Γ of $[a,b]$ with $|\Gamma| < \delta$.*

Proof. We recall the discussion following Example 3. Given M, $M < V$, we must find $\delta > 0$ so that $S_\Gamma > M$ if $|\Gamma| < \delta$. Select $\mu > 0$ such that $M + \mu < V$, and choose a fixed partition $\bar{\Gamma} = \{\bar{x}_j\}_{j=0}^k$ such that $S_{\bar{\Gamma}} > M + \mu$. Using the uniform continuity of f on $[a,b]$, find $\eta > 0$ so that

(i) $|f(x) - f(x')| < \mu/2(k+1)$ if $|x - x'| < \eta$.

Now let Γ be any partition which satisfies

(ii) $|\Gamma| < \eta$,
(iii) $|\Gamma| < \min_j (\bar{x}_j - \bar{x}_{j-1})$.

We claim that $S_\Gamma > M$, from which the theorem will follow by choosing δ to be the smaller of η and $\min_j (\bar{x}_j - \bar{x}_{j-1})$. Write $\Gamma = \{x_i\}_{i=0}^m$ and

$$S_\Gamma = \sum_{i=1}^m |f(x_i) - f(x_{i-1})| = \Sigma' + \Sigma'',$$

where Σ'' is extended over all i such that (x_{i-1},x_i) contains some $\bar{x}_j$. By (iii), any (x_{i-1},x_i) can contain at most one $\bar{x}_j$, and therefore the number of terms of Σ'' is at most $k + 1$. Let $\Gamma \cup \bar{\Gamma}$ denote the partition formed by the union of the points of Γ and $\bar{\Gamma}$. Then $\Gamma \cup \bar{\Gamma}$ is a refinement of both Γ and $\bar{\Gamma}$. Moreover, $S_{\Gamma \cup \bar{\Gamma}} = \Sigma' + \Sigma'''$, where Σ''' is obtained from Σ'' by replacing each term by $|f(x_i) - f(\bar{x}_j)| + |f(\bar{x}_j) - f(x_{i-1})|$, $\bar{x}_j$ being the point of $\bar{\Gamma}$ in (x_{i-1},x_i). By (i) and (ii), each of these two terms is less than $\mu/2(k+1)$, and therefore

$$\Sigma''' < 2(k+1)\frac{\mu}{2(k+1)} = \mu.$$

Hence,

$$\Sigma' = S_{\Gamma \cup \bar{\Gamma}} - \Sigma''' > S_{\Gamma \cup \bar{\Gamma}} - \mu,$$

so that $S_\Gamma > S_{\Gamma \cup \bar{\Gamma}} - \mu$. Since $\Gamma \cup \bar{\Gamma}$ is a refinement of $\bar{\Gamma}$, $S_{\Gamma \cup \bar{\Gamma}} \geq S_{\bar{\Gamma}}$. This gives $S_\Gamma > S_{\bar{\Gamma}} - \mu > M$ and completes the proof.

(2.10) Corollary *If f has a continuous derivative f' on* $[a,b]$, *then*

$$V = \int_a^b |f'|\, dx, \qquad P = \int_a^b \{f'\}^+\, dx, \qquad N = \int_a^b \{f'\}^-\, dx.$$

Proof. By the mean-value theorem,

$$S_\Gamma = \sum_{i=1}^m |f(x_i) - f(x_{i-1})| = \sum_{i=1}^m |f'(\xi_i)|(x_i - x_{i-1})$$

for appropriate $\xi_i \in (x_{i-1},x_i)$, $i = 1, \ldots, m$. Hence, by (2.9),

$$V = \lim_{|\Gamma|\to 0} S_\Gamma = \lim_{|\Gamma|\to 0} \sum_{i=1}^m |f'(\xi_i)|(x_i - x_{i-1}) = \int_a^b |f'(x)|\, dx,$$

by definition of the Riemann integral. Moreover, by (2.6),

$$P = \tfrac{1}{2}[V + f(b) - f(a)] = \tfrac{1}{2}\left[\int_a^b |f'(x)|\,dx + \int_a^b f'(x)\,dx\right]$$

$$= \tfrac{1}{2}\int_a^b [|f'(x)| + f'(x)]\,dx = \int_a^b [f'(x)]^+\,dx.$$

The formula for N follows similarly from the fact that

$$N = \tfrac{1}{2}[V - f(b) + f(a)].$$

For an extension of (2.10), see (7.31) in Chapter 7.

In passing, we note that there are notions of bounded variation for open or partly open intervals, as well as for infinite intervals. Suppose, for example, that (a,b) is a bounded open interval. Let $[a',b'] \subset (a,b)$, and define $\mathring{V}(a,b) = \lim V[a',b']$ as $a' \to a$ and $b' \to b$. If $\mathring{V}(a,b) < +\infty$, we say f is of *bounded variation on* (a,b). Similarly, if f is defined on $(-\infty,+\infty)$, let $V(-\infty,+\infty) = \lim V[a,b]$ as $a \to -\infty$ and $b \to +\infty$. Analogous definitions hold for $[a,b)$, $(a,+\infty)$, $[a,+\infty)$, etc. See Exercise 8.

We may also consider the notion of bounded variation for complex-valued f defined on an interval. The definition is the same as for the real-valued case, and we leave it to the reader to show that a complex-valued f is of bounded variation if and only if both its real and imaginary parts are.

2. Rectifiable Curves

As an application of the notion of bounded variation, we shall discuss its relation to rectifiable curves (initially, those in the plane). A *curve* C in the plane is two parametric equations

(2.11) $$C: \begin{cases} x = \phi(t) \\ y = \psi(t) \end{cases}, \qquad a \le t \le b.$$

The *graph* of C is $\{(x,y) : x = \phi(t),\ y = \psi(t),\ a \le t \le b\}$. The graph may have self-intersections and is not necessarily continuous or bounded. We think of the curve itself as the mapping of $[a,b]$ onto the graph.

Let $\Gamma = \{a = t_0 < t_1 < \cdots < t_m = b\}$ be a partition of $[a,b]$, and consider the corresponding points $P_i = (\phi(t_i), \psi(t_i))$, $i = 0, 1, \ldots, m$, on the graph of C. Draw the polygonal (broken) line connecting P_0 to P_1, P_1 to $P_2, \ldots, P_{m-1}$ to P_m in order, and let

$$l(\Gamma) = \sum_{i=1}^{m} ([\phi(t_i) - \phi(t_{i-1})]^2 + [\psi(t_i) - \psi(t_{i-1})]^2)^{1/2}$$

denote its length. The *length* L of C is defined by the equation

(2.12) $$L = L(C) = \sup_{\Gamma} l(\Gamma).$$

Thus, $0 \le L \le +\infty$. If the graph of C is discontinuous, then as we move along the graph, the length of every missing segment will contribute to L. Moreover, the possibility that the graph may be traversed more than once, that is, that the mapping $t \to (\phi(t), \psi(t))$, $a \le t \le b$, may not be one-to-one, will add to L.

We call C *rectifiable* if $L < +\infty$.

(2.13) Theorem *Let C be a curve defined by (2.11). Then C is rectifiable if and only if both ϕ and ψ are of bounded variation. Moreover,*

$$V(\phi),\ V(\psi) \le L \le V(\phi) + V(\psi).$$

Proof. We will use the simple inequalities

$$|a|,\ |b| \le (a^2 + b^2)^{1/2} \le |a| + |b|$$

for real a and b. Thus, if C is rectifiable and $\Gamma = \{t_i\}$ is any partition of $[a,b]$, the inequality

$$l(\Gamma) = \sum ([\phi(t_i) - \phi(t_{i-1})]^2 + [\psi(t_i) - \psi(t_{i-1})]^2)^{1/2} \le L$$

implies $\sum |\phi(t_i) - \phi(t_{i-1})| \le L$ and $\sum |\psi(t_i) - \psi(t_{i-1})| \le L$. Hence, $V(\phi)$, $V(\psi) \le L$. On the other hand, for any C,

$$l(\Gamma) \le \sum |\phi(t_i) - \phi(t_{i-1})| + \sum |\psi(t_i) - \psi(t_{i-1})| \le V(\phi) + V(\psi).$$

Hence, $L \le V(\phi) + V(\psi)$.

It follows that if $\phi(t)$ is any bounded function which is not of bounded variation on $[a,b]$ (see Example 5 and Exercise 1), then the curve given by $x = y = \phi(t)$, $a \le t \le b$, is not rectifiable, even though its graph lies in a finite segment of the line $y = x$. Thus, the length of the graph of a curve is not necessarily the same as the length of the curve.

In the special case that C is given by a function $y = f(x)$, (2.13) reduces to the simple statement that C is rectifiable if and only if f is of bounded variation.

Curves in $\mathbf{R}^n$ can be treated similarly, and we shall be brief. By a *curve C* in $\mathbf{R}^n$ we mean a system $x_1 = \phi_1(t), \ldots, x_n = \phi_n(t)$, for t in some $[a,b]$. We consider a partition $\Gamma = \{t_i\}_{i=0}^m$ of $[a,b]$ and the length $l(\Gamma)$ of the corresponding polygonal line:

$$l(\Gamma) = \sum_{i=1}^{m} \overline{P_{i-1}P_i} = \sum_{i=1}^{m} \left(\sum_{j=1}^{n} [\phi_j(t_i) - \phi_j(t_{i-1})]^2 \right)^{1/2}.$$

The quantity $L = \sup l(\Gamma)$ is called the *length* of C, and if $L < +\infty$, C is said to be *rectifiable*. As seen from the definition of $l(\Gamma)$, exactly as in the case $n = 2$, C is rectifiable if and only if each ϕ_j is of bounded variation.

3. The Riemann-Stieltjes Integral

Let f and ϕ be two functions which are defined and finite on a finite interval $[a,b]$. If $\Gamma = \{a = x_0 < x_1 < \cdots < x_m = b\}$ is a partition of $[a,b]$, we arbitrarily select intermediate points $\{\xi_i\}_{i=1}^m$ satisfying $x_{i-1} \le \xi_i \le x_i$, and write

$$R_\Gamma = \sum_{i=1}^{m} f(\xi_i)[\phi(x_i) - \phi(x_{i-1})]. \tag{2.14}$$

R_Γ is called a *Riemann-Stieltjes sum* for Γ, and of course depends on the points ξ_i, the functions f and ϕ and the interval $[a,b]$, although we shall usually not display this dependence in our notation.

If

$$I = \lim_{|\Gamma|\to 0} R_\Gamma \tag{2.15}$$

exists and is finite, that is, if given $\varepsilon > 0$ there is a $\delta > 0$ such that $|I - R_\Gamma| < \varepsilon$ for any Γ satisfying $|\Gamma| < \delta$, then I is called *the Riemann-Stieltjes integral of f with respect to ϕ on* $[a,b]$, and denoted

$$I = \int_a^b f(x)\, d\phi(x) = \int_a^b f\, d\phi.$$

A necessary and sufficient condition for the existence of $\int_a^b f\, d\phi$ is that given $\varepsilon > 0$, there exist $\delta > 0$ such that $|R_\Gamma - R_{\Gamma'}| < \varepsilon$ if $|\Gamma|, |\Gamma'| < \delta$. See Exercise 11.

We list four preliminary remarks about this integral.

1. If $\phi(x) = x$, $\int_a^b f\, d\phi$ is clearly just the Riemann integral $\int_a^b f\, dx$. In this case, Theorem 5.54 in Chapter 5 gives a necessary and sufficient condition for the existence of the integral.
2. If f is continuous on $[a,b]$ and ϕ is continuously differentiable on $[a,b]$, then $\int_a^b f\, d\phi = \int_a^b f\phi'\, dx$. [See also (7.32).] In fact, by the mean-value theorem,

$$R_\Gamma = \sum f(\xi_i)[\phi(x_i) - \phi(x_{i-1})] = \sum f(\xi_i)\phi'(\eta_i)(x_i - x_{i-1}),$$

   with $x_{i-1} \le \xi_i, \eta_i \le x_i$. Using the uniform continuity of ϕ', we obtain $\lim_{|\Gamma|\to 0} R_\Gamma = \int_a^b f\phi'\, dx$.
3. Let $\phi(x)$ be a *step function*; that is, suppose there are points $a = \alpha_0 < \alpha_1 < \cdots < \alpha_m = b$ such that ϕ is constant on each interval (α_{i-1},α_i). Let

$$\phi(\alpha_i+) = \lim_{x\to\alpha_i+} \phi(x), \qquad i = 0, 1, \ldots, m-1,$$

and

$$\phi(\alpha_i-) = \lim_{x\to\alpha_i-} \phi(x), \qquad i = 1, \ldots, m,$$

denote the limits from the right and left at α_i, and let $d_i = \phi(\alpha_i+) - \phi(\alpha_i-)$, $i = 1, \ldots, m-1$, $d_0 = \phi(\alpha_0+) - \phi(\alpha_0)$, and $d_m = \phi(\alpha_m) - \phi(\alpha_m-)$ denote the jumps of ϕ. Then, for continuous f,

$$\int_a^b f\,d\phi = \sum_{i=0}^{m} f(\alpha_i)\,d_i.$$

The existence of the integral can be verified directly or by appealing to (2.24).

4. In definition (2.15), no condition other than finiteness is imposed on f or ϕ, but we shall see later that the most important applications occur when ϕ is monotone or, more generally, of bounded variation. We note now that if $\int_a^b f\,d\phi$ exists then f and ϕ have no common points of discontinuity. To prove this, suppose that both f and ϕ are discontinuous at $\bar{x}$, $a < \bar{x} < b$. Suppose first that the discontinuity of ϕ is not removable. Then there is a fixed $\eta > 0$ such that, for every $\varepsilon > 0$, there exist points $\bar{x}_1$ and $\bar{x}_2$ with $\bar{x} - \frac{1}{2}\varepsilon < \bar{x}_1 < \bar{x} < \bar{x}_2 < \bar{x} + \frac{1}{2}\varepsilon$ and $|\phi(\bar{x}_2) - \phi(\bar{x}_1)| > \eta$. Let $\Gamma = \{x_i\}$ be a partition of $[a,b]$ with $|\Gamma| < \varepsilon$ such that $x_{i_0-1} = \bar{x}_1$ and $x_{i_0} = \bar{x}_2$ for some i_0. Choose a point $\xi_i \in [x_{i-1},x_i]$ for $i \neq i_0$ and two different points ξ_{i_0} and ξ'_{i_0} in $[x_{i_0-1},x_{i_0}]$. Let R_Γ be the Riemann-Stieltjes sum using ξ_i in each $[x_{i-1},x_i]$, and let R'_Γ be the sum using ξ_i in $[x_{i-1},x_i]$ for $i \neq i_0$ and ξ'_{i_0} in $[x_{i_0-1},x_{i_0}]$. Then, clearly,

$$\begin{aligned} |R_\Gamma - R'_\Gamma| &= |f(\xi_{i_0}) - f(\xi'_{i_0})||\phi(x_{i_0}) - \phi(x_{i_0-1})| \\ &> \eta|f(\xi_{i_0}) - f(\xi'_{i_0})|. \end{aligned}$$

Since f is discontinuous at $\bar{x}$, we can choose ξ_{i_0} and ξ'_{i_0} subject to the restrictions above and such that $|f(\xi_{i_0}) - f(\xi'_{i_0})| > \mu$ for some $\mu > 0$ independent of ε. It follows that $|R_\Gamma - R'_\Gamma|$ exceeds a positive constant independent of ε, contradicting the assumption that $R_\Gamma - R'_\Gamma \to 0$ as $|\Gamma|, |\Gamma'| \to 0$.

If the discontinuity of ϕ at $\bar{x}$ is removable, a similar argument can be given. The main difference is that we consider Γ with $\bar{x}$ as a partitioning point x_{i_0} and argue for either $[x_{i_0-1},\bar{x}]$ or $[\bar{x},x_{i_0+1}]$, depending on the nature of the discontinuity of f at $\bar{x}$. The arguments in the case where $\bar{x}$ is either a or b are similar.

In the theorem which follows, we list some simple properties of the Riemann-Stieltjes integral. The proofs are left as an exercise.

(2.16) Theorem

(i) *If* $\int_a^b f\,d\phi$ *exists, then so do* $\int_a^b cf\,d\phi$ *and* $\int_a^b f\,d(c\phi)$ *for any constant* c, *and*

$$\int_a^b cf\,d\phi = \int_a^b f\,d(c\phi) = c\int_a^b f\,d\phi.$$

(ii) *If* $\int_a^b f_1\,d\phi$ *and* $\int_a^b f_2\,d\phi$ *both exist, so does* $\int_a^b (f_1 + f_2)\,d\phi$, *and*

$$\int_a^b (f_1 + f_2)\,d\phi = \int_a^b f_1\,d\phi + \int_a^b f_2\,d\phi.$$

(iii) *If* $\int_a^b f\,d\phi_1$ *and* $\int_a^b f\,d\phi_2$ *exist, so does* $\int_a^b f\,d(\phi_1 + \phi_2)$, *and*

$$\int_a^b f\,d(\phi_1 + \phi_2) = \int_a^b f\,d\phi_1 + \int_a^b f\,d\phi_2.$$

The additivity of the integral with respect to intervals is given by the following result. See also Exercise 14.

(2.17) Theorem *If* $\int_a^b f\,d\phi$ *exists and* $a < c < b$, *then* $\int_a^c f\,d\phi$ *and* $\int_c^b f\,d\phi$ *both exist and*

$$\int_a^b f\,d\phi = \int_a^c f\,d\phi + \int_c^b f\,d\phi.$$

Proof. In the proof, $R_\Gamma[a,b]$ will denote a Riemann-Stieltjes sum corresponding to a partition Γ of $[a,b]$. To show that $\int_a^c f\,d\phi$ exists, it is enough to show that given $\varepsilon > 0$, there exists $\delta > 0$ so that if Γ_1 and Γ_2 are partitions of $[a,c]$ with $|\Gamma_1|, |\Gamma_2| < \delta$ then

$$|R_{\Gamma_1}[a,c] - R_{\Gamma_2}[a,c]| < \varepsilon. \tag{2.18}$$

Since $\int_a^b f\,d\phi$ exists, there is a $\delta > 0$ so that for any partitions Γ_1' and Γ_2' of $[a,b]$ with $|\Gamma_1'|, |\Gamma_2'| < \delta$, we have

$$|R_{\Gamma_1'}[a,b] - R_{\Gamma_2'}[a,b]| < \varepsilon. \tag{2.19}$$

Let Γ_1 and Γ_2 be partitions of $[a,c]$ with given sets of intermediate points. Complete Γ_1 and Γ_2 to partitions Γ_1' and Γ_2' of $[a,b]$ by adjoining the same points of $[c,b]$; that is, let Γ' be a partition of $[c,b]$ and let $\Gamma_1' = \Gamma_1 \cup \Gamma'$, $\Gamma_2' = \Gamma_2 \cup \Gamma'$. Select a set of intermediate points in $[c,b]$ for Γ', and let the intermediate points of Γ_1' and Γ_2' consist of these together with the sets for Γ_1 and Γ_2, respectively. Then

$$\begin{aligned} R_{\Gamma_1'}[a,b] &= R_{\Gamma_1}[a,c] + R_{\Gamma'}[c,b] \\ R_{\Gamma_2'}[a,b] &= R_{\Gamma_2}[a,c] + R_{\Gamma'}[c,b]. \end{aligned} \tag{2.20}$$

If we now assume that $|\Gamma_1|, |\Gamma_2| < \delta$ and choose Γ' so that $|\Gamma'| < \delta$, then $|\Gamma_1'|, |\Gamma_2'| < \delta$, and (2.18) follows from (2.19) by subtracting the equations in (2.20).

The proof of the existence of $\int_c^b f\,d\phi$ is similar. The fact that

$$\int_a^b f\,d\phi = \int_a^c f\,d\phi + \int_c^b f\,d\phi$$

follows from (2.20). This completes the proof.

The next result is the very useful formula for *integration by parts.*

(2.21) Theorem *If $\int_a^b f\,d\phi$ exists, then so does $\int_a^b \phi\,df$, and*

$$\int_a^b f\,d\phi = [f(b)\phi(b) - f(a)\phi(a)] - \int_a^b \phi\,df.$$

Proof. Let $\Gamma = \{a = x_0 < x_1 < \cdots < x_m = b\}$ and $x_{i-1} \le \xi_i \le x_i$. Then

$$\begin{aligned} R_\Gamma &= \sum_{i=1}^m f(\xi_i)[\phi(x_i) - \phi(x_{i-1})] = \sum_{i=1}^m f(\xi_i)\phi(x_i) - \sum_{i=1}^m f(\xi_i)\phi(x_{i-1}) \\ &= \sum_{i=1}^m f(\xi_i)\phi(x_i) - \sum_{i=0}^{m-1} f(\xi_{i+1})\phi(x_i) \\ &= -\sum_{i=1}^{m-1} \phi(x_i)[f(\xi_{i+1}) - f(\xi_i)] + f(\xi_m)\phi(b) - f(\xi_1)\phi(a) \end{aligned}$$

since $x_m = b$ and $x_0 = a$. If we subtract and add $\phi(a)[f(\xi_1) - f(a)] + \phi(b)[f(b) - f(\xi_m)]$ on the right side of the last equality and cancel like terms, we obtain $R_\Gamma = -T_\Gamma + [f(b)\phi(b) - f(a)\phi(a)]$, where

$$T_\Gamma = \sum_{i=1}^{m-1} \phi(x_i)[f(\xi_{i+1}) - f(\xi_i)] + \phi(a)[f(\xi_1) - f(a)] + \phi(b)[f(b) - f(\xi_m)].$$

Since the ξ_i straddle the x_i (successive ξ_i's may be equal), T_Γ is a Riemann-Stieltjes sum for $\int_a^b \phi\,df$. Observing that the roles of ϕ and f can be interchanged, and taking the limit as $|\Gamma| \to 0$, we see that $\int_a^b f\,d\phi$ exists if and only if $\int_a^b \phi\,df$ exists, and that $\int_a^b f\,d\phi = [f(b)\phi(b) - f(a)\phi(a)] - \int_a^b \phi\,df$. This completes the proof.

Now let f be bounded and ϕ be monotone increasing on $[a,b]$. If $\Gamma = \{x_i\}_{i=0}^m$, let

$$m_i = \inf_{x_{i-1} \le x \le x_i} f(x), \qquad M_i = \sup_{x_{i-1} \le x \le x_i} f(x),$$

$$L_\Gamma = \sum_{i=1}^m m_i[\phi(x_i) - \phi(x_{i-1})],$$

$$U_\Gamma = \sum_{i=1}^m M_i[\phi(x_i) - \phi(x_{i-1})].$$

Since $-\infty < m_i \le M_i < +\infty$ and $\phi(x_i) - \phi(x_{i-1}) \ge 0$, we see that

(2.22) $$L_\Gamma \le R_\Gamma \le U_\Gamma.$$

L_Γ and U_Γ are called the *upper* and *lower Riemann-Stieltjes sums for* Γ, respectively. The behavior of L_Γ and U_Γ is somewhat more predictable than that of R_Γ, as we now show.

(2.23) Lemma *Let f be bounded and ϕ be increasing on $[a,b]$.*

(i) *If Γ' is a refinement of Γ, then $L_{\Gamma'} \ge L_\Gamma$ and $U_{\Gamma'} \le U_\Gamma$.*

(ii) *If Γ_1 and Γ_2 are any two partitions, then $L_{\Gamma_1} \le U_{\Gamma_2}$.*

Proof. To see (i) for upper sums, suppose that Γ' has only one point x' not in Γ. If x' lies between x_{i-1} and x_i of Γ, then $\sup_{[x_{i-1},x']} f(x)$, $\sup_{[x',x_i]} f(x) \le M_i$, so that

$$\sup_{[x_{i-1},x']} f(x)[\phi(x') - \phi(x_{i-1})] + \sup_{[x',x_i]} f(x)[\phi(x_i) - \phi(x')] \le M_i[\phi(x_i) - \phi(x_{i-1})].$$

Hence, $U_{\Gamma'} \le U_\Gamma$. Since Γ' can be obtained by adding one point at a time to Γ, an extension of this reasoning proves (i) for upper sums. The argument for lower sums is similar.

To show (ii), note that $\Gamma_1 \cup \Gamma_2$ is a refinement of both Γ_1 and Γ_2. Hence, by part (i) and (2.22), we obtain $L_{\Gamma_1} \le L_{\Gamma_1 \cup \Gamma_2} \le U_{\Gamma_1 \cup \Gamma_2} \le U_{\Gamma_2}$, which completes the proof.

We now come to an important result which gives sufficient conditions for the existence of $\int f\,d\phi$.

(2.24) Theorem *If f is continuous on $[a,b]$ and ϕ is of bounded variation on $[a,b]$, then $\int_a^b f\,d\phi$ exists. Moreover,*

$$\left| \int_a^b f\,d\phi \right| \le (\sup_{[a,b]} |f|) V[\phi; a,b].$$

Proof. To prove the existence, we may suppose by (2.7) and (2.16)(iii) that ϕ is monotone increasing. Then, by (2.22), $L_\Gamma \le R_\Gamma \le U_\Gamma$, and it is enough to show that $\lim_{|\Gamma|\to 0} L_\Gamma$ and $\lim_{|\Gamma|\to 0} U_\Gamma$ exist and are equal. This is clear if ϕ is constant on $[a,b]$. If ϕ is not constant, let $\Gamma = \{x_i\}$ and note that given $\varepsilon > 0$, the uniform continuity of f implies there exists $\delta > 0$ such that if $|\Gamma| < \delta$, then $M_i - m_i < \varepsilon/[\phi(b) - \phi(a)]$. Hence, if $|\Gamma| < \delta$,

(2.25) $$0 \le U_\Gamma - L_\Gamma = \sum (M_i - m_i)[\phi(x_i) - \phi(x_{i-1})] < \varepsilon.$$

Therefore,

(2.26) $$\lim_{|\Gamma|\to 0} (U_\Gamma - L_\Gamma) = 0,$$

and it is enough to show that $\lim_{|\Gamma|\to 0} U_\Gamma$ exists. This is immediate since otherwise there would exist an $\varepsilon > 0$ and two sequences of partitions, $\{\Gamma_k'\}$ and $\{\Gamma_k''\}$, with norms tending to zero such that $U_{\Gamma_k'} - U_{\Gamma_k''} > \varepsilon$. In view of (2.26), we would then have, for k large enough, $L_{\Gamma_k'} - U_{\Gamma_k''} > \frac{1}{2}\varepsilon > 0$, contradicting the fact that $L_{\Gamma'} \leq U_{\Gamma''}$ for any Γ' and Γ'' [lemma (2.23)].

To complete the proof, note that the inequality $|\int_a^b f\,d\phi| \leq (\sup_{[a,b]}|f|)\, V[\phi; a,b]$ follows from a similar one for R_Γ by taking the limit.

Combining (2.21) and (2.24), we see that $\int_a^b f\,d\phi$ exists if either f or ϕ is continuous and the other is of bounded variation.

(2.27) Theorem (*Mean-value Theorem*) *If f is continuous on $[a,b]$ and ϕ is bounded and increasing on $[a,b]$, there exists $\xi \in [a,b]$ such that*

$$\int_a^b f\,d\phi = f(\xi)[\phi(b) - \phi(a)].$$

Proof. Since ϕ is increasing, we have

$$(\min_{[a,b]} f)[\phi(b) - \phi(a)] \leq R_\Gamma \leq (\max_{[a,b]} f)[\phi(b) - \phi(a)]$$

for any R_Γ. Since $\int_a^b f\,d\phi$ exists [see (2.24)], it must satisfy

$$(\min_{[a,b]} f)[\phi(b) - \phi(a)] \leq \int_a^b f\,d\phi \leq (\max_{[a,b]} f)[\phi(b) - \phi(a)].$$

The result now follows immediately from the intermediate value theorem for continuous functions.

In passing, we note that Riemann-Stieltjes integrals can be defined on open or partly open bounded intervals and on infinite intervals. If f and ϕ are defined on (a,b) for example, let $a < a' < b' < b$ and define

$$\int_a^b f\,d\phi = \lim_{\substack{a'\to a\\ b'\to b}} \int_{a'}^{b'} f\,d\phi,$$

provided the limit exists. Similarly, let

$$\int_{-\infty}^{+\infty} f\,d\phi = \lim_{\substack{a\to -\infty\\ b\to +\infty}} \int_a^b f\,d\phi$$

if the limit exists. Analogous definitions can be given for $[a,b)$, $(a,+\infty)$, $[a,+\infty)$, etc.

4. Further Results About Riemann-Stieltjes Integrals

We will discuss a variant of the definition of $\int_a^b f\,d\phi$ in the case where f is bounded and ϕ is increasing. Note that it then follows from part (ii) of (2.23) that $-\infty < \sup_\Gamma L_\Gamma \le \inf_\Gamma U_\Gamma < +\infty$. It is natural to ask if the existence of $\int_a^b f\,d\phi$ in this case is equivalent to the statement that

(2.28) $$\sup_\Gamma L_\Gamma = \inf_\Gamma U_\Gamma,$$

which we know to be an equivalent definition in the case of Riemann integrals [see (1.20)]. Unfortunately, the answer in general is *no*, as the following example shows. Let $[a,b] = [-1,1]$ and

$$f(x) = \begin{cases} 0 \text{ if } -1 \le x < 0, \\ 1 \text{ if } \quad 0 \le x \le 1, \end{cases}$$

$$\phi(x) = \begin{cases} 0 \text{ if } -1 \le x \le 0. \\ 1 \text{ if } \quad 0 < x \le 1. \end{cases}$$

Since f and ϕ have a common discontinuity, $\int_{-1}^1 f\,d\phi$ does not exist. In fact, if Γ straddles 0, that is, if $x_{i_0-1} < 0 < x_{i_0}$ for some i_0, then $R_\Gamma = f(\xi_{i_0})$ for $x_{i_0-1} \le \xi_{i_0} \le x_{i_0}$. Hence, R_Γ may be 0 or 1, and thus cannot have a limit. On the other hand, it is easy to check that $U_\Gamma = 1$ for any Γ, and that $L_\Gamma = 0$ if Γ straddles 0 and $L_\Gamma = 1$ otherwise. Hence, neither $\lim_{|\Gamma|\to 0} R_\Gamma$ nor $\lim_{|\Gamma|\to 0} L_\Gamma$ exists, but $\sup_\Gamma L_\Gamma = \inf_\Gamma U_\Gamma = 1$.

In the following two theorems, we explore relations between (2.15) and (2.28).

(2.29) Theorem *Let f be bounded and ϕ be monotone increasing on $[a,b]$. If $\int_a^b f\,d\phi$ exists, then $\lim_{|\Gamma|\to 0} L_\Gamma$ and $\lim_{|\Gamma|\to 0} U_\Gamma$ exist, and*

$$\lim_{|\Gamma|\to 0} L_\Gamma = \lim_{|\Gamma|\to 0} U_\Gamma = \sup_\Gamma L_\Gamma = \inf_\Gamma U_\Gamma = \int_a^b f\,d\phi.$$

Proof. Let $I = \int_a^b f\,d\phi$. By hypothesis, given $\varepsilon > 0$, there is a $\delta > 0$ such that $|I - R_\Gamma| < \varepsilon$ for any R_Γ with $|\Gamma| < \delta$. Given $\Gamma = \{x_i\}_{i=0}^m$ with $|\Gamma| < \delta$, choose ξ_i and η_i in $[x_{i-1},x_i]$, $i = 1, \ldots, m$, such that

$$0 \le M_i - f(\xi_i) < \frac{\varepsilon}{\phi(b) - \phi(a)} \quad \text{and} \quad 0 \le f(\eta_i) - m_i < \frac{\varepsilon}{\phi(b) - \phi(a)}.$$

Let $R'_\Gamma = \sum f(\xi_i)[\phi(x_i) - \phi(x_{i-1})]$ and $R''_\Gamma = \sum f(\eta_i)[\phi(x_i) - \phi(x_{i-1})]$. Then $|I - R'_\Gamma| < \varepsilon$, $|I - R''_\Gamma| < \varepsilon$,

$$0 \le U_\Gamma - R'_\Gamma \le \sum \frac{\varepsilon}{\phi(b) - \phi(a)} [\phi(x_i) - \phi(x_{i-1})] = \varepsilon,$$

and

$$0 \le R''_\Gamma - L_\Gamma \le \sum \frac{\varepsilon}{\phi(b) - \phi(a)} [\phi(x_i) - \phi(x_{i-1})] = \varepsilon.$$

Combining inequalities, we obtain

$$|U_\Gamma - I| \le |U_\Gamma - R'_\Gamma| + |R'_\Gamma - I| < \varepsilon + \varepsilon = 2\varepsilon$$

and

$$|L_\Gamma - I| \le |L_\Gamma - R''_\Gamma| + |R''_\Gamma - I| < \varepsilon + \varepsilon = 2\varepsilon.$$

Hence, $\lim_{|\Gamma|\to 0} U_\Gamma = \lim_{|\Gamma|\to 0} L_\Gamma = I$. Since, by (2.23), $L_\Gamma \le \sup_\Gamma L_\Gamma \le \inf_\Gamma U_\Gamma \le U_\Gamma$, the theorem follows.

(2.30) Theorem *Let f be bounded, and let ϕ be monotone increasing and continuous on $[a,b]$. Then $\lim_{|\Gamma|\to 0} L_\Gamma$ and $\lim_{|\Gamma|\to 0} U_\Gamma$ exist, and*

$$\lim_{|\Gamma|\to 0} L_\Gamma = \sup_\Gamma L_\Gamma, \qquad \lim_{|\Gamma|\to 0} U_\Gamma = \inf_\Gamma U_\Gamma.$$

In particular, if in addition $\sup_\Gamma L_\Gamma = \inf_\Gamma U_\Gamma$, then $\int_a^b f\,d\phi$ exists, and

$$\sup_\Gamma L_\Gamma = \inf_\Gamma U_\Gamma = \int_a^b f\,d\phi.$$

Proof. The proof is similar to that of (2.9). It is enough to show that $\lim_{|\Gamma|\to 0} L_\Gamma = \sup_\Gamma L_\Gamma$ and $\lim_{|\Gamma|\to 0} U_\Gamma = \inf_\Gamma U_\Gamma$ since the last assertion of the theorem will then follow by (2.22). We will give the argument for the upper sums; the one for the lower sums is similar. Let $\inf_\Gamma U_\Gamma = U$. Given $\varepsilon > 0$, we must find $\delta > 0$ such that $U_\Gamma < U + \varepsilon$ if $|\Gamma| < \delta$. Choose $\bar{\Gamma} = \{\bar{x}_j\}_{j=0}^k$ such that $U_{\bar{\Gamma}} < U + \frac{1}{2}\varepsilon$, and let $M = \sup_{[a,b]} |f|$. By the uniform continuity of ϕ, there exists $\eta > 0$ such that

$$|\phi(x) - \phi(x')| < \frac{\varepsilon}{4(k+1)M}$$

if $|x - x'| < \eta$.

Now let $\Gamma = \{x_i\}_{i=0}^m$ be any partition for which $|\Gamma| < \eta$ and $|\Gamma| < \min_j (\bar{x}_j - \bar{x}_{j-1})$. It is enough to show that $U_\Gamma < U + \varepsilon$. Write

$$U_\Gamma = \sum_{i=1}^m M_i[\phi(x_i) - \phi(x_{i-1})] = \Sigma' + \Sigma'',$$

where Σ'' is as in the proof of (2.9). Then $U_{\Gamma\cup\bar{\Gamma}} = \Sigma' + \Sigma'''$, where Σ''' is obtained from Σ'' by replacing each of the terms $M_i[\phi(x_i) - \phi(x_{i-1})]$ by

$$\text{(2.31)} \qquad \sup_{[x_{i-1},\bar{x}_j]} f(x)[\phi(\bar{x}_j) - \phi(x_{i-1})] + \sup_{[\bar{x}_j,x_i]} f(x)[\phi(x_i) - \phi(\bar{x}_j)],$$

$\bar{x}_j$ being the point of $\bar{\Gamma}$ in (x_{i-1},x_i). Hence $U_\Gamma - U_{\Gamma\cup\bar{\Gamma}} = \Sigma'' - \Sigma'''$. At least one of $\sup_{[x_{i-1},\bar{x}_j]} f$ and $\sup_{[\bar{x}_j,x_i]} f$ equals M_i. If it is the first, the difference between $M_i[\phi(x_i) - \phi(x_{i-1})]$ and (2.31) is easily seen to be

$$(M_i - \sup_{[\bar{x}_j,x_i]} f)[\phi(x_i) - \phi(\bar{x}_j)].$$

If it is the second, the difference is

$$(M_i - \sup_{[x_{i-1},\bar{x}_j]} f)[\phi(\bar{x}_j) - \phi(x_{i-1})].$$

In either case, the difference is at most $2M\varepsilon/4(k+1)M = \varepsilon/2(k+1)$ in absolute value. Hence, $U_\Gamma - U_{\Gamma\cup\bar{\Gamma}} \le (k+1)\varepsilon/2(k+1) = \frac{1}{2}\varepsilon$. Moreover, $U_{\Gamma\cup\bar{\Gamma}} \le U_{\bar{\Gamma}} < U + \frac{1}{2}\varepsilon$. Therefore, $U_\Gamma < U + \frac{1}{2}\varepsilon + \frac{1}{2}\varepsilon = U + \varepsilon$, and the theorem follows.

Exercises

1. Let $f(x) = x \sin(1/x)$ for $0 < x \le 1$ and $f(0) = 0$. Show that f is bounded and continuous on $[0,1]$, but that $V[f; 0,1] = +\infty$.
2. Prove theorem (2.1).
3. If $[a',b']$ is a subinterval of $[a,b]$, show that $P[a',b'] \le P[a,b]$ and $N[a',b'] \le N[a,b]$.
4. Let $\{f_k\}$ be a sequence of functions of bounded variation on $[a,b]$. If $V[f_k; a,b] \le M < +\infty$ for all k and if $f_k \to f$ pointwise on $[a,b]$, show that f is of bounded variation and that $V[f; a,b] \le M$. Give an example of a convergent sequence of functions of bounded variation whose limit is not of bounded variation.
5. Suppose f is finite on $[a,b]$ and of bounded variation on every interval $[a+\varepsilon, b]$, $\varepsilon > 0$, with $V[f; a+\varepsilon, b] \le M < +\infty$. Show that $V[f; a,b] < +\infty$. Is $V[f; a,b] \le M$? If not, what additional assumption will make it so?
6. Let $f(x) = x^2 \sin(1/x)$ for $0 < x \le 1$ and $f(0) = 0$. Show that $V[f; 0,1] < +\infty$. [Examine the graph of f, or use Exercise 5 and (2.10).]
7. Suppose f is of bounded variation on $[a,b]$. If f is continuous on $[a,b]$, show that $V(x)$, $P(x)$, and $N(x)$ are also continuous. (If $\Gamma = \{x_i\}$, note that $V[x_{i-1},x_i] - |f(x_{i-1}) - f(x_i)| \le V[a,b] - S_\Gamma$.)
8. The main results about functions of bounded variation on a closed interval remain true for open or partly open intervals and for infinite intervals. Prove, for example, that if f is of bounded variation on $(-\infty,+\infty)$, then f is the difference of two increasing bounded functions.
9. Let C be a curve with parametric equations $x = \phi(t)$ and $y = \psi(t)$, $a \le t \le b$.
(a) If ϕ and ψ are of bounded variation and continuous, show that $L = \lim_{|\Gamma|\to 0} l(\Gamma)$.
(b) If ϕ and ψ are continuously differentiable, show that $L = \int_a^b([\phi'(t)]^2 + [\psi'(t)]^2)^{1/2}\,dt$.
10. If $\lambda_1 < \lambda_2 < \cdots < \lambda_m$ is a finite sequence and $-\infty < s < +\infty$, write $\sum_k a_k e^{-s\lambda_k}$ as a Riemann-Stieltjes integral. [Take $f(x) = e^{-sx}$, ϕ to be an appropriate step function, and $[a,b]$ to contain all the λ_k in its interior.]
11. Show that $\int_a^b f\,d\phi$ exists if and only if given $\varepsilon > 0$, there exists $\delta > 0$ such that $|R_\Gamma - R_{\Gamma'}| < \varepsilon$ if $|\Gamma|, |\Gamma'| < \delta$.

12. Prove that the conclusion of (2.30) is valid if the assumption that ϕ is continuous is replaced by the assumption that f and ϕ have no common discontinuities. (Instead of the uniform continuity of ϕ, use the fact that either f or ϕ is continuous at each point $\bar{x}_j$ of $\bar{\Gamma}$.)
13. Prove theorem (2.16).
14. Give an example which shows that for $a < c < b$, $\int_a^c f\, d\phi$ and $\int_c^b f\, d\phi$ may both exist but $\int_a^b f\, d\phi$ may not. Compare (2.17). [Take $[a,b] = [-1,1]$, $c = 0$, and f and ϕ as in the example following (2.28).]
15. Suppose f is continuous and ϕ is of bounded variation on $[a,b]$. Show that the function $\psi(x) = \int_a^x f\, d\phi$ is of bounded variation on $[a,b]$. If g is continuous on $[a,b]$, show that $\int_a^b g\, d\psi = \int_a^b gf\, d\phi$.
16. Suppose that ϕ is of bounded variation on $[a,b]$ and that f is bounded and continuous except for a finite number of jump discontinuities in $[a,b]$. If ϕ is continuous at each discontinuity of f, show that $\int_a^b f\, d\phi$ exists.
17. If ϕ is of bounded variation on $(-\infty,+\infty)$, f is continuous on $(-\infty,+\infty)$, and $\lim_{|x|\to+\infty} f(x) = 0$, show that $\int_{-\infty}^{+\infty} f\, d\phi$ exists.
18. Let $f(z) = \sum_{k=0}^{\infty} a_k z^k$ be a power series. Show that if $\sum |a_k| < +\infty$, then $f(z)$ is of bounded variation on every radius of the circle $|z| = 1$. [If for example the radius is $0 \le x \le 1$ and the a_k are real, then $f(x) = \sum a_k^+ x^k - \sum a_k^- x^k$.]

Chapter 3

Lebesgue Measure and Outer Measure

In this chapter, we will define and study the Lebesgue measure of sets in $\mathbf{R}^n$. This will be the foundation for the theory of integration to be developed later. We shall base the presentation on the notion of the outer measure of a set.

1. Lebesgue Outer Measure; the Cantor Set

We consider closed n-dimensional intervals $I = \{\mathbf{x} : a_j \le x_j \le b_j, j = 1, \dots, n\}$ and their volumes $v(I) = \prod_{j=1}^{n} (b_j - a_j)$. (See p. 7.) To define the outer measure of an arbitrary subset E of $\mathbf{R}^n$, cover E by a *countable* collection S of intervals I_k, and let

$$\sigma(S) = \sum_{I_k \in S} v(I_k).$$

The *Lebesgue outer measure* (*or exterior measure*) *of* E, denoted $|E|_e$, is defined by

$$|E|_e = \inf \sigma(S), \tag{3.1}$$

where the infimum is taken over all such covers S of E. Thus, $0 \le |E|_e \le +\infty$.

(3.2) Theorem *For an interval I, $|I|_e = v(I)$.*

Proof. Since I covers itself, we have $|I|_e \le v(I)$. To show the opposite inequality, suppose that $S = \{I_k\}_{k=1}^{\infty}$ is a cover of I. Given $\varepsilon > 0$, let I_k^* be an interval containing I_k in its interior such that $v(I_k^*) \le (1 + \varepsilon)v(I_k)$. Then $I \subset \bigcup_k (I_k^*)^\circ$, and since I is compact, the Heine-Borel theorem implies there is an integer N such that $I \subset \bigcup_{k=1}^{N} I_k^*$. Clearly, $v(I) \le \sum_{k=1}^{N} v(I_k^*)$. Hence, $v(I) \le (1 + \varepsilon) \sum_{k=1}^{N} v(I_k) \le (1 + \varepsilon)\sigma(S)$. Since ε can be chosen arbitrarily small, it follows that $v(I) \le \sigma(S)$, and therefore, that $v(I) \le |I|_e$. This completes the proof.

Note that the boundary of any interval has outer measure zero.

The following two theorems state simple but basic properties of outer measure.

(3.3) Theorem *If $E_1 \subset E_2$, then $|E_1|_e \le |E_2|_e$.*

The proof follows immediately from the fact that any cover of E_2 is also a cover of E_1.

(3.4) Theorem *If $E = \bigcup E_k$ is a countable union of sets, then $|E|_e \le \sum |E_k|_e$.*

Proof. We may assume that $|E_k|_e < +\infty$ for each $k = 1, 2, \ldots$, since otherwise the conclusion is obvious. Fix $\varepsilon > 0$. Given k, choose intervals $I_j^{(k)}$ such that $E_k \subset \bigcup_j I_j^{(k)}$ and $\sum_j v(I_j^{(k)}) < |E_k|_e + \varepsilon 2^{-k}$. Since $E \subset \bigcup_{j,k} I_j^{(k)}$, we have $|E|_e \le \sum_{j,k} v(I_j^{(k)}) = \sum_k \sum_j v(I_j^{(k)})$. Therefore,

$$|E|_e \le \sum_k (|E_k|_e + \varepsilon 2^{-k}) = \sum_k |E_k|_e + \varepsilon,$$

and the result follows by letting $\varepsilon \to 0$.

We see in particular that any subset of a set with outer measure zero has outer measure zero, and that the countable union of sets with outer measure zero has outer measure zero. Since any set consisting of a single point clearly has outer measure zero, it follows that *any countable subset of* $\mathbf{R}^n$ *has outer measure zero*. For example, the set consisting of all points each of whose coordinates is rational has outer measure zero, even though it is dense in $\mathbf{R}^n$.

There are sets with outer measure zero which are not countable. As an illustration, we will construct a subset of the real line with outer measure zero which is perfect, and therefore uncountable, by (1.9). Variants of the construction and analogues for $\mathbf{R}^n$, $n > 1$, are given in the exercises.

Consider the closed interval [0,1]. The first stage of the construction is to subdivide [0,1] into thirds and remove the interior of the middle third; that is, remove the open interval $(\frac{1}{3},\frac{2}{3})$. Each successive step of the construction is essentially the same. Thus, at the second stage, we subdivide each of the remaining two intervals $[0,\frac{1}{3}]$ and $[\frac{2}{3},1]$ into thirds and remove the interiors, $(\frac{1}{9},\frac{2}{9})$ and $(\frac{7}{9},\frac{8}{9})$, of their middle thirds. We continue the construction for each of the remaining intervals. The sets removed in the first three successive stages are indicated below by darkened intervals:

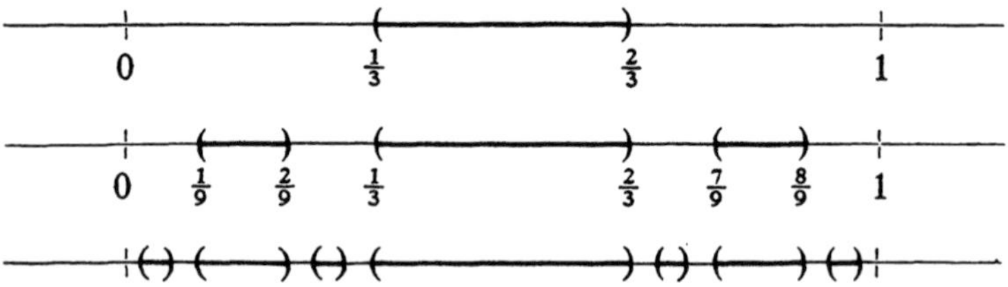

The subset of [0,1] which remains after infinitely many such operations is called the *Cantor set* C: thus, if C_k denotes the union of the intervals left at the kth stage, then

$$C = \bigcap_{k=1}^{\infty} C_k. \tag{3.5}$$

Since each C_k is closed, it follows from theorem (1.7) that C is closed. Note that C_k consists of 2^k closed intervals, each of length 3^{-k}, and that C contains the endpoints of all these intervals. Any point of C belongs to an interval in C_k for every k and is therefore a limit point of the endpoints of the intervals. This proves that C is perfect. Finally, since C is covered by the intervals in any C_k, we have $|C|_e \le 2^k 3^{-k}$ for each k. Therefore, $|C|_e = 0$.

We now introduce a function associated with the Cantor set which will be useful later. If $D_k = [0,1] - C_k$, then D_k consists of the $2^k - 1$ intervals I_j^k (ordered from left to right) removed in the first k stages of construction of the Cantor set. Let f_k be the continuous function on [0,1] which satisfies $f_k(0) = 0$, $f_k(1) = 1$, $f_k(x) = j2^{-k}$ on I_j^k, $j = 1, \ldots, 2^k - 1$, and which is linear on each interval of C_k. The graphs of f_1 and f_2 are shown in the following illustration:

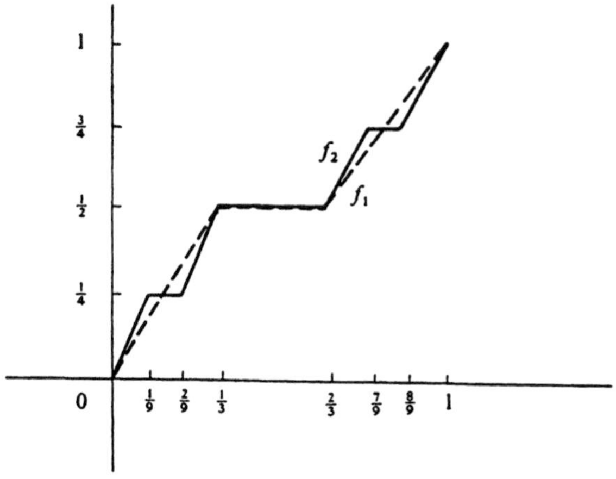

By construction, each f_k is monotone increasing, $f_{k+1} = f_k$ on I_j^k, $j = 1, 2, \ldots, 2^k - 1$, and $|f_k - f_{k+1}| < 2^{-k}$. Hence, $\sum (f_k - f_{k+1})$ converges uniformly on [0,1], and therefore, $\{f_k\}$ converges uniformly on [0,1]. Let $f = \lim_{k\to\infty} f_k$. Then $f(0) = 0$, $f(1) = 1$, f is monotone increasing and continuous on [0,1], and f is constant on every interval removed in constructing C. This f is called the *Cantor-Lebesgue function.*

The next two theorems give useful relations between the outer measure of a set and the outer measures of open sets and G_δ sets (see p. 6) which contain it.

(3.6) Theorem *Let $E \subset \mathbf{R}^n$. Then given $\varepsilon > 0$, there exists an open set G such that $E \subset G$ and $|G|_e \le |E|_e + \varepsilon$. Hence,*

$$|E|_e = \inf |G|_e, \tag{3.7}$$

where the infimum is taken over all open sets G containing E.

Proof. Given $\varepsilon > 0$, choose intervals I_k such that $E \subset \bigcup_{k=1}^\infty I_k$ and $\sum_{k=1}^\infty v(I_k) \le |E|_e + \frac{1}{2}\varepsilon$. Let I_k^* be an interval containing I_k in its interior $(I_k^*)^\circ$ such that $v(I_k^*) \le v(I_k) + \varepsilon 2^{-k-1}$. If $G = \bigcup (I_k^*)^\circ$, then G is open and contains E. Furthermore,

$$|G|_e \le \sum_{k=1}^\infty v(I_k^*) \le \sum_{k=1}^\infty v(I_k) + \varepsilon \sum_{k=1}^\infty 2^{-k-1} \le |E|_e + \tfrac{1}{2}\varepsilon + \tfrac{1}{2}\varepsilon = |E|_e + \varepsilon.$$

This completes the proof.

(3.8) Theorem *If $E \subset \mathbf{R}^n$, there exists a set H of type G_δ such that $E \subset H$ and $|E|_e = |H|_e$.*

Proof. By (3.6), there is for every positive integer k an open set $G_k \supset E$ such that $|G_k|_e \le |E|_e + 1/k$. If $H = \bigcap_{k=1}^\infty G_k$, then H is of type G_δ and $H \supset E$. Moreover, for every k, $|E|_e \le |H|_e \le |G_k|_e \le |E|_e + 1/k$. Thus, $|E|_e = |H|_e$.

The significance of (3.8) is that the most general set in $\mathbf{R}^n$ can be included in a set of relatively simple type, namely G_δ, with the same outer measure.

In defining the notion of outer measure, we used intervals I with edges parallel to the coordinate axes. The question arises whether the outer measure of a set depends on the position of the (orthogonal) coordinate axes. The answer is *no*, and to prove this we will simultaneously consider the usual coordinate system in $\mathbf{R}^n$ and a fixed rotation of this system. Notions pertaining to the rotated system will be denoted by primes. Thus, I' denotes an interval with edges parallel to the rotated coordinate axes, and $|E|_e'$ denotes the outer measure of a subset E relative to these rotated intervals:

$$|E|_e' = \inf \sum v(I_k'), \tag{3.9}$$

where the infimum is taken over all coverings of E by rotated intervals I_k'. The volume of an interval is clearly unchanged by rotation. (See p. 7.)

(3.10) Theorem $|E|_e' = |E|_e$ *for every $E \subset \mathbf{R}^n$.*

Proof. We first claim that given I' and $\varepsilon > 0$, there exist $\{I_l\}$ such that $I' \subset \bigcup I_l$ and $\sum v(I_l) \le v(I') + \varepsilon$. To see this, let I_1' be an interval containing I' in its interior such that $v(I_1') \le v(I') + \varepsilon$. By theorem (1.11), the interior of

I_1' can be written as a union of nonoverlapping intervals I_l. Hence, $I' \subset \bigcup I_l$. Moreover, since the I_l are nonoverlapping and $\bigcup_{l=1}^{N} I_l \subset I_1'$ for every positive integer N, we have $\sum_{l=1}^{N} v(I_l) \leq v(I_1')$. Therefore, $\sum_{l=1}^{\infty} v(I_l) \leq v(I_1') \leq v(I') + \varepsilon$, which proves the claim. A parallel result is that given I and $\varepsilon > 0$, there exist $\{I_l'\}$ such that $I \subset \bigcup I_l'$ and $\sum v(I_l') \leq v(I) + \varepsilon$.

Let E be any subset of $\mathbf{R}^n$. Given ε, choose $\{I_k\}_{k=1}^{\infty}$ such that $E \subset \bigcup I_k$ and $\sum v(I_k) \leq |E|_e + \frac{1}{2}\varepsilon$. For each k, choose $\{I_{k,l}'\}$ such that $I_k \subset \bigcup_l I_{k,l}'$ and $\sum_l v(I_{k,l}') \leq v(I_k) + \varepsilon 2^{-k-1}$. Thus,

$$\sum_{k,l} v(I_{k,l}') \leq \sum_k v(I_k) + \tfrac{1}{2}\varepsilon \leq |E|_e + \varepsilon.$$

Since $E \subset \bigcup_{k,l} I_{k,l}'$, we obtain $|E|_e' \leq |E|_e + \varepsilon$. Hence, $|E|_e' \leq |E|_e$, and by symmetry, $|E|_e' = |E|_e$.

For a related result, see Exercise 22.

2. Lebesgue Measurable Sets

A subset E of $\mathbf{R}^n$ is said to be *Lebesgue measurable*, or simply *measurable*, if given $\varepsilon > 0$, there exists an open set G such that

$$E \subset G \text{ and } |G - E|_e < \varepsilon.$$

If E is measurable, its outer measure is called its *Lebesgue measure*, or simply its *measure*, and denoted $|E|$:

(3.11) $$|E| = |E|_e, \text{ for measurable } E.$$

The condition that E be measurable should not be confused with the conclusion of (3.6), which states that there exists an open G containing E such that $|G|_e \leq |E|_e + \varepsilon$. In general, since $G = E \cup (G - E)$ when $E \subset G$, we only have $|G|_e \leq |E|_e + |G - E|_e$, and we cannot conclude from $|G|_e \leq |E|_e + \varepsilon$ that $|G - E|_e < \varepsilon$.

We now list two simple examples of measurable sets. A nonmeasurable set will be constructed in (3.38).

Example 1. Every open set is measurable.

This is immediate from the definition.

Example 2. Every set of outer measure zero is measurable.

For suppose that $|E|_e = 0$. Then given $\varepsilon > 0$, there is by (3.6) an open G containing E with $|G| < |E|_e + \varepsilon = \varepsilon$. Hence,

$$|G - E|_e \leq |G| < \varepsilon.$$

(3.12) Theorem *The union $E = \bigcup E_k$ of a countable number of measurable sets is measurable, and*

$$|E| \leq \sum |E_k|.$$

Proof. Let $\varepsilon > 0$. For each $k = 1, 2, \ldots$, choose an open set G_k such that $E_k \subset G_k$ and $|G_k - E_k|_e < \varepsilon 2^{-k}$. Then $G = \bigcup G_k$ is open and $E \subset G$. Moreover, since $G - E \subset \bigcup (G_k - E_k)$, we have

$$|G - E|_e \leq |\bigcup (G_k - E_k)|_e \leq \sum |G_k - E_k|_e < \varepsilon.$$

This proves that E is measurable. The fact that $|E| \leq \sum |E_k|$ follows from (3.4).

(3.13) Corollary *An interval I is measurable, and $|I| = v(I)$.*

Proof. I is the union of its interior and its boundary. Since its boundary has measure zero, the fact that I is measurable follows from (3.12) and the results of Examples 1 and 2. Theorem (3.2) shows that $|I| = v(I)$.

(3.14) Theorem *Every closed set is measurable.*

In order to prove this, we will use (1.11) and the next two lemmas.

(3.15) Lemma *If $\{I_k\}_{k=1}^N$ is a finite collection of nonoverlapping intervals, then $\bigcup I_k$ is measurable and $|\bigcup I_k| = \sum |I_k|$.*

Proof. The fact that $\bigcup I_k$ is measurable follows from (3.13). The rest of the lemma is a minor extension of (3.2), and its proof is left as an exercise.

We recall from Chapter 1 that the distance between two sets E_1 and E_2 is defined as $d(E_1,E_2) = \inf \{|x_1 - x_2| : x_1 \in E_1, x_2 \in E_2\}$.

(3.16) Lemma *If $d(E_1,E_2) > 0$, then $|E_1 \cup E_2|_e = |E_1|_e + |E_2|_e$.*

Proof. By (3.4), $|E_1 \cup E_2|_e \leq |E_1|_e + |E_2|_e$. To prove the opposite inequality, suppose $\varepsilon > 0$, and choose intervals $\{I_k\}$ such that $E_1 \cup E_2 \subset \bigcup I_k$ and $\sum |I_k| \leq |E_1 \cup E_2|_e + \varepsilon$. We may assume that the diameter of each I_k is less than $d(E_1,E_2)$. (Otherwise, we divide each I_k into a finite number of subintervals with this property.) Hence, $\{I_k\}$ splits into two subsequences $\{I_k'\}$ and $\{I_k''\}$, the first of which covers E_1 and the second, E_2. Clearly,

$$|E_1|_e + |E_2|_e \leq \sum |I_k'| + \sum |I_k''| = \sum |I_k| \leq |E_1 \cup E_2|_e + \varepsilon.$$

Therefore, $|E_1|_e + |E_2|_e \leq |E_1 \cup E_2|_e$, which completes the proof.

A special case of this result will be used in the proof of (3.14). If E_1 and E_2 are compact and disjoint, then $d(E_1,E_2) > 0$ by Exercise 1(l), p. 13; therefore, $|E_1 \cup E_2|_e = |E_1|_e + |E_2|_e$ if E_1 and E_2 are compact and disjoint.

Proof of theorem (3.14). Suppose first that F is compact. Given $\varepsilon > 0$, choose an open set G such that $F \subset G$ and $|G| < |F|_e + \varepsilon$. Since $G - F$ is open, theorem (1.11) implies there are nonoverlapping closed intervals I_k, $k = 1, 2, \ldots$, such that $G - F = \bigcup I_k$. Therefore, $|G - F|_e \leq \sum |I_k|$, and it suffices to show that $\sum |I_k| \le \varepsilon$. We have $G = F \cup (\bigcup_{k=1}^{\infty} I_k) \supset F \cup (\bigcup_{k=1}^{N} I_k)$ for every positive integer N. Therefore,

$$|G| \ge \left| F \cup \left(\bigcup_{k=1}^{N} I_k \right) \right|_e = |F|_e + \left| \bigcup_{k=1}^{N} I_k \right|_e$$

by (3.16), F and $\bigcup_{k=1}^{N} I_k$ being disjoint and compact. Since $|\bigcup_{k=1}^{N} I_k|_e = \sum_{k=1}^{N} |I_k|$ by (3.15), we obtain $\sum_{k=1}^{N} |I_k| \le |G| - |F|_e < \varepsilon$ for every N, so that $\sum_{k=1}^{\infty} |I_k| \le \varepsilon$, as desired. This proves the result in the case where F is compact.

To complete the proof, let F be any closed subset of $\mathbf{R}^n$ and write $F = \bigcup F_k$, where $F_k = F \cap \{\mathbf{x} : |\mathbf{x}| \le k\}$, $k = 1, 2, \ldots$. Each F_k is compact and, therefore, measurable. Hence, F is measurable by (3.12).

(3.17) Theorem *The complement of a measurable set is measurable.*

Proof. Let E be measurable. For each positive integer k, choose an open set G_k such that $E \subset G_k$ and $|G_k - E|_e < 1/k$. Since CG_k is closed, it is measurable by (3.14). Let $H = \bigcup_k CG_k$. Then H is measurable and $H \subset CE$. Write $CE = H \cup Z$, where $Z = CE - H$. Then $Z \subset CE - CG_k = G_k - E$, and therefore, $|Z|_e < 1/k$ for every k. Hence, $|Z|_e = 0$ and, in particular, Z is measurable. Thus, CE is measurable since it is the union of two measurable sets.

The following two theorems are corollaries of (3.12) and (3.17).

(3.18) Theorem *The intersection $E = \bigcap_k E_k$ of a countable number of measurable sets is measurable.*

Proof. Since E_k is measurable, CE_k is measurable by (3.17). However, $CE = C(\bigcap_k E_k) = \bigcup_k CE_k$, and hence, CE is measurable. Therefore, by another application of (3.17), E is measurable.

(3.19) Theorem *If E_1 and E_2 are measurable, then $E_1 - E_2$ is measurable.*

Proof. Since $E_1 - E_2 = E_1 \cap CE_2$, the result follows from (3.17) and (3.18).

As a consequence of (3.12) and (3.17) and their corollaries (3.18) and (3.19), it follows that the class of measurable subsets of $\mathbf{R}^n$ is closed under the set-theoretic operations of taking complements, countable unions, and countable intersections. Such a class of sets is called a σ-algebra; that is, a nonempty collection Σ of sets E is called a *σ-algebra of sets* if it satisfies the following two conditions:

(i) $CE \in \Sigma$ if $E \in \Sigma$.
(ii) $\bigcup_k E_k \in \Sigma$ if $E_k \in \Sigma$, $k = 1, 2, \ldots$.

Note that it follows from the relation $C(\bigcap_k E_k) = \bigcup_k CE_k$ that $\bigcap_k E_k \in \Sigma$ if Σ is a σ-algebra and $E_k \in \Sigma$, $k = 1, 2, \ldots$. Moreover, it is easy to see that the entire space and the empty set belong to any σ-algebra.

The following result is just a reformulation of (3.12) and (3.17).

(3.20) Theorem *The collection of measurable subsets of* $\mathbf{R}^n$ *is a* σ*-algebra.*

Note, for example, that if E_k, $k = 1, 2, \ldots$, are measurable, then so are $\limsup E_k$ and $\liminf E_k$ since

$$\limsup E_k = \bigcap_{j=1}^{\infty} \bigcup_{k=j}^{\infty} E_k \quad \text{and} \quad \liminf E_k = \bigcup_{j=1}^{\infty} \bigcap_{k=j}^{\infty} E_k.$$

If $\mathscr{C}_1$ and $\mathscr{C}_2$ are two collections of sets, we say that $\mathscr{C}_1$ *is contained in* $\mathscr{C}_2$ if every set in $\mathscr{C}_1$ is also in $\mathscr{C}_2$. If $\mathscr{F}$ is a family of σ-algebras Σ, we define $\bigcap_{\Sigma \in \mathscr{F}} \Sigma$ to be the collection of all sets E which belong to every Σ in $\mathscr{F}$. It is easy to check that $\bigcap_{\Sigma \in \mathscr{F}} \Sigma$ is itself a σ-algebra, and is contained in every Σ in $\mathscr{F}$.

Given a collection $\mathscr{C}$ of subsets of $\mathbf{R}^n$, consider the family $\mathscr{F}$ of all σ-algebras which contain $\mathscr{C}$, and let $\mathscr{E} = \bigcap_{\Sigma \in \mathscr{F}} \Sigma$. Then $\mathscr{E}$ is the *smallest* σ*-algebra containing* $\mathscr{C}$; that is, $\mathscr{E}$ is a σ-algebra containing $\mathscr{C}$, and if Σ is any other σ-algebra containing $\mathscr{C}$, then Σ contains $\mathscr{E}$.

The smallest σ-algebra of subsets of $\mathbf{R}^n$ containing all the open subsets of $\mathbf{R}^n$ is called the *Borel* σ*-algebra* $\mathscr{B}$ of $\mathbf{R}^n$, and the sets in $\mathscr{B}$ are called *Borel subsets of* $\mathbf{R}^n$. Sets of type F_σ, G_δ, $F_{\sigma\delta}$, $G_{\delta\sigma}$ (see p. 6), etc., are Borel sets.

(3.21) Theorem *Every Borel set is measurable.*

Proof. Let $\mathscr{M}$ be the collection of measurable subsets of $\mathbf{R}^n$. By (3.20), $\mathscr{M}$ is a σ-algebra. Since every open set belongs to $\mathscr{M}$, and $\mathscr{B}$ is the smallest σ-algebra containing the open sets, $\mathscr{B}$ is contained in $\mathscr{M}$.

3. Two Properties of Lebesgue Measure

The next two theorems give properties of Lebesgue measure which are of fundamental importance. To prove them, we first need the following characterization of measurability in terms of closed sets.

(3.22) Lemma *A set* E *in* $\mathbf{R}^n$ *is measurable if and only if given* $\varepsilon > 0$, *there exists a closed set* $F \subset E$ *such that* $|E - F|_e < \varepsilon$.

Proof. E is measurable if and only if CE is measurable, that is, if and only if given $\varepsilon > 0$, there exists an open G such that $CE \subset G$ and $|G - CE|_e < \varepsilon$. Such G exists if and only if the set $F = CG$ is closed, $F \subset E$, and $|E - F|_e < \varepsilon$ (since $G - CE = E - F$).

(3.23) Theorem *If $\{E_k\}$ is a countable collection of disjoint measurable sets, then*

$$\Big|\bigcup_k E_k\Big| = \sum_k |E_k|.$$

Proof. First, suppose each E_k is bounded. Given $\varepsilon > 0$ and $k = 1, 2, \dots$, use (3.22) to choose a closed $F_k \subset E_k$ with $|E_k - F_k| < \varepsilon 2^{-k}$. Then $|E_k| \le |F_k| + \varepsilon 2^{-k}$ by (3.12). Since the E_k are bounded and disjoint, the F_k are compact and disjoint. Therefore, by (3.16), $|\bigcup_{k=1}^m F_k| = \sum_{k=1}^m |F_k|$ for every m. The fact that $\bigcup_{k=1}^m F_k \subset \bigcup_{k=1}^\infty E_k$ then implies $\sum_{k=1}^m |F_k| \le |\bigcup_{k=1}^\infty E_k|$. Hence,

$$\Big|\bigcup_{k=1}^{\infty} E_k\Big| \ge \sum_{k=1}^{\infty} |F_k| \ge \sum_{k=1}^{\infty} (|E_k| - \varepsilon 2^{-k}) = \sum_{k=1}^{\infty} |E_k| - \varepsilon,$$

so that $|\bigcup_{k=1}^\infty E_k| \ge \sum_{k=1}^\infty |E_k|$. Since the opposite inequality is always true, the theorem follows in this case.

For the general case, let I_j, $j = 1, 2, \dots$, be a sequence of intervals increasing to $\mathbf{R}^n$, and define $S_1 = I_1$ and $S_j = I_j - I_{j-1}$ for $j \ge 2$. Then the sets $E_{k,j} = E_k \cap S_j$, $k,j = 1, 2, \dots$, are bounded, disjoint, and measurable; $E_k = \bigcup_j E_{k,j}$ and $\bigcup_k E_k = \bigcup_{k,j} E_{k,j}$. Therefore, by the case already established, we have

$$\Big|\bigcup_k E_k\Big| = \Big|\bigcup_{k,j} E_{k,j}\Big| = \sum_{k,j} |E_{k,j}| = \sum_k \Big(\sum_j |E_{k,j}|\Big) = \sum_k |E_k|.$$

(3.24) Corollary *If $\{I_k\}$ is a sequence of nonoverlapping intervals, then $|\bigcup I_k| = \sum |I_k|$.*

Proof. It is clear that $|\bigcup I_k| \le \sum |I_k|$. On the other hand, the $\mathring{I}_k$ being disjoint, we have $|\bigcup I_k| \ge |\bigcup \mathring{I}_k| = \sum |\mathring{I}_k| = \sum |I_k|$. Thus, $|\bigcup I_k| = \sum |I_k|$.

(3.25) Corollary *Suppose E_1 and E_2 are measurable, $E_2 \subset E_1$, and $|E_2| < +\infty$. Then $|E_1 - E_2| = |E_1| - |E_2|$.*

Proof. Since $E_1 = E_2 \cup (E_1 - E_2)$, $|E_1| = |E_2| + |E_1 - E_2|$ by (3.23). Since $|E_2| < +\infty$, the corollary follows.

The second basic property of Lebesgue measure concerns its behavior for monotone sequences of sets.

(3.26) Theorem *Let $\{E_k\}_{k=1}^\infty$ be a sequence of measurable sets.*

(i) *If $E_k \nearrow E$, then* $\lim_{k\to\infty} |E_k| = |E|$.
(ii) *If $E_k \searrow E$ and $|E_k| < +\infty$ for some k, then* $\lim_{k\to\infty} |E_k| = |E|$.

Proof. (i) We may assume that $|E_k| < +\infty$ for all k; otherwise both $\lim_{k\to\infty} |E_k|$ and $|E|$ are infinite. Write

$$E = E_1 \cup (E_2 - E_1) \cup \cdots \cup (E_k - E_{k-1}) \cup \cdots.$$

Since the terms in this union are measurable and disjoint, we have by (3.23) that

$$|E| = |E_1| + |E_2 - E_1| + \cdots + |E_k - E_{k-1}| + \cdots.$$

By (3.25),

$$|E| = |E_1| + (|E_2| - |E_1|) + \cdots + (|E_k| - |E_{k-1}|) + \cdots = \lim_{k\to\infty} |E_k|.$$

(ii) We may clearly assume that $|E_1| < +\infty$. Write

$$E_1 = E \cup (E_1 - E_2) \cup \cdots \cup (E_k - E_{k+1}) \cup \cdots.$$

Since the terms on the right are disjoint measurable sets, and since each E_k has finite measure, we have

$$\begin{aligned} |E_1| &= |E| + (|E_1| - |E_2|) + \cdots + (|E_k| - |E_{k+1}|) + \cdots \\ &= |E| + |E_1| - \lim_{k\to\infty} |E_k|. \end{aligned}$$

Therefore, $|E| = \lim_{k\to\infty} |E_k|$, which completes the proof.

The restriction in (ii) that $|E_k| < +\infty$ for some k is necessary, as the following example shows. Let E_k be the complement of the ball with center 0 and radius k. Then $|E_k| = +\infty$ for all k and $E_k \searrow \emptyset$, the empty set. Therefore, $\lim_{k\to\infty} |E_k| = +\infty$, while $|\emptyset| = 0$.

Although we are interested almost exclusively in measurable sets, proving the measurability of a given set is occasionally difficult in practice, and it may be desirable to apply theorems about outer measure. A particularly useful result is the following modification of part (i) of (3.26). (See also Exercise 21.)

(3.27) Theorem *If $E_k \nearrow E$, then $\lim_{k\to\infty} |E_k|_e = |E|_e$.*

Proof. For each k, let H_k be a measurable set (e.g., a set of type G_δ) such that $E_k \subset H_k$ and $|H_k| = |E_k|_e$. For $m = 1,2,\ldots$, let $V_m = \bigcap_{k=m}^{\infty} H_k$. Since the V_m are measurable and increase to $V = \bigcup V_m$, it follows from (3.26) that $\lim_{m\to\infty} |V_m| = |V|$. Since $E_m \subset V_m \subset H_m$, we have $|E_m|_e \le |V_m| \le |H_m| = |E_m|_e$. Hence, $|V_m| = |E_m|_e$, $\lim_{m\to\infty} |E_m|_e$ exists, and $\lim_{m\to\infty} |E_m|_e = |V|$. However, $V = \bigcup V_m \supset \bigcup E_m = E$, and therefore, $\lim_{m\to\infty} |E_m|_e \ge |E|_e$. The opposite inequality is obvious since $E_m \subset E$, and the theorem follows.

4. Characterizations of Measurability

Lemma (3.22) was a characterization of measurability in terms of closed subsets of a set. The next three theorems give some other characterizations. The first one states that the most general measurable set differs from a Borel set by a set of measure zero.

(3.28) Theorem

(i) *E is measurable if and only if $E = H - Z$, where H is of type G_δ and $|Z| = 0$.*

(ii) *E is measurable if and only if $E = H \cup Z$, where H is of type F_σ and $|Z| = 0$.*

Proof. If E has the representation expressed in either (i) or (ii), it is measurable since H and Z are.

Conversely, to prove the necessity in (i), suppose that E is measurable. For each $k = 1, 2, \ldots$, choose an open set G_k such that $E \subset G_k$ and $|G_k - E| < 1/k$. Set $H = \bigcap_k G_k$. Then H is of type G_δ, $E \subset H$, and $H - E \subset G_k - E$ for every k. Hence, $|H - E| = 0$, and (i) is proved.

The necessity of (ii) follows from that of (i) by taking complements: if E is measurable, so is CE, and therefore $CE = \bigcap G_k - Z$, where the G_k are open and $|Z| = 0$. Hence, $E = (\bigcup CG_k) \cup Z$, which completes the proof.

(3.29) Theorem *Suppose that $|E|_e < +\infty$. Then E is measurable if and only if given $\varepsilon > 0$, $E = (S \cup N_1) - N_2$, where S is a finite union of non-overlapping intervals and $|N_1|_e, |N_2|_e < \varepsilon$.*

The proof is left as an exercise.

Our final characterization of measurability states that the measurable sets are those which split every set (measurable or not) into pieces that are additive with respect to outer measure. This characterization will be used in Chapter 11 to construct measures in abstract spaces.

(3.30) Theorem (*Carathéodory*) *E is measurable if and only if for every set A*

$$|A|_e = |A \cap E|_e + |A - E|_e.$$

Proof. Suppose that E is measurable. Given A, let H be a set of type G_δ such that $A \subset H$ and $|A|_e = |H|$. Since $H = (H \cap E) \cup (H - E)$, and since $H \cap E$ and $H - E$ are measurable and disjoint, $|H| = |H \cap E| + |H - E|$. Therefore, $|A|_e = |H \cap E| + |H - E| \geq |A \cap E|_e + |A - E|_e$. Since the opposite inequality is clearly true, we obtain $|A|_e = |A \cap E|_e + |A - E|_e$.

Conversely, suppose that E satisfies the stated condition for every A. In case $|E|_e < +\infty$, choose a G_δ set H such that $E \subset H$ and $|H| = |E|_e$. Then $H = E \cup (H - E)$, and by hypothesis, $|H| = |E|_e + |H - E|_e$. Therefore, $|H - E|_e = 0$; so the set $Z = H - E$ is measurable, and consequently, E is measurable.

In case $|E|_e = +\infty$, let B_k be the ball with center 0 and radius k, $k = 1, 2, \ldots$, and let $E_k = E \cap B_k$. Then each E_k has finite outer measure and $E = \bigcup E_k$. Let H_k be a set of type G_δ containing E_k with $|H_k| = |E_k|_e$. By hypothesis, $|H_k| = |H_k \cap E|_e + |H_k - E|_e \geq |E_k|_e + |H_k - E|_e$. Therefore, $|H_k - E| = 0$. Let $H = \bigcup H_k$. Then H is measurable, $E \subset H$, and $H - E = \bigcup (H_k - E)$. In particular, $H - E$ has measure zero, and since $E = H - (H - E)$, E is measurable. This completes the proof.

As a special case of Carathéodory's theorem, we obtain the following result.

(3.31) Corollary *If E is a measurable subset of a set A, then $|A|_e = |E| + |A - E|_e$. Hence, if $|E| < +\infty$, $|A - E|_e = |A|_e - |E|$.*

We can now prove a stronger version of (3.8).

(3.32) Theorem *Let E be a subset of $\mathbf{R}^n$. There exists a set H of type G_δ such that $E \subset H$ and for any* measurable *set M, $|E \cap M|_e = |H \cap M|$.*

If $M = \mathbf{R}^n$, this reduces to (3.8).

Proof. Consider first the case when $|E|_e < +\infty$. Let H be a set of type G_δ such that $E \subset H$ and $|E|_e = |H|$. If M is measurable, then by Carathéodory's theorem $|E|_e = |E \cap M|_e + |E - M|_e$ and $|H| = |H \cap M| + |H - M|$. Therefore,

$$|E \cap M|_e + |E - M|_e = |H \cap M| + |H - M|.$$

Since all these terms are finite, and since the inclusion $E - M \subset H - M$ implies that $|E - M|_e \le |H - M|$, we have $|E \cap M|_e \ge |H \cap M|$. The opposite inequality is also true since $E \cap M \subset H \cap M$, and the theorem follows in this case.

In case $|E|_e = +\infty$, we write $E = \bigcup E_k$ with $|E_k|_e < +\infty$ and $E_k \nearrow E$. For example, E_k could be the intersection of E with the ball of radius k centered at the origin. By the case already considered, there is a set U_k of type G_δ such that $E_k \subset U_k$ and $|E_k \cap M|_e = |U_k \cap M|$ for any measurable M. Let $H_k = \bigcap_{m=k}^{\infty} U_m$. Then H_k is measurable, $H_k \nearrow H = \bigcup H_k$, and $E_k \subset H_k \subset U_k$. Hence, $E_k \cap M \subset H_k \cap M \subset U_k \cap M$, and therefore, $|E_k \cap M|_e = |H_k \cap M|$ for measurable M. Since $E_k \nearrow E$ and $H_k \nearrow H$, we have $E_k \cap M \nearrow E \cap M$ and $H_k \cap M \nearrow H \cap M$. By (3.27), $|E \cap M|_e = |H \cap M|$ for measurable M.

Note that our set H is of type $G_{\delta\sigma}$. To obtain a set of type G_δ, use (3.28) to write $H = H_1 - Z$, where H_1 is of type G_δ and $|Z| = 0$. Then $E \subset H_1$. Moreover, $H_1 \cap M = (H \cap M) \cup (Z \cap M)$, and since $|Z| = 0$, we have $|H_1 \cap M| = |H \cap M|$, so that $|E \cap M|_e = |H_1 \cap M|$. This completes the proof.

5. Lipschitz Transformations of $\mathbf{R}^n$

Theorem (3.10) showed that the notion of outer measure is independent of the orientation of the coordinate axes. Since measurability and measure are defined in terms of outer measure, it follows that these too are independent of rotation of the axes. We wish to study the effect of other transformations of $\mathbf{R}^n$ on the class of measurable sets; that is, we seek a condition on a transformation $T : \mathbf{R}^n \to \mathbf{R}^n$ such that the image $TE = \{\mathbf{y} : \mathbf{y} = T\mathbf{x}, \mathbf{x} \in E\}$ of every measurable set E is measurable. We note that a continuous transformation may not preserve measurability: see Exercise 17.

A transformation $\mathbf{y} = T\mathbf{x}$ of $\mathbf{R}^n$ into itself is called a *Lipschitz transformation* if there is a constant c such that

$$|T\mathbf{x} - T\mathbf{x}'| \le c|\mathbf{x} - \mathbf{x}'|.$$

If $y_j = f_j(\mathbf{x})$, $j = 1, \ldots, n$, are the coordinate functions representing T (see p. 11), it follows that T is Lipschitz if and only if each f_j satisfies a Lipschitz condition $|f_j(\mathbf{x}) - f_j(\mathbf{x}')| \leq c_j|\mathbf{x} - \mathbf{x}'|$. For example, a linear transformation of $\mathbf{R}^n$ is clearly a Lipschitz transformation. More generally, a mapping $y_j = f_j(\mathbf{x})$, $j = 1, \ldots, n$, for which each f_j has bounded first partial derivatives in $\mathbf{R}^n$ is a Lipschitz mapping.

(3.33) Theorem *If* $\mathbf{y} = T\mathbf{x}$ *is a Lipschitz transformation of* $\mathbf{R}^n$, *then* T *maps measurable sets into measurable sets.*

Proof. We will first show that a continuous transformation sends sets of type F_σ into sets of type F_σ. A continuous T maps compact sets into compact sets by (1.17); therefore, since any closed set can be written as a countable union of compact sets, T maps closed sets into sets of type F_σ. [Here, we have used the relation $T(\bigcup E_k) = \bigcup TE_k$, which holds for any T and $\{E_k\}$.] It follows that T preserves the class of F_σ sets.

We will next show that a Lipschitz transformation T sends sets of measure zero into sets of measure zero. Since $|T\mathbf{x} - T\mathbf{x}'| \leq c|\mathbf{x} - \mathbf{x}'|$, the image of a set with diameter of d has diameter at most cd. It follows by covering with cubes that there is a constant c' such that $|TI| \leqslant c'|I|$ for any interval I. (TI is measurable since I is closed.) If $|Z| = 0$ and $\varepsilon > 0$, choose intervals $\{I_k\}$ covering Z such that $\sum |I_k| < \varepsilon$. Since $TZ \subset \bigcup TI_k$, we have $|TZ|_e \leqslant \sum |TI_k| \leqslant c' \sum |I_k| < c'\varepsilon$. Hence, $|TZ| = 0$.

If E is a measurable set, we use (3.28) to write $E = H \cup Z$, where H is of type F_σ and $|Z| = 0$. Since $TE = TH \cup TZ$, TE is measurable as the union of measurable sets. This completes the proof.

For an extension of (3.33) when $n = 1$, see Exercise 10 of Chapter 7.

In the special case that T is linear, we will derive a formula for the measure of TE. It may be helpful to think of T as an $n \times n$ matrix and of $T\mathbf{x}$ as the vector resulting from the matrix action of T on $\mathbf{x}$. If P is a parallelepiped (see p. 7), arguments like those used in proving (3.2) and (3.10) show that

$$|P| = v(P) \tag{3.34}$$

(see Exercise 16). Hence, by p. 7, $|P|$ is the absolute value of the $n \times n$ determinant whose rows are the edges of P.

(3.35) Theorem *Let* T *be a linear transformation of* $\mathbf{R}^n$, *and let* E *be measurable. Then* $|TE| = \delta|E|$, *where* δ *is the absolute value of the determinant of* T.

Proof. Let $\delta = |\det T|$. Then for any interval I, $|TI| = \delta|I|$ by p. 7. If E is any subset of $\mathbf{R}^n$ and $\varepsilon > 0$, choose intervals $\{I_k\}$ covering E with $\sum |I_k| < |E|_e + \varepsilon$. Then $|TE|_e \leq \sum |TI_k| = \delta \sum |I_k| < \delta(|E|_e + \varepsilon)$. Therefore,

$$|TE|_e \leq \delta|E|_e. \tag{3.36}$$

To see that $|TE| = \delta|E|$ for measurable E, choose an open set G containing E with $|G - E| < \varepsilon$. Using (1.11), write G as a union of nonoverlapping intervals $\{I_k\}$. By (3.36), we may assume $\delta > 0$. Then since the TI_k are nonoverlapping parallelepipeds, $|TG| = \sum |TI_k| = \delta \sum |I_k| = \delta|G|$. Now, since $E \subset G$, we have $\delta|E| \le \delta|G| = |TG|$. Also,

$$|TG| \le |TE| + |T(G - E)| \le |TE| + \delta\varepsilon,$$

by (3.36). Combining inequalities, we get $\delta|E| \le |TE| + \delta\varepsilon$, so that $\delta|E| \le |TE|$. Hence, by (3.36), $\delta|E| = |TE|$.

6. A Nonmeasurable Set

We will now construct a nonmeasurable subset of $\mathbf{R}^1$; the construction in $\mathbf{R}^n$, $n > 1$, is similar. We will need the axiom of choice in the following form.

Zermelo's Axiom: *Consider a family of arbitrary nonempty disjoint sets indexed by a set A, $\{E_\alpha : \alpha \in A\}$. Then there exists a set consisting of exactly one element from each E_α, $\alpha \in A$.*

We also need the following lemma.

(3.37) Lemma *Let E be a measurable subset of $\mathbf{R}^1$ with $|E| > 0$. Then the set of differences $\{d : d = x - y, x \in E, y \in E\}$ contains an interval centered at the origin.*

Proof. Given $\varepsilon > 0$, there exists an open set G such that $E \subset G$ and $|G| < (1 + \varepsilon)|E|$. By (1.11), G can be written as a union of nonoverlapping intervals, $G = \bigcup I_k$. Letting $E_k = E \cap I_k$, it follows that $E = \bigcup E_k$, that the E_k are measurable, and that two different E_k's have at most one point in common. Therefore, $|G| = \sum |I_k|$ and $|E| = \sum |E_k|$. Since $|G| < (1 + \varepsilon)|E|$, we must have $|I_{k_0}| < (1 + \varepsilon)|E_{k_0}|$ for some k_0. Choosing $\varepsilon = \frac{1}{3}$ and letting I and $\mathscr{E}$ denote the sets I_{k_0} and E_{k_0}, respectively, we have $\mathscr{E} \subset I$ and $|\mathscr{E}| > \frac{3}{4}|I|$. We claim that if $\mathscr{E}$ is translated by any number d satisfying $|d| < \frac{1}{2}|I|$, the translated set $\mathscr{E}_d$ has points in common with $\mathscr{E}$. Otherwise, since $\mathscr{E} \cup \mathscr{E}_d$ is contained in an interval of length $|I| + |d|$, we would have $|\mathscr{E}| + |\mathscr{E}_d| \le |I| + |d|$, or $2|\mathscr{E}| \le |I| + |d|$. [Here, we have used the fact that $|E_d| = |E|$ (see Exercise 18).] However, the last inequality is false if $|d| < \frac{1}{2}|I|$ since $|\mathscr{E}| > \frac{3}{4}|I|$. This proves the claim, and thus the lemma.

(3.38) Theorem (*Vitali*) *There exist nonmeasurable sets.*

Proof. We define an equivalence relation on the real line by saying x and y are equivalent if $x - y$ is rational. The equivalence classes then have the form $E_x = \{x + r : r \text{ is rational}\}$. Two classes E_x and E_y are either identical or disjoint; therefore, one equivalence class consists of all the rational numbers, and the other distinct classes are disjoint sets of irrational numbers.

The number of distinct equivalence classes is uncountable since each class is countable but the union of all the classes is uncountable (this union being the entire line).

Using Zermelo's axiom, let E be a set consisting of exactly one element from each distinct equivalence class. Since any two points of E must differ by an irrational number, the set $\{d : d = x - y, x \in E, y \in E\}$ cannot contain an interval. By (3.37), it follows that either E is not measurable or $|E| = 0$. Since the union of the translates of E by every rational number is all of $\mathbf{R}^1$, $\mathbf{R}^1$ would have measure zero if E did. We conclude that E is not measurable.

(3.39) Corollary *Any set in* $\mathbf{R}^1$ *with positive outer measure contains a nonmeasurable set.*

Proof. Let A satisfy $|A|_e > 0$, and let E be the nonmeasurable set of (3.38). For rational r, let E_r denote the translate of E by r. Then the E_r are disjoint and $\bigcup E_r = (-\infty, +\infty)$. Thus, $A = \bigcup (A \cap E_r)$ and $|A|_e \le \sum |A \cap E_r|_e$. If $A \cap E_r$ is measurable, then $|A \cap E_r| = 0$ by lemma (3.37). Since $|A|_e > 0$, it follows that there is some r such that $A \cap E_r$ is not measurable. This completes the proof.

Exercises

1. There is an analogue for bases different from 10 of the usual decimal expansion of a number. If b is an integer larger than 1 and $0 < x < 1$, show that there exist integral coefficients c_k, $0 \le c_k < b$, such that $x = \sum_{k=1}^{\infty} c_k b^{-k}$. Show that this expansion is unique unless $x = cb^{-k}$, in which case there are two expansions.
2. When $b = 3$ in Exercise 1, the expansion is called the *triadic* or *ternary expansion of* x. Show that the Cantor set C consists of all x such that x has *some* triadic expansion for which every c_k is either 0 or 2.

   Let $f(x)$ be the Cantor-Lebesgue function: see p. 35. Show that if $x \in C$ and $x = \Sigma\, c_k 3^{-k}$, where each c_k is either 0 or 2, then $f(x) = \Sigma\, (\tfrac{1}{2} c_k) 2^{-k}$.
3. Construct a two-dimensional Cantor set in the unit square $\{(x,y) : 0 \le x,y \le 1\}$ as follows. Subdivide the square into nine equal parts and keep only the four closed corner squares, removing the remaining region (which forms a cross). Then repeat this process in a suitably scaled version for the remaining squares, *ad infinitum*. Show that the resulting set is perfect, has plane measure zero, and equals $C \times C$.
4. Construct a subset of [0,1] in the same manner as the Cantor set by removing from each remaining interval a subinterval of relative length θ, $0 < \theta < 1$.

Show that the resulting set is perfect and has measure zero.

5. Construct a subset of [0,1] in the same manner as the Cantor set, except that at the kth stage, each interval removed has length $\delta 3^{-k}$, $0 < \delta < 1$. Show that the resulting set is perfect, has measure $1 - \delta$, and contains no intervals.
6. Construct a Cantor-type subset of [0,1] by removing from each interval remaining at the kth stage a subinterval of relative length θ_k, $0 < \theta_k < 1$. Show that the remainder has measure zero if and only if $\sum \theta_k = +\infty$. (Use the fact that for $a_k > 0$, $\prod_{k=1}^{\infty} a_k$ converges, in the sense that $\lim_{N\to\infty} \prod_{k=1}^{N} a_k$ exists and is not zero, if and only if $\sum_{k=1}^{\infty} \log a_k$ converges.)
7. Prove (3.15).
8. Show that the Borel σ-algebra $\mathscr{B}$ in $\mathbf{R}^n$ is the smallest σ-algebra containing the closed sets in $\mathbf{R}^n$.
9. If $\{E_k\}_{k=1}^{\infty}$ is a sequence of sets with $\sum |E_k|_e < +\infty$, show that $\limsup E_k$ (and so also $\liminf E_k$) has measure zero.
10. If E_1 and E_2 are measurable, show that $|E_1 \cup E_2| + |E_1 \cap E_2| = |E_1| + |E_2|$.
11. Prove (3.29).
12. If E_1 and E_2 are measurable subsets of $\mathbf{R}^1$, show that $E_1 \times E_2$ is a measurable subset of $\mathbf{R}^2$ and $|E_1 \times E_2| = |E_1||E_2|$. (Interpret $0 \cdot \infty$ as 0.) (HINT: Use a characterization of measurability.)
13. Motivated by (3.7), define the *inner measure* of E by $|E|_i = \sup |F|$, where the supremum is taken over all closed subsets F of E. Show that (i) $|E|_i \leq |E|_e$, and (ii) if $|E|_e < +\infty$, then E is measurable if and only if $|E|_i = |E|_e$. [Use (3.22).]
14. Show that the conclusion of part (ii) of Exercise 13 is false if $|E|_e = +\infty$.
15. If E is measurable and A is any subset of E, show that $|E| = |A|_i + |E - A|_e$. (See Exercise 13 for the definition of $|A|_i$.)
16. Prove (3.34).
17. Give an example which shows that the image of a measurable set under a continuous transformation may not be measurable. (Consider the Cantor-Lebesgue function and the pre-image of an appropriate nonmeasurable subset of its range.)
18. Prove that outer measure is *translation invariant*; that is, if $E_{\mathbf{h}} = \{\mathbf{x} + \mathbf{h} : \mathbf{x} \in E\}$ is the translate of E by $\mathbf{h}$, $\mathbf{h} \in \mathbf{R}^n$, show that $|E_{\mathbf{h}}|_e = |E|_e$. If E is measurable, show that $E_{\mathbf{h}}$ is also measurable. [This fact was used in proving (3.37).]
19. Carry out the details of the construction of a nonmeasurable subset of $\mathbf{R}^n$, $n > 1$.
20. Show that there exist disjoint $E_1, E_2, \ldots, E_k, \ldots$ such that $|\bigcup E_k|_e < \sum |E_k|_e$ with strict inequality. (Let E be a nonmeasurable subset of [0,1] whose rational translates are disjoint. Consider the translates of E by all rational numbers r, $0 < r < 1$, and use Exercise 18.)
21. Show that there exist sets $E_1, E_2, \ldots, E_k, \ldots$ such that $E_k \searrow E$, $|E_k|_e < +\infty$, and $\lim_{k\to\infty} |E_k|_e > |E|_e$ with strict inequality.

22. Show that the outer measure of a set is unchanged if in the definition of outer measure we use coverings of the set by parallelepipeds with a fixed orientation (i.e., with edges parallel to a fixed set of n linearly independent vectors), rather than coverings by intervals.

23. Let Z be a subset of $\mathbf{R}^1$ with measure zero. Show that the set $\{x^2 : x \in Z\}$ also has measure zero.

24. Let $0.\alpha_1\alpha_2 \cdots$ be the dyadic development of any x in $[0,1]$, that is, $x = \alpha_1 2^{-1} + \alpha_2 2^{-2} + \cdots$, $\alpha_i = 0,1$. Let $k_1, k_2, \ldots$ be a fixed permutation of the positive integers $1, 2, \ldots$, and consider the transformation T which sends $x = 0.\alpha_1\alpha_2 \cdots$ to $Tx = 0.\alpha_{k_1}\alpha_{k_2} \cdots$. If E is a measurable subset of $[0,1]$, show that its image TE is also measurable and that $|TE| = |E|$. [Consider first the special cases of E a dyadic interval $[s2^{-k}, (s+1)2^{-k}]$, $s = 0, 1, \ldots, 2^k - 1$, and then of E an open set (which is a countable union of nonoverlapping dyadic intervals). Also show that if E has small measure, then so has TE.]

25. Construct a measurable subset E of $[0,1]$ such that for every subinterval I, both $E \cap I$ and $I - E$ have positive measure. [Take a Cantor-type subset of $[0,1]$ with positive measure (see Exercise 5), and on each subinterval of the complement of this set, construct another such set, and so on. The measures can be arranged so that the union of all the sets has the desired property.]

26. Construct a continuous function on $[0,1]$ which is not of bounded variation on any subinterval. [The construction follows the pattern of the Cantor-Lebesgue function with some modifications. At the first stage, for example, make the corresponding function increase to $2/3$ (rather than $1/2$) in $(0,1/3)$, then make it decrease by $1/3$ in $(1/3,2/3)$, and then increase again $2/3$ in $(2/3,1)$. The construction at other stages is similar, depending on whether the preceding function was increasing or decreasing in the subinterval under consideration.]

27. Construct a continuous function of bounded variation on $[0,1]$ which is not monotone in any subinterval. (The construction is like that in the preceding exercise, except that the approximating functions are less steep. For example, at the first stage, let the function increase to $1/2 + \varepsilon$, then decrease by 2ε, and then increase again by $1/2 + \varepsilon$. Choose the ε's at each stage so that their sum converges.)

Chapter 4

Lebesgue Measurable Functions

We will use the concept of Lebesgue measure to introduce a rich class of functions and a method of integrating these functions. In this chapter, we describe the class of functions.

Let f be a real-valued function defined on a set E in $\mathbf{R}^n$, that is, $-\infty \le f(\mathbf{x}) \le +\infty$, $\mathbf{x} \in E$. Then f is called a *Lebesgue measurable function* on E, or simply a *measurable function*, if for every finite a, the set

$$\{\mathbf{x} \in E : f(\mathbf{x}) > a\}$$

is a measurable subset of $\mathbf{R}^n$. In what follows, we shall often use the abbreviation $\{f > a\}$ for $\{\mathbf{x} \in E : f(\mathbf{x}) > a\}$. Since

$$E = \{f = -\infty\} \cup \left(\bigcup_{k=1}^{\infty} \{f > -k\} \right),$$

the measurability of E is equivalent to that of $\{f = -\infty\}$ if we assume that f is measurable. For simplicity, we shall always assume that $\{f = -\infty\}$ is measurable, and so consider only measurable functions defined on measurable sets.

As a varies, the behavior of the sets $\{f > a\}$ describes how the values of f are distributed. Intuitively, it is clear that the smoother f is, the smaller the variety of such sets will be. For example, if $E = \mathbf{R}^n$ and f is continuous in $\mathbf{R}^n$, then $\{f > a\}$ is always open. A function f defined on a Borel set E is said to be *Borel measurable* if $\{f > a\}$ is a Borel set for every a. Thus, every Borel measurable function is measurable.

One further comment will be helpful later. Let $\mathcal{M}$ denote the class of measurable subsets of $\mathbf{R}^n$. Much of the development of measurable functions given below depends only on the σ-algebra structure of $\mathcal{M}$ and the properties of Lebesgue measure. Thus, a measurable function inherits its elementary properties from those of measurable sets. It is reasonable to expect, therefore, that many of the methods of this chapter can be used to develop results in more general settings, for spaces other than $\mathbf{R}^n$ and σ-algebras other than $\mathcal{M}$. This will be done in Chapter 10. To save too much repetition there, it will be helpful if the reader notices which properties of $\mathcal{M}$ and Lebesgue measure are used in the proofs here. These will be discussed at the end of the chapter.

1. Elementary Properties of Measurable Functions

(4.1) Theorem *Let f be a real-valued function defined on a (measurable) set E. Then f is measurable if and only if any of the following statements holds for every finite a:*

(i) *$\{f \geq a\}$ is measurable.*
(ii) *$\{f < a\}$ is measurable.*
(iii) *$\{f \leq a\}$ is measurable.*

Proof. Since $\{f \geq a\} = \bigcap_{k=1}^{\infty} \{f > a - 1/k\}$, the measurability of f implies (i). Since $\{f < a\}$ is the complement of $\{f \geq a\}$, it follows that (i) implies (ii). Since $\{f \leq a\} = \bigcap_{k=1}^{\infty} \{f < a + 1/k\}$, we see that (ii) implies (iii). Finally, since $\{f > a\}$ is the complement of $\{f \leq a\}$, it follows that f is measurable if (iii) holds.

The proof of the following is left as an exercise.

(4.2) Corollary *If f is measurable, then $\{f > -\infty\}, \{f < +\infty\}, \{f = +\infty\}$, $\{a \leqslant f \leqslant b\}, \{f = a\}$, etc., are all measurable. Moreover, f is measurable if and only if $\{a < f < +\infty\}$ is measurable for every finite a.*

Also, observe that $\{a < f \leq b\} = \{f > a\} - \{f > b\}$.

Let f be defined in E. If S is a subset of $\mathbf{R}^1$, the *inverse image* of S under f is defined by

$$f^{-1}(S) = \{\mathbf{x} \in E : f(\mathbf{x}) \in S\}.$$

(4.3) Theorem *f is measurable if and only if for every open set G in $\mathbf{R}^1$, the inverse image $f^{-1}(G)$ is a measurable subset of $\mathbf{R}^n$.*

Proof. If $G = (a, +\infty)$, then $f^{-1}(G) = \{a < f < +\infty\}$. Hence, if $f^{-1}(G)$ is measurable for every such G, f must be measurable. To prove the converse, suppose that f is measurable and let G be any open subset of $\mathbf{R}^1$. By (1.10), G can be written $G = \bigcup_k (a_k, b_k)$. But $f^{-1}((a_k, b_k))$ equals $\{a_k < f < b_k\}$, and is therefore measurable. Since $f^{-1}(G) = \bigcup_k f^{-1}((a_k, b_k))$, it follows that $f^{-1}(G)$ is measurable too.

We observe that the proof above also shows that f is Borel measurable if and only if $f^{-1}(G)$ is Borel measurable for every open $G \subset \mathbf{R}^1$.

(4.4) Theorem *Let A be a dense subset of $\mathbf{R}^1$. Then f is measurable if $\{f > a\}$ is measurable for all $a \in A$.*

Proof. Given any real a, choose a sequence $\{a_k\}$ in A which converges to

a from above: $a_k \in A$, $a_k \geq a$, $\lim_{k\to\infty} a_k = a$. Then $\{f > a\} = \bigcup \{f > a_k\}$, and the theorem follows.

A property is said to hold *almost everywhere in E* or, in abbreviated form, *a.e.*, if it holds in E except in some subset of E with measure zero. For example, the statement "$f = 0$ a.e. in E" means that $f(\mathbf{x}) = 0$ in E, with the possible exception of those $\mathbf{x}$ in some subset Z of E with $|Z| = 0$.

The next few theorems give some simple properties of the class of measurable functions.

(4.5) Theorem *If f is measurable and if $g = f$ a.e., then g is measurable and* $|\{g > a\}| = |\{f > a\}|$.

Proof. If $Z = \{f \neq g\}$, then $\{g > a\} \cup Z = \{f > a\} \cup Z$. Therefore, f being measurable, $\{g > a\} \cup Z$ is measurable, and since this differs from $\{g > a\}$ by a set of measure zero, g is measurable. Finally,

$$|\{g > a\}| = |\{g > a\} \cup Z| = |\{f > a\} \cup Z| = |\{f > a\}|.$$

In view of the last theorem, it is natural to extend the definition of measurability to include functions which are defined only a.e. in E, by saying that such an f is measurable on E if it is measurable on the subset of E where it is defined. Note also that if f is measurable on E, then it is measurable on any measurable $E_1 \subset E$ since $\{\mathbf{x} \in E_1 : f(\mathbf{x}) > a\} = \{\mathbf{x} \in E : f(\mathbf{x}) > a\} \cap E_1$.

If ϕ and f are real-valued measurable functions defined on $\mathbf{R}^1$ and $\mathbf{R}^n$, respectively, their composition $\phi(f(\mathbf{x}))$ may not be measurable (see Exercise 5). If ϕ is continuous, however, we have the following result.

(4.6) Theorem *Let ϕ be continuous on $\mathbf{R}^1$ and let f be finite a.e. in E, so that, in particular, $\phi(f)$ is defined a.e. in E. Then $\phi(f)$ is measurable if f is.*

Proof. We may assume that f is finite everywhere in E. We will use the fact that since ϕ is continuous, the inverse image $\phi^{-1}(G)$ of an open set G is open. [The reader who is not already familiar with this fact can find related material in both (4.15) and Exercise 10.] By (4.3), it is enough to show that for every open G in $\mathbf{R}^1$, $\{\mathbf{x} : \phi(f(\mathbf{x})) \in G\}$ is measurable. However, $\{\mathbf{x} : \phi(f(\mathbf{x})) \in G\} = f^{-1}(\phi^{-1}(G))$, and since $\phi^{-1}(G)$ is open and f is measurable, $f^{-1}(\phi^{-1}(G))$ is measurable by (4.3).

Remark: The cases which arise most frequently are $\phi(t) = |t|$, $|t|^p (p > 0)$, e^{ct}, etc. Thus,

$$|f|, \ |f|^p (p > 0), \ e^{cf},$$

are measurable if f is (even if we do not assume that f is finite a.e., as is easily seen). Another special case worth mentioning is that of

$$f^+ = \max\{f, 0\}, \qquad f^- = -\min\{f, 0\}.$$

It is enough to observe that the functions x^+ and x^- are continuous.

(4.7) Theorem *If f and g are measurable, then $\{f > g\}$ is measurable.*

Proof. Let $\{r_k\}$ be the rational numbers. Then

$$\{f > g\} = \bigcup_k \{f > r_k > g\} = \bigcup_k (\{f > r_k\} \cap \{g < r_k\}),$$

and the theorem follows.

(4.8) Theorem *If f is measurable and λ is any real number, then $f + \lambda$ and λf are measurable.*

The proof is left as an exercise.

We note that the sum $f + g$ of two functions f and g is well-defined wherever it is not of the form $+\infty + (-\infty)$ or $-\infty + \infty$. In the next theorem, we assume for simplicity that $f + g$ is well-defined everywhere. (For extensions, see Exercise 6.)

(4.9) Theorem *If f and g are measurable, so is $f + g$.*

Proof. Since g is measurable, so is $a - g$ for any real a, by (4.8). Since $\{f + g > a\} = \{f > a - g\}$, the result follows from (4.7).

A corollary of (4.8) and (4.9) is that a finite linear combination $\lambda_1 f_1 + \cdots + \lambda_N f_N$ of measurable functions $f_1, \ldots, f_N$ is measurable. Thus, the class of measurable functions on a set E forms a vector space.

In the theorem which follows we consider products of functions. In addition to the familiar conventions about products of infinities, we adopt the convention $0 \cdot \pm\infty = \pm\infty \cdot 0 = 0$.

(4.10) Theorem *If f and g are measurable, so is fg. If $g \neq 0$ a.e., f/g is measurable.*

Proof. By (4.6) and the remark following it, $f^2 (= |f|^2)$ is measurable if f is. Hence, if f and g are measurable and finite, the formula $fg = [(f + g)^2 - (f - g)^2]/4$ implies that fg is measurable. The proof when f and g can be infinite and the proof of the second statement of the theorem are left as exercises.

(4.11) Theorem *If $\{f_k(\mathbf{x})\}_{k=1}^{\infty}$ is a sequence of measurable functions, then $\sup_k f_k(\mathbf{x})$ and $\inf_k f_k(\mathbf{x})$ are measurable.*

Proof. Since $\inf_k f_k = -\sup_k (-f_k)$, it is enough to prove the result for $\sup_k f_k$. But this follows easily from the fact that $\{\sup_k f_k > a\} = \bigcup_k \{f_k > a\}$.

As a special case of the preceding theorem, we see that if $f_1, \ldots, f_N$ are measurable, then so are $\max_k f_k$ and $\min_k f_k$. In particular, if f is measurable, then so are $f^+ = \max\{f, 0\}$ and $f^- = -\min\{f, 0\}$, a fact we have already observed [see the remark following (4.6).]

(4.12) Theorem *If $\{f_k\}$ is a sequence of measurable functions, then*

$\limsup_{k\to\infty} f_k$ and $\liminf_{k\to\infty} f_k$ are measurable. In particular, if $\lim_{k\to\infty} f_k$ exists a.e., it is measurable.

Proof. Since

$$\limsup_{k\to\infty} f_k = \inf_j \{\sup_{k\geq j} f_k\}, \qquad \liminf_{k\to\infty} f_k = \sup_j \{\inf_{k\geq j} f_k\},$$

the first statement follows from (4.11). The second statement is then a corollary of (4.5) since wherever $\lim_{k\to\infty} f_k$ exists, it equals $\limsup_{k\to\infty} f_k$.

The *characteristic function*, or *indicator function*, $\chi_E(\mathbf{x})$, of a set E is defined by

$$\chi_E(\mathbf{x}) = \begin{cases} 1 \text{ if } \mathbf{x} \in E \\ 0 \text{ if } \mathbf{x} \notin E \end{cases}.$$

Clearly, χ_E is measurable if and only if E is measurable. χ_E is an example of what is called a simple function: a *simple function* is one which assumes only a finite number of values, all of which are finite. If f is a simple function taking (distinct) values $a_1, \dots, a_N$ on (disjoint) sets $E_1, \dots, E_N$, then

$$f(\mathbf{x}) = \sum_{k=1}^{N} a_k \chi_{E_k}(\mathbf{x}).$$

We leave it as an exercise to show that such an f is measurable if and only if $E_1, \dots, E_N$ are measurable.

(4.13) Theorem

(i) *Every function f can be written as the limit of a sequence $\{f_k\}$ of simple functions.*

(ii) *If $f \geq 0$, the sequence can be chosen to increase to f, that is, chosen such that $f_k \leq f_{k+1}$ for every k.*

(iii) *If the function f in either* (i) *or* (ii) *is measurable, then the f_k can be chosen to be measurable.*

Proof. We will prove (ii) first. Thus, suppose that $f \geq 0$. For each k, $k = 1, 2, \dots$, subdivide the values of f which fall in $[0,k]$ by partitioning $[0,k]$ into subintervals $[(j-1)2^{-k}, j2^{-k}]$, $j = 1, \dots, k2^k$. Let

$$f_k(\mathbf{x}) = \begin{cases} \dfrac{j-1}{2^k} & \text{if } \dfrac{j-1}{2^k} \leq f(\mathbf{x}) < \dfrac{j}{2^k}, \quad j = 1, \dots, k2^k. \\ k & \text{if } f(\mathbf{x}) \geq k. \end{cases}$$

Each f_k is a simple function defined everywhere in the domain of f. Clearly, $f_k \leq f_{k+1}$ since in passing from f_k to f_{k+1}, each subinterval $[(j-1)2^{-k}, j2^{-k}]$ is divided in half. Moreover, $f_k \to f$ since $0 \leq f - f_k \leq 2^{-k}$ for sufficiently large k wherever f is finite, and $f_k = k \to +\infty$ wherever $f = +\infty$. This proves (ii).

To prove (i), apply the result of (ii) to each of the nonnegative functions f^+ and f^-, obtaining increasing sequences $\{f'_k\}$ and $\{f''_k\}$ of simple functions such that $f'_k \to f^+$ and $f''_k \to f^-$. Then $f'_k - f''_k$ is simple and $f'_k - f''_k \to f^+ - f^- = f$.

Finally, it is enough to prove (iii) for $f \geq 0$ since otherwise we may consider f^+ and f^-. In this case, however,

$$f_k = \sum_{j=1}^{k2^k} \frac{j-1}{2^k} \chi_{\{(j-1)/2^k \leq f < j/2^k\}} + k\chi_{\{f \geq k\}}.$$

This is measurable if f is since all the sets involved are measurable.

Note that if f is bounded, the simple functions above will converge uniformly to f.

2. Semicontinuous Functions

We now study classes of functions f whose continuity properties on a set E can be characterized by the nature of $\{f > a\}$ or $\{f < a\}$. Let f be defined on E, and let $\mathbf{x}_0$ be a limit point of E which lies in E. Then f is said to be *upper semicontinuous at* $\mathbf{x}_0$ if

$$\limsup_{\mathbf{x}\to\mathbf{x}_0;\mathbf{x}\in E} f(\mathbf{x}) \leq f(\mathbf{x}_0).$$

We will usually abbreviate this by saying that f is *usc at* $\mathbf{x}_0$. Note that if $f(\mathbf{x}_0) = +\infty$, then f is automatically usc at $\mathbf{x}_0$; otherwise, the statement that f is usc at $\mathbf{x}_0$ means that given $M > f(\mathbf{x}_0)$, there exists $\delta > 0$ such that $f(\mathbf{x}) < M$ for all $\mathbf{x} \in E$ which lie in the ball $|\mathbf{x} - \mathbf{x}_0| < \delta$. Intuitively, this means that near $\mathbf{x}_0$ the values of f do not exceed $f(\mathbf{x}_0)$ by a fixed amount.

Similarly, f is said to be *lower semicontinuous at* $\mathbf{x}_0$, or *lsc at* $\mathbf{x}_0$, if

$$\liminf_{\mathbf{x}\to\mathbf{x}_0;\mathbf{x}\in E} f(\mathbf{x}) \geq f(\mathbf{x}_0).$$

Thus, if $f(\mathbf{x}_0) = -\infty$, f is lsc at $\mathbf{x}_0$, while if $f(\mathbf{x}_0) > -\infty$, the definition amounts to saying that given $m < f(\mathbf{x}_0)$, there exists $\delta > 0$ such that $f(\mathbf{x}) > m$ if $\mathbf{x} \in E$ and $|\mathbf{x} - \mathbf{x}_0| < \delta$. Equivalently, f is lsc at $\mathbf{x}_0$ if and only if $-f$ is usc at $\mathbf{x}_0$.

It follows that f is continuous at $\mathbf{x}_0$ if and only if $|f(\mathbf{x}_0)| < +\infty$ and f is both usc and lsc at $\mathbf{x}_0$. As simple examples of functions which are usc everywhere in $\mathbf{R}^1$ but not continuous at some x_0, we have

$$u_1(x) = \begin{cases} 0 & \text{if } x < x_0, \\ 1 & \text{if } x \geq x_0, \end{cases} \qquad u_2(x) = \begin{cases} 0 & \text{if } x \neq x_0. \\ 1 & \text{if } x = x_0. \end{cases}$$

Hence, $-u_1$ and $-u_2$ are lsc everywhere in $\mathbf{R}^1$. The Dirichlet function of p. 16, Example 4, is usc at the rational numbers and lsc at the irrationals.

A function defined on E is called *usc* (*lsc*, *continuous*) *relative to* E if it is usc (lsc, continuous) at every limit point of E which is in E. The next theorem characterizes functions which are semicontinuous relative to a set.

(4.14) Theorem

(i) *A function f is usc relative to E if and only if $\{\mathbf{x} \in E : f(\mathbf{x}) \geq a\}$ is relatively closed [equivalently, $\{\mathbf{x} \in E : f(\mathbf{x}) < a\}$ is relatively open] for all finite a.*

(ii) *A function f is lsc relative to E if and only if $\{\mathbf{x} \in E : f(\mathbf{x}) \leq a\}$ is relatively closed [equivalently, $\{\mathbf{x} \in E : f(\mathbf{x}) > a\}$ is relatively open] for all finite a.*

Proof. Statements (i) and (ii) are equivalent since f is usc if and only if $-f$ is lsc. It is therefore enough to prove (i). Suppose first that f is usc relative to E. Given a, let $\mathbf{x}_0$ be a limit point of $\{\mathbf{x} \in E : f(\mathbf{x}) \geq a\}$ which is in E. Then there exist $\mathbf{x}_k \in E$ such that $\mathbf{x}_k \to \mathbf{x}_0$ and $f(\mathbf{x}_k) \geq a$. Since f is usc at $\mathbf{x}_0$, we have $f(\mathbf{x}_0) \geq \limsup_{k\to\infty} f(\mathbf{x}_k)$. Therefore, $f(\mathbf{x}_0) \geq a$, so that $\mathbf{x}_0 \in \{\mathbf{x} \in E : f(\mathbf{x}) \geq a\}$. This shows that $\{\mathbf{x} \in E : f(\mathbf{x}) \geq a\}$ is relatively closed.

Conversely, let $\mathbf{x}_0$ be a limit point of E which is in E. If f is not usc at $\mathbf{x}_0$, then $f(\mathbf{x}_0) < +\infty$, and there exist M and $\{\mathbf{x}_k\}$ such that $f(\mathbf{x}_0) < M$, $\mathbf{x}_k \in E$, $\mathbf{x}_k \to \mathbf{x}_0$, and $f(\mathbf{x}_k) \geq M$. Hence, $\{\mathbf{x} \in E : f(\mathbf{x}) \geq M\}$ is not relatively closed since it does not contain all its limit points which are in E.

(4.15) Corollary *A finite function f is continuous relative to E if and only if all sets of the form $\{\mathbf{x} \in E : f(\mathbf{x}) \geq a\}$ and $\{\mathbf{x} \in E : f(\mathbf{x}) \leq a\}$ are relatively closed [or, equivalently, all $\{\mathbf{x} \in E : f(\mathbf{x}) > a\}$ and $\{\mathbf{x} \in E : f(\mathbf{x}) < a\}$ are relatively open] for finite a.*

(4.16) Corollary *Let E be measurable, and let f be defined on E. If f is usc (lsc, continuous) relative to E, then f is measurable.*

Proof. Let f be usc relative to E. Since $\{\mathbf{x} \in E : f(\mathbf{x}) \geq a\}$ is relatively closed, it is the intersection of E with a closed set. Hence, it is measurable, and the result follows from (4.1).

The results in (4.14)–(4.16) deserve special attention in certain cases. Suppose, for example, that $E = \mathbf{R}^n$ and f is usc everywhere in $\mathbf{R}^n$. Since $\{f > a\} = \bigcup_{k=1}^{\infty} \{f \geq a + 1/k\}$, it follows from (4.14) that $\{f > a\}$ is of type F_σ. Since an F_σ set is a Borel set, we see that a function which is usc (similarly, lsc or continuous) at every point of $\mathbf{R}^n$ is Borel measurable.

3. Properties of Measurable Functions: Egorov's Theorem and Lusin's Theorem

Our next theorem states in effect that if a sequence of measurable functions converges at each point of a set E, then, with the exception of a subset of E with arbitrarily small measure, the sequence actually converges *uniformly*.

This remarkable result cannot hold, at least in the form just stated, without some further restrictions. For example, if $E = \mathbf{R}^n$ and $f_k = \chi_{\{\mathbf{x}:|\mathbf{x}|<k\}}$, then f_k converges to 1 everywhere but does not converge uniformly outside any bounded set. Again, if the f_k are finite but the limit f is infinite in a set of positive measure, then $|f_k - f|$ is also infinite in this set. The difficulties in these examples can be easily overcome: the missing ingredient in the first case is that $|E| < +\infty$, and in the second, that $|f| < +\infty$ a.e. Adding these restrictions, we obtain the following basic result.

(4.17) Theorem (*Egorov's Theorem*) *Suppose that $\{f_k\}$ is a sequence of measurable functions which converges almost everywhere in a set E of finite measure to a finite limit f. Then given $\varepsilon > 0$, there is a closed subset F of E such that $|E - F| < \varepsilon$ and $\{f_k\}$ converges uniformly to f on F.*

In order to prove this, we need a preliminary result which is interesting in its own right.

(4.18) Lemma *Under the same hypothesis as in Egorov's theorem, given $\varepsilon, \eta > 0$, there is a closed subset F of E and an integer K such that $|E - F| < \eta$ and $|f(\mathbf{x}) - f_k(\mathbf{x})| < \varepsilon$ for $\mathbf{x} \in F$ and $k > K$.*

Proof. Fix $\varepsilon, \eta > 0$. For each m, let $E_m = \{|f - f_k| < \varepsilon \text{ for all } k > m\}$. Thus, $E_m = \bigcap_{k>m} \{|f - f_k| < \varepsilon\}$, so that E_m is measurable. Clearly, $E_m \subset E_{m+1}$. Moreover, since $f_k \to f$ a.e. in E and f is finite, $E_m \nearrow E - Z$, $|Z| = 0$. Hence, by (3.26), $|E_m| \to |E - Z| = |E|$. Since $|E| < +\infty$, it follows that $|E - E_m| \to 0$. Choose m_0 so that $|E - E_{m_0}| < \frac{1}{2}\eta$, and let F be a closed subset of E_{m_0} with $|E_{m_0} - F| < \frac{1}{2}\eta$. Then $|E - F| < \eta$ and $|f - f_k| < \varepsilon$ in F if $k > m_0$.

Proof of Egorov's theorem. Given $\varepsilon > 0$, use (4.18) to select closed $F_m \subset E$, $m \geq 1$, and integers $K_{m,\varepsilon}$ such that $|E - F_m| < \varepsilon 2^{-m}$ and $|f - f_k| < 1/m$ in F_m if $k > K_{m,\varepsilon}$. The set $F = \bigcap_m F_m$ is closed, and since $F \subset F_m$ for all m, f_k converges uniformly to f on F. Finally, $E - F = E - \bigcap F_m = \bigcup (E - F_m)$ and, therefore, $|E - F| \leq \sum |E - F_m| < \varepsilon$. This completes the proof.

See Exercises 13 and 14 for an analogue of Egorov's theorem in the continuous parameter case; i.e., in the case when $f_{\mathbf{y}}(\mathbf{x}) \to f(\mathbf{x})$ as $\mathbf{y} \to \mathbf{y}_0$.

We have observed that a continuous function is measurable. Our next result, Lusin's theorem, gives a continuity property which characterizes measurable functions. In order to state the result, we first make the following definition. A function f defined on a measurable set E has *property* $\mathscr{C}$ on E if given $\varepsilon > 0$, there is a closed set $F \subset E$ such that

(i) $|E - F| < \varepsilon$,

(ii) f is continuous relative to F.

We recall that condition (ii) means that if $\mathbf{x}_0$ and $\{\mathbf{x}_k\}$ belong to F and $\mathbf{x}_k \to \mathbf{x}_0$, then $f(\mathbf{x}_k) \to f(\mathbf{x}_0)$. In case F is bounded (and, therefore, compact), (ii) implies that the restriction of f to F is uniformly continuous [theorem (1.15)].

(4.19) Lemma *A simple measurable function has property $\mathscr{C}$.*

Proof. Suppose that f is a simple measurable function on E, taking distinct values $a_1, \dots, a_N$ on measurable subsets $E_1, \dots, E_N$. Given $\varepsilon > 0$, choose closed $F_j \subset E_j$ with $|E_j - F_j| < \varepsilon/N$. Then the set $F = \bigcup_{j=1}^N F_j$ is closed, and since $E - F = \bigcup E_j - \bigcup F_j \subset \bigcup (E_j - F_j)$, we have $|E - F| \le \sum |E_j - F_j| < \varepsilon$. It remains only to show that f is continuous on F. Note that each F_j is relatively open in F, so the only points of F in a small neighborhood of any point of F_j are points of F_j itself. The continuity on F of f follows from this since f is constant on each F_j.

Property $\mathscr{C}$ is actually equivalent to measurability, as we now show.

(4.20) Theorem (*Lusin's Theorem*) *Let f be defined and finite on a measurable set E. Then f is measurable if and only if it has property $\mathscr{C}$ on E.*

Proof. If f is measurable, then by (4.13) there exist simple measurable f_k, $k = 1, 2, \dots$, which converge to f. By (4.19), each f_k has property $\mathscr{C}$, so given $\varepsilon > 0$, there exist closed $F_k \subset E$ such that $|E - F_k| < \varepsilon 2^{-k-1}$ and f_k is continuous relative to F_k. Assuming for the moment that $|E| < +\infty$, we see by Egorov's theorem that there is a closed $F_0 \subset E$ with $|E - F_0| < \frac{1}{2}\varepsilon$ such that $\{f_k\}$ converges uniformly to f on F_0. If $F = F_0 \cap (\bigcap_{k=1}^\infty F_k)$, then F is closed, each f_k is continuous relative to F, and $\{f_k\}$ converges uniformly to f on F. Hence, f is continuous relative to F by (1.16). Since

$$|E - F| \le |E - F_0| + \sum_{k=1}^{\infty} |E - F_k| < \tfrac{1}{2}\varepsilon + \tfrac{1}{2}\varepsilon = \varepsilon,$$

it follows that f has property $\mathscr{C}$ on E. This proves the necessity of property $\mathscr{C}$ for measurability if $|E| < +\infty$.

If $|E| = +\infty$, write $E = \bigcup_{k=1}^\infty E_k$, where E_k is the part of E in the ring $\{\mathbf{x} : k - 1 \le |\mathbf{x}| < k\}$. Since $|E_k| < +\infty$, we may select closed $F_k \subset E_k$ such that $|E_k - F_k| < \varepsilon 2^{-k}$ and f is continuous relative to F_k. If $F = \bigcup_{k=1}^\infty F_k$, it follows that $|E - F| \le \sum |E_k - F_k| < \varepsilon$, and that f is continuous relative to F. A simple argument shows that F is closed, and therefore f has property $\mathscr{C}$ on E.

Conversely, suppose that f has property $\mathscr{C}$ on E. For each k, $k = 1, 2, \dots$, choose a closed $F_k \subset E$ such that $|E - F_k| < 1/k$ and the restriction of f to

F_k is continuous. If $H = \bigcup_{k=1}^{\infty} F_k$, then $H \subset E$ and the set $Z = E - H$ has measure zero. We have

$$\begin{aligned}\{\mathbf{x} \in E : f(\mathbf{x}) > a\} &= \{\mathbf{x} \in H : f(\mathbf{x}) > a\} \cup \{\mathbf{x} \in Z : f(\mathbf{x}) > a\} \\ &= \bigcup_{k=1}^{\infty} \{\mathbf{x} \in F_k : f(\mathbf{x}) > a\} \cup \{\mathbf{x} \in Z : f(\mathbf{x}) > a\}.\end{aligned}$$

Since $\{\mathbf{x} \in Z : f(\mathbf{x}) > a\}$ has measure zero, the measurability of f will follow from that of each $\{\mathbf{x} \in F_k : f(\mathbf{x}) > a\}$. However, since f is continuous relative to F_k and F_k is measurable, $\{\mathbf{x} \in F_k : f(\mathbf{x}) > a\}$ is measurable by (4.16). This completes the proof of Lusin's theorem.

4. Convergence in Measure

Let f and $\{f_k\}$ be measurable functions which are defined and finite a.e. in a set E. Then $\{f_k\}$ is said to *converge in measure* on E to f if for every $\varepsilon > 0$,

$$\lim_{k\to\infty} |\{\mathbf{x} \in E : |f(\mathbf{x}) - f_k(\mathbf{x})| > \varepsilon\}| = 0.$$

We will indicate convergence in measure by writing

$$f_k \xrightarrow{m} f.$$

This concept has many useful applications in analysis. Here we will discuss its relation to ordinary pointwise convergence; the first result is basically a reformulation of (4.18).

(4.21) Theorem *Let f and f_k, $k = 1, 2, \ldots$, be measurable and finite a.e. in E. If $f_k \to f$ a.e. on E and $|E| < +\infty$, then $f_k \xrightarrow{m} f$ on E.*

Proof. Given $\varepsilon, \eta > 0$, let F and K be as defined in lemma (4.18). Then if $k > K$, $\{\mathbf{x} \in E: |f(\mathbf{x}) - f_k(\mathbf{x})| > \varepsilon\} \subset E - F$, and since $|E - F| < \eta$, the result follows.

We recall that the conclusion above may not hold if $|E| = +\infty$, as shown by the example $E = \mathbf{R}^n$, $f_k = \chi_{\{\mathbf{x}:|\mathbf{x}|<k\}}$, and $f = 1$.

Convergence in measure does not imply pointwise convergence a.e., even for sets of finite measure. To see this, take $n = 1$ and let $\{I_k\}$ be a sequence of subintervals of $[0,1]$ satisfying the following conditions:

(i) Each point of $[0,1]$ belongs to infinitely many I_k.
(ii) $\lim_{k\to\infty} |I_k| = 0$.

For example, let the first interval be $[0,1]$, the next two be the two halves of $[0,1]$, the next four be the four quarters, and so on. Then if $f_k = \chi_{I_k}$, we have $f_k \xrightarrow{m} 0$, while f_k diverges at every point of $[0,1]$.

There is, however, the following partial converse to (4.21).

(4.22) Theorem *If $f_k \xrightarrow{m} f$ on E, there is a subsequence f_{k_j} such that $f_{k_j} \to f$ a.e. in E.*

Proof. Since $f_k \xrightarrow{m} f$, given $j = 1, 2, \ldots$, there exists k_j such that

$$\left|\left\{|f - f_k| > \frac{1}{j}\right\}\right| < \frac{1}{2^j}$$

for $k \geq k_j$. We may assume that $k_j \nearrow$. Let $E_j = \{|f - f_{k_j}| > 1/j\}$, and $H_m = \bigcup_{j=m}^{\infty} E_j$. Then $|E_j| < 2^{-j}$, $|H_m| \leq \sum_{j=m}^{\infty} 2^{-j} = 2^{-m+1}$, and $|f - f_{k_j}| \leq 1/j$ in $E - E_j$. Thus, if $j \geq m$, $|f - f_{k_j}| \leq 1/j$ in $E - H_m$, so that $f_{k_j} \to f$ in $E - H_m$. Since $|H_m| \to 0$, it follows that $f_{k_j} \to f$ a.e. in E. This completes the proof.

The following theorem gives a Cauchy criterion for convergence in measure.

(4.23) Theorem *A necessary and sufficient condition that $\{f_k\}$ converge in measure on E is that for each $\varepsilon > 0$,*

$$\lim_{k,l \to \infty} |\{\mathbf{x} \in E : |f_k(\mathbf{x}) - f_l(\mathbf{x})| > \varepsilon\}| = 0.$$

Proof. The necessity follows from the formula

$$\{|f_k - f_l| > \varepsilon\} \subset \{|f_k - f| > \tfrac{1}{2}\varepsilon\} \cup \{|f_l - f| > \tfrac{1}{2}\varepsilon\}$$

and the fact that the measures of the sets on the right tend to zero as $k,l \to \infty$ if $f_k \xrightarrow{m} f$.

To prove the converse, choose N_j, $j = 1, 2, \ldots$, so that if $k,l \geq N_j$, then $|\{|f_k - f_l| > 2^{-j}\}| < 2^{-j}$. We may assume that $N_j \nearrow$. Then $|f_{N_{j+1}} - f_{N_j}| \leq 2^{-j}$ except for a set E_j, $|E_j| < 2^{-j}$. Let $H_i = \bigcup_{j=i}^{\infty} E_j$. Then

$$|f_{N_{j+1}}(\mathbf{x}) - f_{N_j}(\mathbf{x})| \leq 2^{-j} \text{ for } j \geq i \text{ and } \mathbf{x} \notin H_i.$$

It follows that $\sum (f_{N_{j+1}} - f_{N_j})$ converges uniformly outside H_i, and therefore, that $\{f_{N_j}\}$ converges uniformly outside H_i. Since $|H_i| \leq \sum_{j \geq i} 2^{-j} = 2^{-i+1}$, we obtain that $\{f_{N_j}\}$ converges a.e. in E and, letting $f = \lim f_{N_j}$, that $f_{N_j} \xrightarrow{m} f$ on E. In order to show that $f_k \xrightarrow{m} f$ on E, note that

$$\{|f_k - f| > \varepsilon\} \subset \{|f_k - f_{N_j}| > \tfrac{1}{2}\varepsilon\} \cup \{|f_{N_j} - f| > \tfrac{1}{2}\varepsilon\}$$

for any N_j. To show that the measure of the set on the left is less than a prescribed $\eta > 0$ for all sufficiently large k, select N_j so that the first term on the right has measure less than $\frac{1}{2}\eta$ for all large k (here, we use the Cauchy condition), and so that the measure of the second term on the right is also less than $\frac{1}{2}\eta$. This completes the proof.

As pointed out at the beginning of the chapter, many of the results above depend on only a few basic properties of Lebesgue measurable sets and

Lebesgue measure. This is especially true for the elementary properties of measurable functions [theorems (4.1)–(4.12)] and the section about convergence in measure, which use only the fact that the class of measurable subsets of $\mathbf{R}^n$ is a σ-algebra, and in (4.5), that subsets of a set of measure zero are measurable. Egorov's theorem uses two additional facts: the fundamental result (3.26) concerning monotone sequences of sets, and (3.22) about the approximability of measurable sets by closed sets. Actually, even (3.22), which is a topological property of Lebesgue measure, is not needed in the proof of Egorov's theorem if instead of requiring that F be closed, we merely require that it be measurable.

The rest of the chapter, namely, the material on semicontinuous functions and Lusin's theorem, uses somewhat more restrictive topological properties of $\mathbf{R}^n$ and Lebesgue measure. For example, about $\mathbf{R}^n$, we have used metric properties, and about Lebesgue measure, we have used (3.22) [e.g., in (4.19)] and the fact that Borel sets are measurable [e.g., in (4.16)].

Exercises

1. Prove corollary (4.2) and theorem (4.8).
2. Let f be a simple function, taking its distinct values on disjoint sets $E_1, \dots, E_N$. Show that f is measurable if and only if $E_1, \dots, E_N$ are measurable.
3. Theorem (4.3) can be used to define measurability for vector-valued (e.g., complex-valued) functions. Suppose, for example, that f and g are real-valued and defined in $\mathbf{R}^n$, and let $F(\mathbf{x}) = (f(\mathbf{x}), g(\mathbf{x}))$. Then F is said to be measurable if $F^{-1}(G)$ is measurable for every open $G \subset \mathbf{R}^2$. Prove that F is measurable if and only if both f and g are measurable in $\mathbf{R}^n$.
4. Let f be defined and measurable in $\mathbf{R}^n$. If T is a nonsingular linear transformation of $\mathbf{R}^n$, show that $f(T\mathbf{x})$ is measurable. [If $E_1 = \{\mathbf{x} : f(\mathbf{x}) > a\}$ and $E_2 = \{\mathbf{x} : f(T\mathbf{x}) > a\}$, show that $E_2 = T^{-1}E_1$.]
5. Give an example to show that $\phi(f(\mathbf{x}))$ may not be measurable if ϕ and f are measurable. (Let F be the Cantor-Lebesgue function and let f be its inverse, suitably defined. Let ϕ be the characteristic function of a set of measure zero whose image under F is not measurable.)
6. Let f and g be measurable functions on E.
   (a) If f and g are finite a.e. in E, show that $f + g$ is measurable no matter how we define it at the points when it has the form $+\infty + (-\infty)$ or $-\infty + \infty$.
   (b) Show that fg is measurable without restriction on the finiteness of f and g. Show that $f + g$ is measurable if it is defined to have the same value at every point where it has the form $+\infty + (-\infty)$ or $-\infty + \infty$. (Note that a function h defined on E is measurable if and only if both $\{h = +\infty\}$ and $\{h = -\infty\}$ are measurable and the restriction of h to the subset of E where h is finite is measurable.)
7. Let f be usc and less than $+\infty$ on a compact set E. Show that f is bounded

above on E. Show also that f assumes its maximum on E, i.e., that there exists $\mathbf{x}_0 \in E$ such that $f(\mathbf{x}_0) \geq f(\mathbf{x})$ for all $\mathbf{x} \in E$.

8. (a) Let f and g be two functions which are usc at $\mathbf{x}_0$. Show that $f + g$ is usc at $\mathbf{x}_0$. Is $f - g$ usc at $\mathbf{x}_0$? When is fg usc at $\mathbf{x}_0$?
   (b) If $\{f_k\}$ is a sequence of functions which are usc at $\mathbf{x}_0$, show that $\inf_k f_k(\mathbf{x})$ is usc at $\mathbf{x}_0$.
   (c) If $\{f_k\}$ is a sequence of functions which are usc at $\mathbf{x}_0$ and which converge uniformly near $\mathbf{x}_0$, show that $\lim f_k$ is usc at $\mathbf{x}_0$.
9. (a) Show that the limit of a decreasing (increasing) sequence of functions usc (lsc) at $\mathbf{x}_0$ is usc (lsc) at $\mathbf{x}_0$. In particular, the limit of a decreasing (increasing) sequence of functions continuous at $\mathbf{x}_0$ is usc (lsc) at $\mathbf{x}_0$.
   (b) Let f be usc and less than $+\infty$ on $[a,b]$. Show that there exist continuous f_k on $[a,b]$ such that $f_k \searrow f$.
10. (a) If f is defined and continuous on E, show that $\{a < f < b\}$ is relatively open, and that $\{a \leq f \leq b\}$ and $\{f = a\}$ are relatively closed.
   (b) Let f be a finite function on $\mathbf{R}^n$. Show that f is continuous on $\mathbf{R}^n$ if and only if $f^{-1}(G)$ is open for every open G in $\mathbf{R}^1$, or if and only if $f^{-1}(F)$ is closed for every closed F in $\mathbf{R}^1$.
11. Let f be defined on $\mathbf{R}^n$ and let $B(\mathbf{x})$ denote the open ball $\{\mathbf{y} : |\mathbf{x} - \mathbf{y}| < r\}$ with center $\mathbf{x}$ and fixed radius r. Show that the function $g(\mathbf{x}) = \sup \{f(\mathbf{y}) : \mathbf{y} \in B(\mathbf{x})\}$ is lsc and that the function $h(\mathbf{x}) = \inf \{f(\mathbf{y}) : \mathbf{y} \in B(\mathbf{x})\}$ is usc on $\mathbf{R}^n$. Is the same true for the *closed* ball $\{\mathbf{y} : |\mathbf{x} - \mathbf{y}| \leq r\}$?
12. If $f(x)$, $x \in \mathbf{R}^1$, is continuous at almost every point of an interval $[a,b]$, show that f is measurable on $[a,b]$. Generalize this to functions defined in $\mathbf{R}^n$. [For a constructive proof, use the subintervals of a sequence of partitions to define a sequence of simple measurable functions converging to f a.e. in $[a,b]$. Use (4.12). See also the proof of (5.54) in Chapter 5.]
13. One difficulty encountered in trying to extend the proof of Egorov's Theorem to the continuous parameter case $f_y(x) \to f(x)$ as $y \to y_0$ is showing that the analogues of the sets E_m in (4.18) are measurable. This difficulty can often be overcome in individual cases. Suppose, for example, that $f(x,y)$ is defined and continuous in the square $0 \leq x \leq 1$, $0 < y \leq 1$, and that $f(x) = \lim_{y \to 0} f(x,y)$ exists and is finite for x in a measurable subset E of $[0,1]$. Show that if ε and δ satisfy $0 < \varepsilon, \delta < 1$, the set $E_{\varepsilon\delta} = \{x \in E : |f(x,y) - f(x)| \leq \varepsilon$ for all $y < \delta\}$ is measurable. [If y_k, $k = 1, 2, \ldots,$ is a dense subset of $(0,\delta)$, show that $E_{\varepsilon\delta} = \bigcap_k \{x \in E : |f(x,y_k) - f(x)| \leq \varepsilon\}$.]
14. Let $f(x,y)$ be as in Exercise 13. Show that given $\varepsilon > 0$, there exists a closed $F \subset E$ with $|E - F| < \varepsilon$ such that $f(x,y)$ converges uniformly for $x \in F$ to $f(x)$ as $y \to 0$. [Follow the proof of Egorov's Theorem, using the sets $E_{\varepsilon, 1/m}$ defined in Exercise 13 for the sets E_m of (4.18).]
15. Let $\{f_k\}$ be a sequence of measurable functions defined on a measurable E with $|E| < +\infty$. If $|f_k(\mathbf{x})| \leq M_\mathbf{x} < +\infty$ for all k for each $\mathbf{x} \in E$, show that given $\varepsilon > 0$, there is a closed $F \subset E$ and a finite M such that $|E - F| < \varepsilon$ and $|f_k(\mathbf{x})| \leq M$ for all k and all $\mathbf{x} \in F$.

16. Prove that $f_k \xrightarrow{m} f$ on E if and only if given $\varepsilon > 0$, there exists K such that $|\{|f - f_k| > \varepsilon\}| < \varepsilon$ if $k > K$. Give an analogous Cauchy criterion.

17. Suppose that $f_k \xrightarrow{m} f$ and $g_k \xrightarrow{m} g$ on E. Show that $f_k + g_k \xrightarrow{m} f + g$ on E and, if $|E| < +\infty$, that $f_k g_k \xrightarrow{m} fg$ on E. If, in addition, $g_k \to g$ on E, $g \neq 0$ a.e., and $|E| < +\infty$, show that $f_k/g_k \xrightarrow{m} f/g$ on E. [For the product f_k/g_k, write $f_k g_k - fg = (f_k - f)(g_k - g) + f(g_k - g) + g(f_k - f)$. Consider each term separately, using the fact that a function which is finite on E, $|E| < +\infty$, is bounded outside a subset of E with small measure.]

18. If f is measurable on E, define $\omega_f(a) = |\{f > a\}|$ for $-\infty < a < +\infty$. If $f_k \nearrow f$, show that $\omega_{f_k} \nearrow \omega_f$. If $f_k \xrightarrow{m} f$, show that $\omega_{f_k} \to \omega_f$ at each point of continuity of ω_f. [For the second part, show that if $f_k \xrightarrow{m} f$, then $\limsup_{k\to\infty} \omega_{f_k}(a) \leq \omega_f(a - \varepsilon)$ and $\liminf_{k\to\infty} \omega_{f_k}(a) \geq \omega_f(a + \varepsilon)$ for every $\varepsilon > 0$.]

19. Let $f(x,y)$ be a function defined on the unit square $0 \leq x \leq 1$, $0 \leq y \leq 1$ which is continuous in each variable separately. Show that f is a measurable function of (x,y).

20. If f is measurable on $[a,b]$, show that given $\varepsilon > 0$, there is a continuous g on $[a,b]$ such that $|\{x : f(x) \neq g(x)\}| < \varepsilon$. (See Exercise 18 of Chapter 1.)

Chapter 5

The Lebesgue Integral

1. Definition of the Integral of a Nonnegative Function

There are several equivalent ways to define the Lebesgue integral and develop its main properties. The approach we have chosen is based on the notion that the integral of a nonnegative f should represent the volume of the region under the graph of f.

We start then with a *nonnegative* function f, $0 \le f \le +\infty$, defined on a measurable subset E of $\mathbf{R}^n$. Let

$$\begin{aligned}\Gamma(f,E) &= \{(\mathbf{x},f(\mathbf{x})) \in \mathbf{R}^{n+1} : \mathbf{x} \in E, f(\mathbf{x}) < +\infty\},\\ R(f,E) &= \{(\mathbf{x},y) \in \mathbf{R}^{n+1} : \mathbf{x} \in E,\ 0 \le y \le f(\mathbf{x}) \text{ if } f(\mathbf{x}) < +\infty\\ &\qquad \text{and} \quad 0 \le y < +\infty \quad \text{if} \quad f(\mathbf{x}) = +\infty\}.\end{aligned}$$

$\Gamma(f,E)$ is called the *graph of f over E* and $R(f,E)$ the *region under f over E*.

If $R(f,E)$ is measurable (as a subset of $\mathbf{R}^{n+1}$), its measure $|R(f,E)|_{(n+1)}$ is called the *Lebesgue integral of f over E*, and we write

$$|R(f,E)|_{(n+1)} = \int_E f(\mathbf{x})\, d\mathbf{x}.$$

Usually, one of the abbreviations

$$\int_E f\, d\mathbf{x} \quad \text{or} \quad \int_E f$$

is used, and at times the lengthy notation

$$\int \underset{E}{\cdots} \int f(x_1, \ldots, x_n)\, dx_1 \cdots dx_n$$

is convenient. We stress that the definition applies only to nonnegative f; a definition for functions which are not nonnegative will be given later. We also note that the existence of the integral is equivalent to the measurability of $R(f,E)$ and does not require the finiteness of $|R(f,E)|_{(n+1)}$. The next theorem is of basic importance.

(5.1) Theorem *Let f be a nonnegative function defined on a measurable set E. Then* $\int_E f$ *exists if and only if f is measurable.*

We will show here only that the integral exists if f is measurable, postponing a proof of the converse until Chapter 6, (6.11). We need several lemmas, the first of which proves the theorem for functions which are constant on E. In this case, $R(f,E)$ is a *cylinder set*; that is, it has the form $\{(\mathbf{x},y) : \mathbf{x} \in E, 0 \le y \le a\}$.

(5.2) Lemma *Let E be a subset of* $\mathbf{R}^n$, $0 \le a \le +\infty$, *and define* $E_a = \{(\mathbf{x},y) : \mathbf{x} \in E, 0 \le y \le a\}$ *for finite a and* $E_\infty = \{(\mathbf{x},y) : \mathbf{x} \in E, 0 \le y < +\infty\}$. *If E is measurable (as a subset of* $\mathbf{R}^n$*), then* E_a *is measurable (as a subset of* $\mathbf{R}^{n+1}$*) and* $|E_a|_{(n+1)} = a|E|_{(n)}$.*

Proof. The result follows from a series of simple observations. First, assume that a is finite. If $|E| = 0$ or if E is an interval which is either closed, partly open, or open, the result is clear. Next, if E is an open set, then by (1.11), it can be written as a disjoint union of partly open intervals, $E = \bigcup I_k$. Therefore, $E_a = \bigcup I_{k,a}$, and since the $I_{k,a}$ are measurable and disjoint, E_a is measurable and $|E_a| = \sum |I_{k,a}| = \sum a|I_k| = a|E|$.

Let E be of type G_δ, $E = \bigcap_{k=1}^\infty G_k$, with $|G_1| < +\infty$. We may assume that $G_k \searrow E$ by writing $E = G_1 \cap (G_1 \cap G_2) \cap (G_1 \cap G_2 \cap G_3) \cap \cdots$. Therefore, by (3.26), $|G_k| \to |E|$ as $k \to \infty$. Moreover, $G_{k,a}$ is measurable, $|G_{k,a}| = a|G_k|$, and $G_{k,a} \searrow E_a$. Therefore, E_a is measurable and $|E_a| = \lim_{k\to\infty} |G_{k,a}| = a \lim_{k\to\infty} |G_k| = a|E|$.

If E is any measurable set with $|E| < +\infty$, then by (3.28), $E = H - Z$, where $|Z| = 0$, H is a set of type G_δ, $H = \bigcap_{k=1}^\infty G_k$, and $|G_1| < +\infty$. Since $E_a = H_a - Z_a$, we see that E_a is measurable and $|E_a| = |H_a| = a|H| = a|E|$. Finally, if $|E| = +\infty$, the result follows by writing E as the countable union of disjoint measurable sets with finite measure. This completes the proof in case a is finite.

If $a = +\infty$, choose a sequence $\{a_k\}$ of finite numbers increasing to $+\infty$. The conclusion then follows easily from the fact that $E_{a_k} \nearrow E_\infty$.

(5.3) Lemma *If f is a nonnegative measurable function on E,* $0 \le |E| \le +\infty$, *then* $\Gamma(f,E)$ *has measure zero.*

Proof. Given $\varepsilon > 0$ and $k = 0, 1, \ldots,$ let $E_k = \{k\varepsilon \le f < (k+1)\varepsilon\}$. The E_k are disjoint and measurable, and their union is the subset of E where f is finite. Hence, $\Gamma(f,E) = \bigcup \Gamma(f,E_k)$. Since $|\Gamma(f,E_k)|_e \le \varepsilon|E_k|$ by (5.2), we obtain

$$|\Gamma(f,E)|_e \le \sum |\Gamma(f,E_k)|_e \le \varepsilon \sum |E_k| \le \varepsilon|E|.$$

*Here and below, $0 \cdot \infty$ should be interpreted as 0.

If $|E| < +\infty$, this implies that $\Gamma(f,E)$ has measure zero. If $|E| = +\infty$, write E as the countable union of disjoint measurable sets with finite measure. Then $\Gamma(f,E)$ is the countable union of sets of measure zero, and the lemma follows.

Proof of the sufficiency in (5.1). Let f be nonnegative and measurable on E. By (4.13), there exist simple measurable $f_k \nearrow f$. Therefore, $R(f_k,E) \cup \Gamma(f,E) \nearrow R(f,E)$, and since $\Gamma(f,E)$ is measurable (with measure zero), it is enough to show that each $R(f_k,E)$ is measurable. Fix k and suppose that the distinct values of f_k are $a_1, \dots, a_N$, taken on measurable sets $E_1, \dots, E_N$, respectively. Then $R(f_k,E) = \bigcup_{j=1}^N E_{j,a_j}$. Therefore, $R(f_k,E)$ is measurable by (5.2), and the proof is complete.

(5.4) Corollary *If f is a nonnegative measurable function, taking values $a_1, \dots, a_N$ on disjoint sets $E_1, \dots, E_N$, respectively, and if $E = \bigcup E_j$, then*

$$\int_E f = \sum_{j=1}^N a_j |E_j|.$$

Proof. Clearly $R(f,E) = \bigcup_{j=1}^N E_{j,a_j}$. Since the E_j are measurable and disjoint, so are the E_{j,a_j}. Therefore, $\int_E f = \sum_{j=1}^N |E_{j,a_j}|$, and the corollary follows from the fact that $|E_{j,a_j}| = a_j|E_j|$.

2. Properties of the Integral

(5.5) Theorem

(i) *If f and g are measurable and if $0 \le g \le f$ on E, then $\int_E g \le \int_E f$. In particular, $\int_E (\inf f) \le \int_E f$.*
(ii) *If f is nonnegative and measurable on E and if $\int_E f$ is finite, then $f < +\infty$ a.e. in E.*
(iii) *Let E_1 and E_2 be measurable and $E_1 \subset E_2$. If f is nonnegative and measurable on E_2, then $\int_{E_1} f \le \int_{E_2} f$.*

Proof. Parts (i) and (iii) follow from the relations $R(g,E) \subset R(f,E)$ and $R(f,E_1) \subset R(f,E_2)$, respectively. To prove (ii), we may assume that $|E| > 0$. If $f = +\infty$ in a subset E_1 of E with positive measure, then by (i) and (iii), we have $\int_E f \ge \int_{E_1} f \ge \int_{E_1} a = a|E_1|$, no matter how large a is. This contradicts the finiteness of $\int_E f$.

(5.6) Theorem (*Monotone Convergence Theorem for Nonnegative Functions*) *If $\{f_k\}$ is a sequence of nonnegative measurable functions such that $f_k \nearrow f$ on E, then*

$$\int_E f_k \to \int_E f.$$

Proof. By (4.12), f is measurable. Since $R(f_k,E) \cup \Gamma(f,E) \nearrow R(f,E)$ and $\Gamma(f,E)$ has measure zero, the result follows from (3.26).

(5.7) Theorem *Suppose that f is nonnegative and measurable on E, and that E is the countable union of disjoint measurable sets E_j, $E = \bigcup E_j$. Then*

$$\int_E f = \sum \int_{E_j} f.$$

Proof. The sets $R(f,E_j)$ are disjoint and measurable. Since $R(f,E) = \bigcup R(f,E_j)$, the result follows from (3.23).

The next four theorems are corollaries of the results just proved. The first one provides an alternate definition of the integral which will be useful in Chapter 10 as a motivation for defining integration with respect to abstract measures.

(5.8) Theorem *Let f be nonnegative and measurable on E. Then*

$$\int_E f = \sup \sum_j [\inf_{x \in E_j} f(x)]|E_j|,$$

where the supremum is taken over all decompositions $E = \bigcup_j E_j$ of E into the union of a finite number of disjoint measurable sets E_j.

The reader will observe that the formula resembles the definition of the Riemann integral if the E_j are taken to be subintervals.

Proof. If $E = \bigcup_{j=1}^N E_j$ is such a decomposition, consider the measurable function g taking values $a_j = \inf_{y \in E_j} f(y)$ on E_j, $j = 1, \ldots, N$. Since $0 \le g \le f$, we have by (5.4) and (5.5) that $\sum_{j=1}^N a_j|E_j| \le \int_E f$. Therefore,

$$\sup \sum_j (\inf_{E_j} f)|E_j| \le \int_E f.$$

To prove the opposite inequality, consider for $k = 1, 2, \ldots,$ the sets $\{E_j^{(k)}\}$, $j = 0, 1, \ldots, k2^k$, defined by

$$E_j^{(k)} = \left\{\frac{j-1}{2^k} \le f < \frac{j}{2^k}\right\}, j = 1, \ldots, k2^k;\ E_0^{(k)} = \{f \ge k\},$$

and the corresponding measurable functions

$$f_k = \sum_j (\inf_{E_j^{(k)}} f)\chi_{E_j^{(k)}}.$$

[Compare the simple functions in (4.13)(ii).] Then $0 \le f_k \nearrow f$, and by the monotone convergence theorem, $\int_E f_k \to \int_E f$. Since $\int_E f_k = \sum_j (\inf_{E_j^{(k)}} f)|E_j^{(k)}|$, it follows that

$$\sup \sum_j (\inf_{E_j} f)|E_j| \ge \int_E f,$$

which completes the proof.

(5.9) Theorem *Let f be nonnegative on E. If $|E| = 0$, then $\int_E f = 0$.*

This can be proved in many ways; for example, it follows immediately from the last result.

We can now slightly strengthen the statement of part (i) of (5.5).

(5.10) Theorem *If f and g are nonnegative and measurable on E and if $g \leqslant f$ a.e. in E, then $\int_E g \leqslant \int_E f$.*

In particular, if f and g are nonnegative and measurable on E and if $f = g$ a.e. in E, then $\int_E f = \int_E g$.

Proof. Write $E = A \cup Z$, where A and Z are disjoint and $Z = \{g > f\}$. Then $|Z| = 0$. Therefore, by (5.7) and (5.9), $\int_E f = \int_A f + \int_Z f = \int_A f$. Since the same is true for g, and since $f \geq g$ everywhere on A, the result follows.

In defining $\int_E f$, we assumed that f was defined everywhere in E. In view of (5.10), $\int_E f$ is unchanged if we modify f in a set of measure zero. Hence, we may consider integrals $\int_E f$ where f is defined only a.e. in E, by completing the definition of f arbitrarily in the set Z of measure zero where it is undefined. As a result of (5.9), this amounts to defining $\int_E f$ to be $\int_{E-Z} f$. Similarly, we may extend the definition of the integral and the results above to measurable functions which are nonnegative only a.e. in E.

(5.11) Theorem *Let f be nonnegative and measurable on E. Then $\int_E f = 0$ if and only if $f = 0$ a.e. in E.*

Proof. If $f = 0$ a.e. in E, then $\int_E f = 0$ by (5.10). Conversely, suppose that f is nonnegative and measurable on E and that $\int_E f = 0$. For $\alpha > 0$, we have by (5.4) and (5.5) that

$$\alpha|\{\mathbf{x} \in E : f(\mathbf{x}) > \alpha\}| \leq \int_{\{\mathbf{x}\in E: f(\mathbf{x})>\alpha\}} f \leq \int_E f = 0.$$

Therefore, $\{\mathbf{x} \in E : f(\mathbf{x}) > \alpha\}$ has measure zero for every $\alpha > 0$. Since the set where $f > 0$ is the union of those where $f > 1/k$, it follows that $f = 0$ a.e. in E.

The proof above also establishes the following useful result.

(5.12) Corollary (*Tchebyshev's Inequality*) *Let f be nonnegative and measurable on E. If $\alpha > 0$, then*

$$|\{\mathbf{x} \in E : f(\mathbf{x}) > \alpha\}| \leq \frac{1}{\alpha}\int_E f.$$

The significance of this inequality is that it estimates the "size" of f in terms of the integral of f.

The next two theorems establish the linear properties of the integral for nonnegative functions.

(5.13) Theorem *If f is nonnegative and measurable, and if c is any nonnegative constant, then*

$$\int_E cf = c\int_E f.$$

Proof. If f is simple, then so is cf, and the theorem follows in this case from the formula for integrating simple functions [see (5.4)]. For arbitrary measurable $f \geq 0$, choose simple measurable f_k with $0 \leq f_k \nearrow f$. Then $0 \leq cf_k \nearrow cf$ and

$$\int_E cf = \lim_{k\to\infty}\int_E cf_k = \lim_{k\to\infty} c\int_E f_k = c\int_E f.$$

(5.14) Theorem *If f and g are nonnegative and measurable, then*

$$\int_E (f+g) = \int_E f + \int_E g.$$

Proof. First suppose that f and g are simple: $f = \sum_{i=1}^N a_i\chi_{A_i}$ and $g = \sum_{j=1}^M b_j\chi_{B_j}$. Then $f+g$ is also simple, taking values $a_i + b_j$ on $A_i \cap B_j$: $f + g = \sum_{i,j}(a_i + b_j)\chi_{A_i\cap B_j}$. Thus,

$$\int_E (f+g) = \sum_{i,j}(a_i+b_j)|A_i \cap B_j| = \sum_i a_i \sum_j |A_i\cap B_j| + \sum_j b_j \sum_i |A_i \cap B_j|$$
$$= \sum a_i|A_i| + \sum b_j|B_j| = \int_E f + \int_E g.$$

For general nonnegative measurable f and g, choose simple measurable f_k and g_k such that $0 \leq f_k \nearrow f$ and $0 \leq g_k \nearrow g$. Then $f_k + g_k$ is simple and $0 \leq f_k + g_k \nearrow f + g$. Therefore,

$$\int_E (f+g) = \lim_{k\to\infty}\int_E (f_k+g_k) = \lim_{k\to\infty}\left(\int_E f_k + \int_E g_k\right) = \int_E f + \int_E g,$$

which completes the proof.

(5.15) Corollary *Suppose that f and ϕ are measurable on E, $0 \leq f \leq \phi$, and $\int_E f$ is finite. Then*

$$\int_E (\phi - f) = \int_E \phi - \int_E f.$$

Proof. By (5.14), we have $\int_E f + \int_E (\phi - f) = \int_E \phi$. Since $\int_E f$ is finite, the result follows by subtraction.

(5.16) Theorem *If f_k, $k = 1, 2, \ldots$, are nonnegative and measurable, then*

$$\int_E \left(\sum_{k=1}^\infty f_k\right) = \sum_{k=1}^\infty \int_E f_k.$$

Proof. The functions F_N defined by $F_N = \sum_{k=1}^{N} f_k$ are nonnegative and measurable, and increase to $\sum_{k=1}^{\infty} f_k$. Hence,

$$\int_E \left(\sum_{k=1}^{\infty} f_k\right) = \lim_{N\to\infty} \int_E F_N = \lim_{N\to\infty} \sum_{k=1}^{N} \int_E f_k = \sum_{k=1}^{\infty} \int_E f_k.$$

Note that the preceding theorem is essentially a corollary of the monotone convergence theorem (5.6). Conversely, the monotone convergence theorem can be deduced from this result. Verification is left to the reader.

The monotone convergence theorem gives a sufficient condition for interchanging the operations of integration and passage to the limit: $\int_E \lim f_k = \lim \int_E f_k$. It is an important problem to find other conditions under which this is true. First, we show that some restriction other than the mere convergence of f_k to f is necessary. Let E be the interval $[0,1]$, and for $k = 1, 2, \ldots,$ let f_k be defined as follows: when $0 \leq x \leq 1/k$, the graph of f_k consists of the sides of the isosceles triangle with altitude k and base $[0,1/k]$; when $1/k \leq x \leq 1, f_k(x) = 0$.

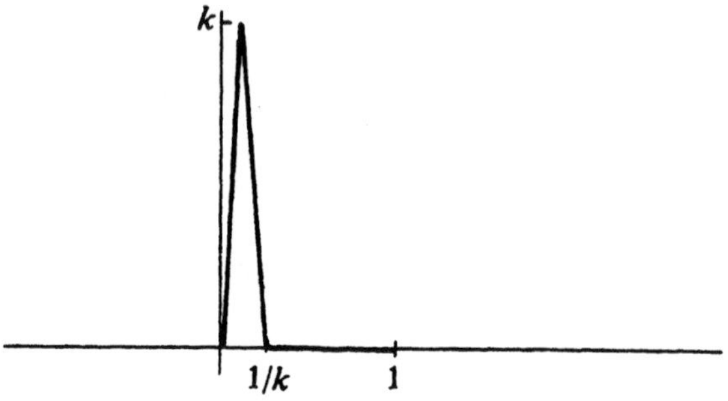

Clearly, $f_k \to 0$ on $[0,1]$, but $\int_0^1 f_k = \frac{1}{2}(1/k)(k) = \frac{1}{2}$ for all k. Hence, $\lim \int_0^1 f_k \neq \int_0^1 \lim f_k$.

(5.17) Theorem (*Fatou's Lemma*) *If $\{f_k\}$ is a sequence of nonnegative measurable functions on E, then*

$$\int_E (\liminf_{k\to\infty} f_k) \leq \liminf_{k\to\infty} \int_E f_k.$$

Proof. First note that the integral on the left exists since its integrand is nonnegative and measurable. Next, let $g_k = \inf\{f_k, f_{k+1}, \ldots\}$ for each k. Then $g_k \nearrow \liminf f_k$ and $0 \leq g_k \leq f_k$. Therefore, by (5.6) and (5.10),

$$\int_E g_k \to \int_E (\liminf f_k), \qquad \int_E g_k \leq \int_E f_k,$$

so that

$$\int_E (\liminf f_k) = \lim \int_E g_k \le \liminf \int_E f_k.$$

(5.18) Corollary *Let f_k, $k = 1, 2, \ldots$, be nonnegative and measurable on E, and let $f_k \to f$ a.e. in E. If $\int_E f_k \le M$ for all k, then $\int_E f \le M$.*

Proof. By Fatou's lemma, $\int_E (\liminf f_k) \le M$. Since $\liminf f_k = \lim f_k = f$ a.e. in E, the conclusion follows.

We now prove a basic result about term-by-term integration of convergent sequences.

(5.19) Theorem (*Lebesgue's Dominated Convergence Theorem for Nonnegative Functions*) *Let $\{f_k\}$ be a sequence of nonnegative measurable functions on E such that $f_k \to f$ a.e. in E. If there exists a measurable function ϕ such that $f_k \le \phi$ a.e. for all k and if $\int_E \phi$ is finite, then*

$$\int_E f_k \to \int_E f.$$

Proof. By Fatou's lemma,

$$\int_E f = \int_E \liminf f_k \le \liminf \int_E f_k,$$

and the theorem will follow if we show that

$$\int_E f \ge \limsup \int_E f_k.$$

To prove this inequality, apply Fatou's lemma to the nonnegative functions $\phi - f_k$, obtaining

$$\int_E \liminf (\phi - f_k) \le \liminf \int_E (\phi - f_k).$$

Since $f_k \to f$ a.e., the integrand on the left equals $\phi - f$ a.e., so that the integral on the left is $\int_E \phi - \int_E f$ by (5.15). The right-hand integral equals

$$\liminf\left(\int_E \phi - \int_E f_k\right) = \int_E \phi - \limsup \int_E f_k.$$

Combining formulas and cancelling $\int_E \phi$, we obtain the inequality $\int_E f \ge \limsup \int_E f_k$, as desired.

3. The Integral of an Arbitrary Measurable f

Let f be any measurable function defined on a set E. Then $f = f^+ - f^-$ and, by the remark following (4.11), f^+ and f^- are measurable. Therefore, the integrals $\int_E f^+(\mathbf{x})\,d\mathbf{x}$ and $\int_E f^-(\mathbf{x})\,d\mathbf{x}$ exist and are nonnegative, possibly having value $+\infty$. *Provided at least one of these integrals is finite*, we define

$$\int_E f(\mathbf{x})\,d\mathbf{x} = \int_E f^+(\mathbf{x})\,d\mathbf{x} - \int_E f^-(\mathbf{x})\,d\mathbf{x},$$

and say that the integral $\int_E f(\mathbf{x})\,d\mathbf{x}$ *exists*. If $f \geq 0$, then $f = f^+$, and this definition agrees with the previous one. As in the case when $f \geq 0$, we will use the abbreviations $\int_E f\,d\mathbf{x}$ and $\int_E f$.

The definition clearly applies if f is defined only a.e. in E, as in the case when $f \geq 0$ (see p. 68). For the sake of simplicity, we shall usually assume that f is defined everywhere in E.

If $\int_E f$ exists then, of course, $-\infty \leq \int_E f \leq +\infty$. If $\int_E f$ exists *and is finite*, we say that f is *Lebesgue integrable*, or simple *integrable*, on E and write $f \in L(E)$. Thus,

$$L(E) = \left\{f : \int_E f \text{ is finite}\right\}.$$

If $\int_E f$ exists, then

$$\left|\int_E f\right| \leq \int_E f^+ + \int_E f^- = \int_E (f^+ + f^-),$$

by (5.14). Since $f^+ + f^- = |f|$, we obtain the inequality

(5.20)
$$\left|\int_E f\,d\mathbf{x}\right| \leq \int_E |f|\,d\mathbf{x}.$$

(5.21) Theorem *Let f be measurable on E. Then f is integrable over E if and only if $|f|$ is.*

Proof. It follows immediately from (5.20) that $f \in L(E)$ if $|f| \in L(E)$. If $f \in L(E)$, then the difference $\int_E f^+ - \int_E f^-$ is finite, and therefore, since at least one of $\int_E f^+$ or $\int_E f^-$ is finite, both must be finite. Hence, their sum is finite. Since this sum is $\int_E (f^+ + f^-) = \int_E |f|$, it follows that $|f| \in L(E)$.

The simple properties of $\int_E f$ for general f follow from the results already established for $f \geq 0$. As a first example, we have the following theorem.

(5.22) Theorem *If $f \in L(E)$, then f is finite a.e. in E.*

Proof. If $f \in L(E)$, then $|f| \in L(E)$, and the result follows from (5.5)(ii).

(5.23) Theorem

(i) *If both $\int_E f$ and $\int_E g$ exist and if $f \leq g$ a.e. in E, then $\int_E f \leq \int_E g$. In particular, if $f = g$ a.e. in E, then $\int_E f = \int_E g$.*
(ii) *If $\int_{E_2} f$ exists and E_1 is a measurable subset of E_2, then $\int_{E_1} f$ exists.*

Proof.

(i) The fact that $f \le g$ a.e. implies that $0 \le f^+ \le g^+$ and $0 \le g^- \le f^-$ a.e. in E. By (5.10), we then have $\int_E f^+ \le \int_E g^+$ and $\int_E f^- \ge \int_E g^-$, and (i) follows by subtracting these inequalities.

(ii) If $\int_{E_2} f$ exists, at least one of $\int_{E_2} f^+$ or $\int_{E_2} f^-$ is finite. If $E_1 \subset E_2$, then by (5.5)(iii), at least one of $\int_{E_1} f^+$ or $\int_{E_1} f^-$ is finite. Therefore, $\int_{E_1} f$ exists.

(5.24) Theorem *If $\int_E f$ exists and $E = \bigcup_k E_k$ is the countable union of disjoint measurable sets E_k, then*

$$\int_E f = \sum_k \int_{E_k} f.$$

Proof. Each $\int_{E_k} f$ exists by (5.23)(ii). We have

$$\int_E f = \int_E f^+ - \int_E f^- = \sum \int_{E_k} f^+ - \sum \int_{E_k} f^-$$

by (5.7). Since at least one of these sums is finite, we obtain

$$\int_E f = \sum \left(\int_{E_k} f^+ - \int_{E_k} f^- \right) = \sum \int_{E_k} f.$$

(5.25) Theorem *If $|E| = 0$ or if $f = 0$ a.e. in E, then $\int_E f = 0$.*

Proof. The theorem follows by applying (5.9) or (5.11) to f^+ and f^-.

The next few results deal with the linear properties of the integral.

(5.26) Lemma *If $\int_E f$ is defined, then so is $\int_E (-f)$, and $\int_E (-f) = -\int_E f$.*

Proof. Since $(-f)^+ = f^-$ and $(-f)^- = f^+$, and at least one of $\int_E f^-$ or $\int_E f^+$ is finite, we have $\int_E (-f) = \int_E f^- - \int_E f^+ = -\int_E f$.

(5.27) Theorem *If $\int_E f$ exists and c is any real constant, then $\int_E (cf)$ exists and*

$$\int_E (cf) = c \int_E f.$$

Proof. If $c \ge 0$, $(cf)^+ = cf^+$ and $(cf)^- = cf^-$. Therefore, by (5.13), $\int_E (cf)^+ = c \int_E f^+$ and $\int_E (cf)^- = c \int_E f^-$. It follows that $\int_E (cf)$ exists and $\int_E (cf) = c(\int_E f^+ - \int_E f^-) = c \int_E f$. If $c = -1$, the theorem reduces to (5.26). For any $c \le 0$, we have $c = (-1)(|c|)$, and the result follows from the cases $c = -1$ and $c \ge 0$.

(5.28) Theorem *If $f, g \in L(E)$, then $f + g \in L(E)$ and*

$$\int_E (f + g) = \int_E f + \int_E g.$$

Proof. Since $|f + g| \leq |f| + |g|$, we have from (5.23)(i) and (5.14) that

$$\int_E |f + g| \leq \int_E (|f| + |g|) = \int_E |f| + \int_E |g| < +\infty.$$

Hence $f + g \in L(E)$.

To prove the rest of the theorem, first suppose that each of f, g, and $f + g$ has constant sign on E. There are then six possibilities: (1) $f \geq 0$, $g \geq 0$ (so that $f + g \geq 0$); (2) $f \geq 0$, $g < 0$, $f + g \geq 0$; (3) $f \geq 0$, $g < 0$, $f + g < 0$; (4) $f < 0$, $g \geq 0$, $f + g \geq 0$; (5) $f < 0$, $g \geq 0$, $f + g < 0$; (6) $f < 0$, $g < 0$ (so that $f + g < 0$). Note that these possibilities are mutually exclusive. [Note also that in some cases the inequalities can be refined: for example, in case 2, f can never be zero.] The result in case 1 is just (5.14). Cases 2–6 are all similar, and we shall consider only case 2. Then $f \geq 0$, $-g > 0$, $f + g \geq 0$, and since by (5.22) each function is finite a.e., $(f + g) + (-g) = f$ a.e. Hence, by (5.23)(i) and (5.14), we have $\int_E (f + g) + \int_E (-g) = \int_E f$. The result in case 2 now follows from (5.26) and the fact that all the integrals involved are finite.

For arbitrary f and g in $L(E)$, we subdivide E into at most six measurable sets, $E_1, \ldots, E_6$, where possibilities (1), ..., (6) hold, respectively. Since E_i and E_j are disjoint for $i \neq j$, we have

$$\int_E (f + g) = \sum_{j=1}^{6} \int_{E_j} (f + g) = \sum_{j=1}^{6} \left(\int_{E_j} f + \int_{E_j} g \right) = \int_E f + \int_E g.$$

This completes the proof.

It follows that if $f_k \in L(E)$, $k = 1, \ldots, N$, and if a_k are real constants, then $\sum_{k=1}^{N} a_k f_k \in L(E)$ and

$$\int_E \left(\sum_{k=1}^{N} a_k f_k \right) = \sum_{k=1}^{N} a_k \int_E f_k.$$

(5.29) Corollary *Let f and ϕ be measurable on E, $f \geqslant \phi$ a.e. and $\phi \in L(E)$. Then*

$$\int_E (f - \phi) = \int_E f - \int_E \phi.$$

Proof. First, note that $\int_E f$ exists since $f^- \leqslant \phi^-$ a.e. implies that $\int_E f^-$ is finite. Next, $\int_E (f - \phi)$ exists since $f - \phi \geqslant 0$ a.e. If $f \in L(E)$, the corollary follows from (5.28). If $f \notin L(E)$, then since $f^- \in L(E)$, we must have $\int_E f = +\infty$. The fact that $\phi \in L(E)$ implies that $f - \phi \notin L(E)$, so that $\int_E (f - \phi) = +\infty$, since $f - \phi \geqslant 0$ a.e. This completes the proof.

In Chapter 8, we will study conditions on f and g which imply that $fg \in L(E)$. For now, we have the following simple result.

(5.30) Theorem *If $f \in L(E)$, g is measurable on E, and there exists a finite constant M such that $|g| \leq M$ a.e. in E, then $fg \in L(E)$.*

Proof. Since $|fg| \le M|f|$ a.e. in E, we have by (5.10) and (5.27) that $\int_E |fg| \le \int_E M|f| = M\int_E |f|$. Hence, $fg \in L(E)$.

(5.31) **Corollary** *If $f \in L(E)$, $f \geqslant 0$ a.e., and there exist finite constants α and β such that $\alpha \leqslant g \leqslant \beta$ a.e. in E, then*

$$\alpha \int_E f \le \int_E fg \le \beta \int_E f.$$

Proof. By (5.30), $fg \in L(E)$. Since $f \geqslant 0$ a.e., we have $\alpha f \leqslant fg \leqslant \beta f$ a.e. in E, and the conclusion follows by integrating.

We now study conditions which imply that $\int_E f_k \to \int_E f$ if $f_k \to f$ in E. Most of the results are extensions of those we derived for nonnegative functions.

(5.32) **Theorem** (***Monotone Convergence Theorem***) *Let $\{f_k\}$ be a sequence of measurable functions on E.*

(i) *If $f_k \nearrow f$ a.e. on E and there exists $\phi \in L(E)$ such that $f_k \geqslant \phi$ a.e. on E for all k, then $\int_E f_k \to \int_E f$.*

(ii) *If $f_k \searrow f$ a.e. on E and there exists $\phi \in L(E)$ such that $f_k \leqslant \phi$ a.e. on E for all k, then $\int_E f_k \to \int_E f$.*

Proof. To prove (i), we may assume by (5.25) that $f_k \nearrow f$ and $f_k \geqslant \phi$ everywhere on E. Then $0 \leqslant f_k - \phi \nearrow f - \phi$ on E, so that by (5.6), $\int_E (f_k - \phi) \to \int_E (f - \phi)$. Therefore, by (5.29), $\int_E f_k - \int_E \phi \to \int_E f - \int_E \phi$, and since $\phi \in L(E)$, the result follows.

We can deduce (ii) from (i) by considering the functions $-f_k$. Details are left to the reader.

(5.33) **Theorem** (***Uniform Convergence Theorem***) *Let $f_k \in L(E)$, $k = 1, 2, \ldots,$ and let $\{f_k\}$ converge uniformly to f on E, $|E| < +\infty$. Then $f \in L(E)$ and $\int_E f_k \to \int_E f$.*

Proof. Since $|f| \le |f_k| + |f - f_k|$ and $\{f_k\}$ converges uniformly to f on E, we have $|f| \le |f_k| + 1$ on E if k is sufficiently large. Since $|E| < +\infty$, it follows that $f \in L(E)$. From (5.28) and (5.20), we obtain

$$\left|\int_E f - \int_E f_k\right| = \left|\int_E (f - f_k)\right| \le \int_E |f - f_k|.$$

The last integral is bounded by $(\sup_{x \in E} |f(x) - f_k(x)|)|E|$, which by hypothesis tends to 0 as $k \to \infty$. This proves the theorem.

(5.34) **Theorem** (***Fatou's Lemma***) *Let $\{f_k\}$ be a sequence of measurable functions on E. If there exists $\phi \in L(E)$ such that $f_k \geqslant \phi$ a.e. on E for all k, then*

$$\int_E (\liminf_{k\to\infty} f_k) \le \liminf_{k\to\infty} \int_E f_k.$$

Proof. The result follows by first applying (5.17) to the sequence $\{f_k - \phi\}$ of nonnegative functions, and then using (5.29). Details are left to the reader.

(5.35) Corollary *Let $\{f_k\}$ be a sequence of measurable functions on E. If there exists $\phi \in L(E)$ such that $f_k \leqslant \phi$ a.e. on E for all k, then*

$$\int_E (\limsup_{k\to\infty} f_k) \geq \limsup_{k\to\infty} \int_E f_k.$$

Proof. This follows from Fatou's lemma since $-f_k \geqslant -\phi$ a.e. and $\liminf\,(-f_k) = -\limsup f_k$.

(5.36) Theorem (*Lebesgue's Dominated Convergence Theorem*) *Let $\{f_k\}$ be a sequence of measurable functions on E such that $f_k \to f$ a.e. in E. If there exists $\phi \in L(E)$ such that $|f_k| \leq \phi$ a.e. in E for all k, then $\int_E f_k \to \int_E f$.*

Proof. By hypothesis, $-\phi \leq f_k \leq \phi$ a.e. in E. Therefore, $0 \leq f_k + \phi \leq 2\phi$ a.e. in E. Since $2\phi \in L(E)$, we conclude from (5.19) that $\int_E (f_k + \phi) \to \int_E (f + \phi)$. Since ϕ, f, and all the f_k are integrable on E, the result follows from (5.28).

The following special case of the dominated convergence theorem is often useful if E has finite measure.

(5.37) Corollary (*Bounded Convergence Theorem*) *Let $\{f_k\}$ be a sequence of measurable functions on E such that $f_k \to f$ a.e. in E. If $|E| < +\infty$ and there is a finite constant M such that $|f_k| \leq M$ a.e. in E, then $\int_E f_k \to \int_E f$.*

In later chapters, we will consider the integrals of complex-valued functions. Here we mention only the definition. (See p. 125 for some further remarks.) If $f = f_1 + if_2$ with f_1 and f_2 real-valued, we define

$$\int_E f = \int_E f_1 + i \int_E f_2.$$

(For the measurability of such f, see p. 61, Exercise 3.) Many basic properties of the ordinary integral are valid in this case.

4. A Relation Between Riemann-Stieltjes and Lebesgue Integrals; the L^p Spaces, $0 < p < \infty$

It turns out that there is a remarkably simple and useful representation of Lebesgue integrals over subsets of $\mathbf{R}^n$ in terms of Riemann-Stieltjes integrals (over subsets of $\mathbf{R}^1$, of course). In order to establish this relation, we must first study the function

$$\omega(\alpha) = \omega_{f,E}(\alpha) = |\{\mathbf{x} \in E : f(\mathbf{x}) > \alpha\}|,$$

where f is a measurable function on E and $-\infty < \alpha < +\infty$. We call $\omega_{f,E}$ *the distribution function of f on E.*

Some properties of ω were given in Exercise 18, Chapter 4. Clearly, it is not affected by changing f in a set of measure zero, and it is decreasing. As $\alpha \to +\infty$,

$$\{\mathbf{x} \in E : f(\mathbf{x}) > \alpha\} \searrow \{\mathbf{x} \in E : f(\mathbf{x}) = +\infty\};$$

hence, if we assume that f is finite a.e. in E, then by (3.26)(ii)

$$\lim_{\alpha \to +\infty} \omega(\alpha) = 0,$$

unless $\omega(\alpha) \equiv +\infty$. Similarly,

$$\lim_{\alpha \to -\infty} \omega(\alpha) = |E|.$$

We will assume from now on that $|E| < +\infty$. This insures that ω is bounded, that $\lim_{\alpha \to +\infty} \omega(\alpha) = 0$, and that ω is of bounded variation on $(-\infty, +\infty)$ with variation equal to $|E|$. The assumption is made only to simplify the properties of ω, and is not entirely necessary (see, e.g., Exercise 16); in fact, the case $|E| = +\infty$ is often important.

In the results below, f is a measurable function which is finite a.e. in E, $|E| < +\infty$, and we write

$$\omega(\alpha) = \omega_{f,E}(\alpha), \qquad \{f > \alpha\} = \{\mathbf{x} \in E : f(\mathbf{x}) > \alpha\}, \text{ etc.}$$

(5.38) Lemma *If $\alpha < \beta$, then $|\{\alpha < f \le \beta\}| = \omega(\alpha) - \omega(\beta)$.*

Proof. We have $\{f > \beta\} \subset \{f > \alpha\}$ and $\{\alpha < f \le \beta\} = \{f > \alpha\} - \{f > \beta\}$. Since $|\{f > \beta\}| < +\infty$, the lemma follows from (3.25).

Given α, let

$$\omega(\alpha+) = \lim_{\varepsilon \searrow 0} \omega(\alpha + \varepsilon), \qquad \omega(\alpha-) = \lim_{\varepsilon \searrow 0} \omega(\alpha - \varepsilon)$$

denote the limits of ω from the right and left at α.

(5.39) Lemma

(a) $\omega(\alpha+) = \omega(\alpha)$; *that is, ω is continuous from the right.*
(b) $\omega(\alpha-) = |\{f \ge \alpha\}|$.

Proof. If $\varepsilon_k \searrow 0$, then $\{f > \alpha + \varepsilon_k\} \nearrow \{f > \alpha\}$ and $\{f > \alpha - \varepsilon_k\} \searrow \{f \ge \alpha\}$. Since these sets have finite measures, it follows from (3.26) that $\omega(\alpha + \varepsilon_k) \to \omega(\alpha)$ and $\omega(\alpha - \varepsilon_k) \to |\{f \ge \alpha\}|$. This completes the proof.

We now know that ω is a decreasing function which is continuous from the right. It may have jump discontinuities, with jumps $\omega(\alpha-) - \omega(\alpha)$, and intervals of constancy. These possibilities are characterized by the behavior of f stated in the following result.

(5.40) Corollary

(a) $\omega(\alpha-) - \omega(\alpha) = |\{f = \alpha\}|$; *in particular, ω is continuous at α if and only if $|\{f = \alpha\}| = 0$.*

(b) *ω is constant in an open interval (α,β) if and only if $|\{\alpha < f < \beta\}| = 0$, that is, if and only if f takes almost no values between α and β.*

Proof. Since $|\{f \geq \alpha\}| = |\{f > \alpha\}| + |\{f = \alpha\}|$, part (a) follows immediately from (5.39)(b). To prove part (b), note that $|\{\alpha < f < \beta\}| = |\{f > \alpha\}| - |\{f \geq \beta\}| = \omega(\alpha) - \omega(\beta-)$. This is zero if and only if ω is constant in the half-open interval $[\alpha,\beta)$. However, since ω is continuous from the right, it is constant in (α,β) if and only if it is constant in $[\alpha,\beta)$.

The rest of the theorems in this section give relations between Lebesgue and Riemann-Stieltjes integrals. As always, f is measurable and finite a.e. in E, $|E| < +\infty$ and $\omega = \omega_{f,E}$.

(5.41) Theorem *If $a < f(\mathbf{x}) \leq b$ (a and b finite) for $\mathbf{x} \in E$, then*

$$\int_E f = -\int_a^b \alpha \, d\omega(\alpha).$$

Proof. The Lebesgue integral on the left exists and is finite since f is bounded and $|E| < +\infty$. The Riemann-Stieltjes integral on the right exists by (2.24). To show that they are equal, partition the interval $[a,b]$ by $a = \alpha_0 < \alpha_1 < \cdots < \alpha_k = b$, and let $E_j = \{\alpha_{j-1} < f \leq \alpha_j\}$. The E_j are disjoint and $E = \bigcup_{j=1}^k E_j$. Hence, $\int_E f = \sum_{j=1}^k \int_{E_j} f$ and, therefore,

$$\sum_{j=1}^k \alpha_{j-1}|E_j| \leq \int_E f \leq \sum_{j=1}^k \alpha_j |E_j|.$$

By (5.38), $|E_j| = \omega(\alpha_{j-1}) - \omega(\alpha_j) = -[\omega(\alpha_j) - \omega(\alpha_{j-1})]$. Hence, both sums above are Riemann-Stieltjes sums for $-\int_a^b \alpha \, d\omega(\alpha)$. Since these sums must converge to $-\int_a^b \alpha \, d\omega(\alpha)$ as the norm of the partition tends to zero, the conclusion follows.

Theorem (5.41) can be extended to the case when f is not bounded on E as follows.

(5.42) Theorem *Let f be any measurable function on E, and let $E_{ab} = \{\mathbf{x} \in E : a < f(\mathbf{x}) \leq b\}$ (a and b finite). Then*

$$\int_{E_{ab}} f = -\int_a^b \alpha \, d\omega(\alpha).$$

Proof. Let $\omega_{ab}(\alpha) = |\{\mathbf{x} \in E_{ab} : f(\mathbf{x}) > \alpha\}|$. Then ω_{ab} is the distribution function of f on E_{ab}. By (5.41), we have $\int_{E_{ab}} f = -\int_a^b \alpha \, d\omega_{ab}(\alpha)$. We claim that the last integral equals $\int_a^b \alpha \, d\omega(\alpha)$. By taking limits of Riemann-Stieltjes sums which approximate these integrals, we only need to show that

$\omega_{ab}(\alpha) - \omega_{ab}(\beta) = \omega(\alpha) - \omega(\beta)$ for $a \le \alpha < \beta \le b$. By (5.38), this is equivalent to showing that $|\{\mathbf{x} \in E_{ab} : \alpha < f(\mathbf{x}) \le \beta\}| = |\{\mathbf{x} \in E : \alpha < f(\mathbf{x}) \le \beta\}|$ for such α and β. However, by the definition of E_{ab} and the restrictions on α and β, $\{\mathbf{x} \in E_{ab} : \alpha < f(\mathbf{x}) \le \beta\} = \{\mathbf{x} \in E : \alpha < f(\mathbf{x}) \le \beta\}$. This proves the claim and the theorem too.

In both (5.41) and (5.42), the integrals of f are extended over sets where f is bounded. This restriction is removed in the next theorem, where we define (see p. 28)

$$\int_{-\infty}^{+\infty} \alpha \, d\omega(\alpha) = \lim_{\substack{a \to -\infty \\ b \to +\infty}} \int_a^b \alpha \, d\omega(\alpha),$$

if the limit exists.

(5.43) Theorem *If either $\int_E f$ or $\int_{-\infty}^{+\infty} \alpha \, d\omega(\alpha)$ is finite, then the other exists and is finite, and*

$$\int_E f = -\int_{-\infty}^{+\infty} \alpha \, d\omega(\alpha).$$

Proof. By (5.42), $\int_{E_{ab}} f = -\int_a^b \alpha \, d\omega(\alpha)$. If $f \in L(E)$, then $\int_{E_{ab}} f \to \int_E f$ as $a \to -\infty$, $b \to +\infty$, since this holds for both f^+ and f^-. Therefore, $\lim_{a \to -\infty, b \to +\infty} [-\int_a^b \alpha \, d\omega(\alpha)]$ exists and equals $\int_E f$, which proves half of the theorem.

Now suppose that $\int_{-\infty}^{+\infty} \alpha \, d\omega(\alpha)$ exists and is finite. Then $\int_0^{\infty} \alpha \, d\omega(\alpha)$ is finite, and we claim that $\int_E f^+ = -\int_0^{\infty} \alpha \, d\omega(\alpha)$. By (5.42), for $b > 0$, $\int_{E_{0b}} f = -\int_0^b \alpha \, d\omega(\alpha)$. Therefore, as $b \to +\infty$, $\int_{E_{0b}} f \to -\int_0^{\infty} \alpha \, d\omega(\alpha)$. On the other hand, as b increases to $+\infty$, $E_{0b} \nearrow \{0 < f < +\infty\}$. Therefore,

$$\int_{E_{0b}} f = \int_{E_{0b}} f^+ \to \int_{\{0 < f < +\infty\}} f^+ = \int_E f^+,$$

and the claim follows. A similar argument, using the sets E_{a0} with $a \to -\infty$, shows that $\int_E f^- = \int_{-\infty}^0 \alpha \, d\omega(\alpha)$. Since all the integrals are finite, it follows that $\int_E f = \int_E f^+ - \int_E f^- = -\int_{-\infty}^{+\infty} \alpha \, d\omega(\alpha)$.

Two measurable functions f and g defined on E are said to be *equimeasurable*, or *equidistributed*, if

$$\omega_{f,E}(\alpha) = \omega_{g,E}(\alpha) \text{ for all } \alpha.$$

We may intuitively think of two equimeasurable functions as being "rearrangements" of each other. For such functions, we have

$$|\{a < f \le b\}| = |\{a < g \le b\}|, \ |\{f = a\}| = |\{g = a\}|, \text{ etc.}$$

We also have the following corollary.

(5.44) Corollary *If f and g are equimeasurable on E and $f \in L(E)$, then $g \in L(E)$ and*

$$\int_E f = \int_E g.$$

The method used to derive (5.41)–(5.43) illustrates a basic difference between Riemann and Lebesgue integrals. The Riemann integral is defined by a limiting process whose initial step involves partioning the *domain* of f. On the other hand, we saw in (5.41) that the Lebesgue integral can be obtained from a process which partitions the *range* of f. In order to define this process more clearly, let f be a nonnegative measurable function which is finite a.e. in E, $|E| < +\infty$. Let $\Gamma = \{0 = \alpha_0 < \alpha_1 < \cdots\}$ be a partition of the positive ordinate axis by a countable number of points $\alpha_k \to +\infty$, and let $|\Gamma| = \sup_k (\alpha_{k+1} - \alpha_k)$. Let $E_k = \{\alpha_k \le f < \alpha_{k+1}\}$ and $Z = \{f = +\infty\}$. Then the E_k are measurable and disjoint, $|Z| = 0$ and $E = (\bigcup E_k) \cup Z$, so that $|E| = \sum |E_k|$. Let

$$s_\Gamma = \sum \alpha_k |E_k|, \qquad S_\Gamma = \sum \alpha_{k+1} |E_k|.$$

(5.45) Theorem *Let f be a nonnegative measurable function which is finite a.e. in E, $|E| < +\infty$. Then*

$$\int_E f = \lim_{|\Gamma| \to 0} s_\Gamma = \lim_{|\Gamma| \to 0} S_\Gamma.$$

Proof. We may assume that f is finite everywhere, since changing it in a set of measure zero does not affect the expressions above. Given Γ, define functions ϕ_Γ and ψ_Γ by setting $\phi_\Gamma = \alpha_k$ in E_k and $\psi_\Gamma = \alpha_{k+1}$ in E_k, $k = 0, 1, \ldots$. Then $0 \le \phi_\Gamma \le f \le \psi_\Gamma$, $\int_E \phi_\Gamma = s_\Gamma$ and $\int_E \psi_\Gamma = S_\Gamma$. Hence,

$$s_\Gamma \le \int_E f \le S_\Gamma.$$

If $s_\Gamma < +\infty$, then

$$0 \le S_\Gamma - s_\Gamma = \sum (\alpha_{k+1} - \alpha_k)|E_k| \le |\Gamma||E|,$$

so that $S_\Gamma < +\infty$ and $S_\Gamma - s_\Gamma \to 0$ as $|\Gamma| \to 0$. The conclusion of the theorem now follows easily in case $\int_E f < +\infty$. On the other hand, if $\int_E f = +\infty$, then $S_\Gamma = +\infty$, and therefore also $s_\Gamma = +\infty$, which completes the proof.

Theorem (5.45) is the origin of an anecdote which compares the methods that Lebesgue and Riemann might have used to count coins. The story goes that Lebesgue would have been a better bankteller. To see why, imagine coins placed at various points along the x-axis (there may be coins of equal value at different points), and think of $f(x)$ as the value of the coin at x. Suppose that we want to determine the total value of all the coins. In Lebesgue's method, partitioning the ordinate axis and forming the sets E_k corresponds to sorting the coins according to value; computing $|E_k|$ corresponds to counting the number with a given value. Thus, $\sum \alpha_k|E_k| = \int f$ represents the total value. Riemann's method is less efficient; it approximates the total by arbitrarily grouping the coins (partitioning the x-axis), and then summing the products of the number of coins in a given group by the value of any chosen coin in the group.

The relation between Lebesgue and Riemann-Stieltjes integrals can be extended in a useful way to give Riemann-Stieltjes representations for integrals of the form $\int_E \phi(f)$, where f and E are subject to the usual restrictions, and ϕ is assumed to be continuous. This last assumption assures the measurability of $\phi(f)$ by (4.6).

(5.46) Theorem *If $a < f \leq b$ (a and b finite) in E and ϕ is continuous on $[a,b]$, then*

$$\int_E \phi(f) = -\int_a^b \phi(\alpha)\, d\omega(\alpha).$$

Proof. Since ϕ is bounded and E (as always) has finite measure, we see that $\phi(f) \in L(E)$. Since ϕ is continuous, the Riemann-Stieltjes integral exists by (2.24). Write f as the limit of simple measurable f_k with $a < f_k \leq b$ as follows: for $k = 1, 2, \ldots,$ let $a = \alpha_0^{(k)} < \alpha_1^{(k)} < \cdots < \alpha_{m_k}^{(k)} = b$ be partitions of $[a,b]$ with norms tending to zero, and let $f_k(\mathbf{x}) = \alpha_j^{(k)}$ when $\alpha_{j-1}^{(k)} < f(\mathbf{x}) \leq \alpha_j^{(k)}$. Then $\phi(f_k) \to \phi(f)$ in E. Since the $\phi(f_k)$ are uniformly bounded and $|E| < +\infty$, it follows from the bounded convergence theorem that $\int_E \phi(f_k) \to \int_E \phi(f)$. However, $\phi(f_k)$ is simple, taking values $\phi(\alpha_j^{(k)})$ on $\{\alpha_{j-1}^{(k)} < f \leq \alpha_j^{(k)}\}$. Therefore, by (5.38),

$$\int_E \phi(f_k) = -\sum_j \phi(\alpha_j^{(k)})[\omega(\alpha_j^{(k)}) - \omega(\alpha_{j-1}^{(k)})],$$

so that as $k \to \infty$, $\int_E \phi(f_k) \to -\int_a^b \phi(\alpha)\, d\omega(\alpha)$. This completes the proof.

In the next theorem, let

$$\int_{-\infty}^{+\infty} \phi(\alpha)\, d\omega(\alpha) = \lim_{\substack{a\to-\infty \\ b\to+\infty}} \int_a^b \phi(\alpha)\, d\omega(\alpha),$$

if the limit exists (cf. p. 28).

(5.47) Theorem *Let ϕ be continuous on $(-\infty,+\infty)$. If $\phi(f) \in L(E)$, then $\int_{-\infty}^{+\infty} \phi(\alpha)\, d\omega(\alpha)$ exists and*

$$\int_E \phi(f) = -\int_{-\infty}^{+\infty} \phi(\alpha)\, d\omega(\alpha).$$

Proof. Since the proof is similar to that of part of (5.43), we shall be brief. For finite a and b, $a < b$, let E_{ab} and ω_{ab} be as in (5.42). By (5.46), $\int_{E_{ab}} \phi(f) = -\int_a^b \phi(\alpha)\, d\omega_{ab}(\alpha)$. Therefore, as in the proof of (5.42), $\int_{E_{ab}} \phi(f) = -\int_a^b \phi(\alpha)\, d\omega(\alpha)$. The theorem now follows by letting $a \to -\infty$ and $b \to +\infty$.

We remark that if ϕ is continuous *and nonnegative*, then the equality

$$\int_E \phi(f) = -\int_{-\infty}^{+\infty} \phi(\alpha)\, d\omega(\alpha)$$

holds without restriction on the finiteness of either side. To see this, let $a \to -\infty$, $b \to +\infty$ in the equation $\int_{E_{ab}} \phi(f) = -\int_a^b \phi(\alpha)\, d\omega(\alpha)$.

Thus, for any continuous ϕ, we have

$$\int_E |\phi(f)| = -\int_{-\infty}^{+\infty} |\phi(\alpha)|\, d\omega(\alpha).$$

Taking $\phi(\alpha) = |\alpha|^p$, $0 < p < \infty$, it follows that

$$\int_E |f|^p = -\int_{-\infty}^{+\infty} |\alpha|^p\, d\omega(\alpha).$$

If f is nonnegative, we obtain

(5.48) $$\int_E f^p = -\int_0^{\infty} \alpha^p\, d\omega(\alpha).$$

Hence, for any measurable f,

$$\int_E |f|^p = -\int_0^{\infty} \alpha^p\, d\omega_{|f|}(\alpha).$$

Given $\phi \geq 0$, let $L_\phi(E)$ denote the class of measurable f such that $\phi(f) \in L(E)$. If $\phi(\alpha) = |\alpha|^p$, $0 < p < \infty$, the standard notation is

$$L^p(E) = \left\{ f : \int_E |f|^p < +\infty \right\},\ 0 < p < \infty.$$

Note that $L^1(E) = L(E)$. We will systematically study the L^p classes in Chapter 8. For now, we only want to complete (5.48).

First, note that for measurable f there is an L^p version of *Tchebyshev's inequality*:

(5.49) $$\omega(\alpha) \leq \frac{1}{\alpha^p} \int_{\{f > \alpha\}} f^p,\ \alpha > 0.$$

The proof is left as an exercise. Hence, if f is in L^p, $\alpha^p \omega(\alpha)$ remains bounded as $\alpha \to +\infty$. A stronger result is actually true.

(5.50) Lemma *If $f \in L^p(E)$, then*

$$\lim_{\alpha \to +\infty} \alpha^p \omega(\alpha) = 0.$$

Proof. This will be a corollary of (5.49) if we show that $\int_{\{f>\alpha\}} f^p \to 0$ as $\alpha \to +\infty$. We may suppose that α runs through a sequence $\alpha_k \to +\infty$. Let $f_k = f$ wherever $f > \alpha_k$ and $f_k = 0$ elsewhere. Then $\int_{\{f>\alpha_k\}} f^p = \int_E f_k^p$. Since f is finite a.e., $f_k \to 0$ a.e. Moreover, $0 \le f_k^p \le |f|^p \in L(E)$, and the result follows from the dominated convergence theorem.

In the following theorem, we use (5.50) to integrate the Riemann-Stieltjes integral in (5.48) by parts.

(5.51) Theorem *If $f \ge 0$ and $f \in L^p(E)$, then*

$$\int_E f^p = -\int_0^\infty \alpha^p \, d\omega(\alpha) = p\int_0^\infty \alpha^{p-1}\omega(\alpha)\, d\alpha.$$

Proof. The first equality is just (5.48). For the second, if $0 < a < b < +\infty$, we have

$$-\int_a^b \alpha^p \, d\omega(\alpha) = -b^p\omega(b) + a^p\omega(a) + p\int_a^b \alpha^{p-1}\omega(\alpha)\, d\alpha,$$

by (2.21) and the fact that α^p is continuously differentiable on $[a,b]$. Now let $a \to 0$ and $b \to +\infty$. Then $b^p\omega(b) \to 0$ by (5.50), and $a^p\omega(a) \to 0$ since $|E| < +\infty$ (see also Exercise 14). The theorem follows immediately.

For an extension of (5.51), see Exercise 16.

5. Riemann and Lebesgue Integrals

We now study a relation between Lebesgue and Riemann integrals over finite intervals $[a,b]$ of $\mathbf{R}^1$, and give a characterization of those bounded functions which are Riemann integrable. The Lebesgue integral $\int_{[a,b]} f$ will be denoted by $\int_a^b f$, and the Riemann integral by $(R)\int_a^b f$.

(5.52) Theorem *Let f be a bounded function which is Riemann integrable on $[a,b]$. Then $f \in L[a,b]$ and*

$$\int_a^b f = (R)\int_a^b f.$$

Proof. Let $\{\Gamma_k\}$ be a sequence of partitions of $[a,b]$ with norms tending to zero. For each k, define two simple functions as follows: if $x_1^{(k)} < x_2^{(k)} < \cdots$ are the partitioning points of Γ_k, let $l_k(x)$ and $u_k(x)$ be defined in each semi-open interval $[x_i^{(k)}, x_{i+1}^{(k)})$ as the lower and upper bounds of f on $[x_i^{(k)}, x_{i+1}^{(k)}]$, respectively. Then l_k and u_k are uniformly bounded and measurable in $[a,b)$, and if L_k and U_k denote the lower and upper Riemann sums of f corresponding to Γ_k, we have

$$\int_a^b l_k = L_k, \qquad \int_a^b u_k = U_k.$$

Note also that $l_k \le f \le u_k$ and, if we assume that Γ_{k+1} is a refinement of Γ_k, that $l_k \nearrow$ and $u_k \searrow$. Let $l = \lim_{k\to\infty} l_k$ and $u = \lim_{k\to\infty} u_k$. Then l and u are measurable, $l \le f \le u$ and, by the bounded convergence theorem, $L_k \to \int_a^b l$ and $U_k \to \int_a^b u$. But since f is Riemann integrable, L_k and U_k both converge to $(R)\int_a^b f$ by (2.29). Therefore,

$$(R)\int_a^b f = \int_a^b l = \int_a^b u.$$

Since $u - l \ge 0$, (5.11) implies that $l = f = u$ a.e. in $[a,b]$. Therefore, f is measurable and $(R)\int_a^b f = \int_a^b f$, which completes the proof.

Theorem (5.52) says that any function which is Riemann integrable is also Lebesgue integrable, and that the two integrals are equal. There are, of course, bounded functions which are Lebesgue integrable but not Riemann integrable. One such is the Dirichlet function defined for $0 \le x \le 1$ by letting $f(x) = 1$ if x is rational and $f(x) = 0$ if x is irrational. Since $f = 0$ except for a subset of $[0,1]$ of measure zero, its Lebesgue integral is 0. On the other hand, its Riemann integral does not exist since every upper Riemann sum is 1 and every lower Riemann sum is zero.

The practical value of (5.52) is that it allows us to compute the Lebesgue integral of Riemann integrable (e.g., continuous) functions. Using the monotone convergence theorem, we can easily extend (5.52) to include *improper* Riemann integrals of nonnegative functions. Special as it is, the following result is useful in applications.

(5.53) Theorem *Let f be nonnegative on a finite interval $[a,b]$ and Riemann integrable (so, in particular, bounded) over every subinterval $[a + \varepsilon, b]$, $\varepsilon > 0$. Suppose that the improper integral*

$$I = \lim_{\varepsilon\to 0} (R)\int_{a+\varepsilon}^b f$$

exists and is finite. Then $f \in L[a,b]$ and

$$\int_a^b f = I.$$

Proof. Observe that by (5.52), $\int_{a+\varepsilon}^b f = (R)\int_{a+\varepsilon}^b f$, $\varepsilon > 0$. The result then follows from the monotone convergence theorem by letting $\varepsilon \to 0$.

We note in passing that the finiteness of the improper Riemann integral of an f which is not nonnegative does not in general imply that f is integrable (see Exercise 7).

Our final result is a characterization of those bounded functions that are Riemann integrable.

(5.54) Theorem *A bounded function is Riemann integrable on $[a,b]$ if and only if it is continuous a.e. in $[a,b]$.*

Proof. Suppose that f is bounded and Riemann integrable. Let Γ_k, l_k, u_k, etc. be as in the proof of (5.52). Let Z be the set of measure zero outside which $l = f = u$. We claim that if x is not a partitioning point of any Γ_k and if $x \notin Z$, then f is continuous at x. In fact, if f is not continuous at x and x is never a partitioning point, there exists $\varepsilon > 0$, depending on x but not on k, such that $u_k(x) - l_k(x) \geq \varepsilon$. This implies that $u(x) - l(x) \geq \varepsilon$, which is impossible if $x \notin Z$. Therefore, f is continuous a.e. in $[a,b]$.

To prove the converse, let f be a bounded function which is continuous a.e. in $[a,b]$. Let $\{\Gamma_k'\}$ be any sequence of partitions with norms tending to zero, and define the corresponding l_k', u_k', L_k', and U_k' as in (5.52). Note that $\{l_k'\}$ and $\{u_k'\}$ may not be monotone since Γ_{k+1}' may not be a refinement of Γ_k'. However, by the continuity of f, both l_k' and u_k' converge a.e. to f. Hence, by the bounded convergence theorem, $\int_a^b l_k'$ and $\int_a^b u_k'$ both converge to $\int_a^b f$. Since $L_k' = \int_a^b l_k'$ and $U_k' = \int_a^b u_k'$, it follows that the upper and lower Riemann sums converge to the same limit. Therefore, f is Riemann integrable.

Exercises

1. If f is a simple measurable function (not necessarily positive) taking values a_j on E_j, $j = 1, 2, \ldots, N$, show that $\int_E f = \sum_{j=1}^N a_j|E_j|$. [Use (5.24).]
2. Show that the conclusions of (5.32) are not true without the assumption that $\phi \in L(E)$. [In part (ii), for example, take $f_k = \chi_{(k,+\infty)}$.)
3. Let $\{f_k\}$ be a sequence of nonnegative measurable functions defined on E. If $f_k \to f$ and $f_k \leq f$ a.e. on E, show that $\int_E f_k \to \int_E f$.
4. If $f \in L(0,1)$, show that $x^k f(x) \in L(0,1)$ for $k = 1, 2, \ldots$, and $\int_0^1 x^k f(x)\,dx \to 0$.
5. Use Egorov's theorem to prove the bounded convergence theorem.
6. Let $f(x,y)$, $0 \leq x,y \leq 1$, satisfy the following conditions: for each x, $f(x,y)$ is an integrable function of y, and $(\partial f(x,y)/\partial x)$ is a bounded function of (x,y). Show that $(\partial f(x,y)/\partial x)$ is a measurable function of y for each x and
$$\frac{d}{dx}\int_0^1 f(x,y)\,dy = \int_0^1 \frac{\partial}{\partial x} f(x,y)\,dy.$$
7. Give an example of an f which is not integrable, but whose improper Riemann integral exists and is finite.
8. Prove (5.49).
9. If $p > 0$ and $\int_E |f - f_k|^p \to 0$ as $k \to \infty$, show that $f_k \xrightarrow{m} f$ on E (and thus that there is a subsequence $f_{k_j} \to f$ a.e. in E).
10. If $p > 0$, $\int_E |f - f_k|^p \to 0$, and $\int_E |f_k|^p \leq M$ for all k, show that $\int_E |f|^p \leq M$.

11. For which $p > 0$ does $1/x \in L^p(0,1)$? $L^p(1,\infty)$? $L^p(0,\infty)$?

12. Give an example of a bounded continuous f on $(0,\infty)$ such that $\lim_{x\to\infty} f(x) = 0$ but $f \notin L^p(0,\infty)$ for any $p > 0$.

13. (a) Let $\{f_k\}$ be a sequence of measurable functions on E. Show that $\sum f_k$ converges absolutely a.e. in E if $\sum \int_E |f_k| < +\infty$. [Use theorems (5.16) and (5.22).]

    (b) If $\{r_k\}$ denotes the rational numbers in $[0,1]$ and $\{a_k\}$ satisfies $\sum |a_k| < +\infty$, show that $\sum a_k|x - r_k|^{-1/2}$ converges absolutely a.e. in $[0,1]$.

14. Prove the following result (which is obvious if $|E| < +\infty$), describing the behavior of $a^p\omega(a)$ as $a \to 0+$. If $f \in L^p(E)$, then $\lim_{a\to 0+} a^p\omega(a) = 0$. (If $f \geq 0$, $\varepsilon > 0$, choose $\delta > 0$ so that $\int_{\{f\leq\delta\}} f^p < \varepsilon$. Thus, $a^p[\omega(a) - \omega(\delta)] \leq \int_{\{a<f\leq\delta\}} f^p < \varepsilon$ for $0 < a < \delta$. Now let $a \to 0$.)

15. Suppose that f is nonnegative and measurable on E and that ω is finite on $(0,\infty)$. If $\int_0^\infty \alpha^{p-1}\omega(\alpha)\,d\alpha$ is finite, show that $\lim_{a\to 0+} a^p\omega(a) = \lim_{b\to+\infty} b^p\omega(b) = 0$. (Consider $\int_{a/2}^a$ and $\int_{b/2}^b$.)

16. Suppose that f is nonnegative and measurable on E and that ω is finite on $(0,\infty)$. Show that (5.51) holds without any further restrictions [that is, f need not be in $L^p(E)$ and $|E|$ need not be finite] if we interpret $\int_0^\infty \alpha^p\,d\omega(\alpha) = \lim_{\substack{a\to 0+ \\ b\to+\infty}} \int_a^b$. [Use E_{ab} to obtain the relation $\int_E f^p = -\int_0^\infty \alpha^p\,d\omega(\alpha)$. If either $\int_0^\infty \alpha^p\,d\omega(\alpha)$ or $\int_0^\infty \alpha^{p-1}\omega(\alpha)\,d\alpha$ is finite, use (5.50) and the results of Exercises 14 or 15 to integrate by parts.]

17. If $f \geq 0$ and $\omega(\alpha) \leq c(1 + \alpha)^{-p}$ for all $\alpha > 0$, show that $f \in L^r$, $0 < r < p$.

18. If $f \geq 0$, show that $f \in L^p$ if and only if $\sum_{k=-\infty}^{+\infty} 2^{kp}\omega(2^k) < +\infty$. (Use Exercise 16.)

19. Derive analogues of (5.52) and (5.54) for integrals over intervals in $\mathbf{R}^n$, $n > 1$.

20. Let $\mathbf{y} = T\mathbf{x}$ be a nonsingular linear transformation of $\mathbf{R}^n$. If $\int_E f(\mathbf{y})\,d\mathbf{y}$ exists, show that

$$\int_E f(\mathbf{y})\,d\mathbf{y} = |\det T| \int_{T^{-1}E} f(T\mathbf{x})\,d\mathbf{x}.$$

    [The case when $f = \chi_{E_1}$, $E_1 \subset E$, follows from integrating the formula $\chi_{E_1}(T\mathbf{x}) = \chi_{T^{-1}E_1}(\mathbf{x})$ over $T^{-1}E$, and then applying (3.35).]

21. If $\int_A f = 0$ for every measurable subset A of a measurable set E, show that $f = 0$ a.e. in E.

Chapter 6

Repeated Integration

Let $f(x,y)$ be defined in a rectangle

$$I = \{(x,y) : a \le x \le b, c \le y \le d\}.$$

If f is continuous, we have the classical formula

$$\iint_I f(x,y)\,dx\,dy = \int_a^b \left[\int_c^d f(x,y)\,dy \right] dx,$$

and there is an analogous formula for functions of n variables.

The first two sections of this chapter extend this and related results on repeated integration to the case of Lebesgue integrable functions. The last section contains some applications.

1. Fubini's Theorem

We shall use the following notation. Let $\mathbf{x} = (x_1, \ldots, x_n)$ be a point of an n-dimensional interval I_1,

$$I_1 = \{\mathbf{x} = (x_1, \ldots, x_n) : a_i \le x_i \le b_i,\ i = 1, \ldots, n\},$$

and let $\mathbf{y}$ be a point of an m-dimensional interval I_2,

$$I_2 = \{\mathbf{y} = (y_1, \ldots, y_m) : c_j \le y_j \le d_j,\ j = 1, \ldots, m\}.$$

Here I_1 and I_2 may be all of $\mathbf{R}^n$ and $\mathbf{R}^m$, respectively. The Cartesian product $I = I_1 \times I_2$ is an $(n + m)$-dimensional interval consisting of points $(x_1, \ldots, x_n, y_1, \ldots, y_m)$. We shall denote such points by $(\mathbf{x},\mathbf{y})$. A function $f(x_1, \ldots, x_n, y_1, \ldots, y_m)$ defined in I will be written $f(\mathbf{x},\mathbf{y})$, and its integral $\int_I f$ will be denoted by $\iint_I f(\mathbf{x},\mathbf{y})\,d\mathbf{x}\,d\mathbf{y}$.

(6.1) Theorem (*Fubini's Theorem*) *Let* $f(\mathbf{x},\mathbf{y}) \in L(I)$, $I = I_1 \times I_2$. *Then*

(i) *for almost every* $\mathbf{x} \in I_1$, $f(\mathbf{x},\mathbf{y})$ *is measurable and integrable on* I_2 *as a function of* $\mathbf{y}$;

(ii) *as a function of* $\mathbf{x}$, $\int_{I_2} f(\mathbf{x},\mathbf{y})\, d\mathbf{y}$ *is measurable and integrable on* I_1, *and*

$$\iint_I f(\mathbf{x},\mathbf{y})\, d\mathbf{x}\, d\mathbf{y} = \int_{I_1} \left[\int_{I_2} f(\mathbf{x},\mathbf{y})\, d\mathbf{y} \right] d\mathbf{x}.$$

Setting $f = 0$ outside I, we see that it is enough to prove the theorem when $I_1 = \mathbf{R}^n$, $I_2 = \mathbf{R}^m$, and $I = \mathbf{R}^{n+m}$. For simplicity, we then drop I_1, I_2, and I from the notation and write $\int f(\mathbf{x},\mathbf{y})\, d\mathbf{x}$ for $\int_{I_1} f(\mathbf{x},\mathbf{y})\, d\mathbf{x}$, $L(d\mathbf{x})$ for $L(I_1)$, $L(d\mathbf{x}\, d\mathbf{y})$ for $L(I)$, etc.

We will prove the theorem by considering a series of special cases. The first two lemmas below will help in passing from one case to the next. In these lemmas, we say that a function f in $L(d\mathbf{x}\, d\mathbf{y})$ for which Fubini's theorem is true has *property* $\mathcal{F}$.

(6.2) Lemma *A finite linear combination of functions with property* $\mathcal{F}$ *has property* $\mathcal{F}$.

This follows immediately from (4.9) and (5.28).

(6.3) Lemma *Let* $f_1, f_2, \ldots, f_k, \ldots,$ *have property* $\mathcal{F}$. *If* $f_k \nearrow f$ *or* $f_k \searrow f$ *and* $f \in L(d\mathbf{x}\, d\mathbf{y})$, *then* f *has property* $\mathcal{F}$.

Proof. Changing signs if necessary, we may assume that $f_k \nearrow f$. For each k, there exists by hypothesis a set Z_k in $\mathbf{R}^n$ with measure zero such that $f_k(\mathbf{x},\mathbf{y}) \in L(d\mathbf{y})$ if $\mathbf{x} \notin Z_k$. Let $Z = \bigcup_k Z_k$, so that Z has $\mathbf{R}^n$-measure zero. If $\mathbf{x} \notin Z$, then $f_k(\mathbf{x},\mathbf{y}) \in L(d\mathbf{y})$ for all k, and therefore, by the monotone convergence theorem applied to $\{f_k(\mathbf{x},\mathbf{y})\}$ as functions of $\mathbf{y}$,

$$h_k(\mathbf{x}) = \int f_k(\mathbf{x},\mathbf{y})\, d\mathbf{y} \nearrow h(\mathbf{x}) = \int f(\mathbf{x},\mathbf{y})\, d\mathbf{y} \qquad (\mathbf{x} \notin Z).$$

By assumption, $h_k(\mathbf{x}) \in L(d\mathbf{x})$, $f_k \in L(d\mathbf{x}\, d\mathbf{y})$ and $\iint f_k(\mathbf{x},\mathbf{y})\, d\mathbf{x}\, d\mathbf{y} = \int h_k(\mathbf{x})\, d\mathbf{x}$. Therefore, using the monotone convergence theorem again, we obtain that $\iint f(\mathbf{x},\mathbf{y})\, d\mathbf{x}\, d\mathbf{y} = \int h(\mathbf{x})\, d\mathbf{x}$. Since $f \in L(d\mathbf{x}\, d\mathbf{y})$, it follows that $h \in L(d\mathbf{x})$, which implies that h is finite a.e. This completes the proof.

The next three lemmas prove special cases of Fubini's theorem.

(6.4) Lemma *If* E *is a set of type* G_δ, *viz.*, $E = \bigcap_{k=1}^{\infty} G_k$, *and if* G_1 *has finite measure, then* χ_E *has property* $\mathcal{F}$.

Proof. Case 1. Suppose that E is a bounded open interval in $\mathbf{R}^{n+m}$: $E = J_1 \times J_2$, where J_1 and J_2 are bounded open intervals in $\mathbf{R}^n$ and $\mathbf{R}^m$, respectively. Then $|E| = |J_1||J_2|$, where $|J_1|$ and $|J_2|$ denote the measures of J_1 and J_2 in $\mathbf{R}^n$ and $\mathbf{R}^m$. For every $\mathbf{x}$, $\chi_E(\mathbf{x},\mathbf{y})$ is measurable as a function of $\mathbf{y}$. If $h(\mathbf{x}) = \int \chi_E(\mathbf{x},\mathbf{y})\, d\mathbf{y}$, then $h(\mathbf{x}) = |J_2|$ for $\mathbf{x} \in J_1$, and $h(\mathbf{x}) = 0$ otherwise. Therefore, $\int h(\mathbf{x})\, d\mathbf{x} = |J_1||J_2|$. But also, $\iint \chi_E(\mathbf{x},\mathbf{y})\, d\mathbf{x}\, d\mathbf{y} = |E| = |J_1||J_2|$, and the lemma is proved in this case.

Case 2. Suppose that E is any set (of type G_δ or not) on the boundary of an interval in $\mathbf{R}^{n+m}$. Then for almost every $\mathbf{x}$, the set $\{\mathbf{y} : (\mathbf{x},\mathbf{y}) \in E\}$ has $\mathbf{R}^m$-measure zero. Therefore, if $h(\mathbf{x}) = \int \chi_E(\mathbf{x},\mathbf{y})\, d\mathbf{y}$, it follows that $h(\mathbf{x}) = 0$ a.e. Hence, $\int h(\mathbf{x})\, d\mathbf{x} = 0$. But also, $\int\int \chi_E(\mathbf{x},\mathbf{y})\, d\mathbf{x}\, d\mathbf{y} = |E| = 0$.

Case 3. Suppose next that E is a partly open interval in $\mathbf{R}^{n+m}$. Then E is the union of its interior and a subset of its boundary. It follows from cases 1 and 2 and (6.2) that χ_E has property $\mathscr{F}$.

Case 4. Let E be an open set in $\mathbf{R}^{n+m}$ with finite measure. Write $E = \bigcup I_j$, where the I_j are disjoint, partly open intervals. If $E_k = \bigcup_{j=1}^k I_j$, then $\chi_{E_k} = \sum_{j=1}^k \chi_{I_j}$, so that χ_{E_k} has property $\mathscr{F}$ by case 3 and (6.2). Since $\chi_{E_k} \nearrow \chi_E$, χ_E has property $\mathscr{F}$ by (6.3).

Case 5. Let E satisfy the hypothesis of (6.4). We may assume that $G_k \searrow E$ by considering the open sets $G_1, G_1 \cap G_2, G_1 \cap G_2 \cap G_3$, etc. Then $\chi_{G_k} \searrow \chi_E$, and the lemma follows from case 4 and (6.3).

(6.5) Lemma *If Z is a subset of $\mathbf{R}^{n+m}$ with measure zero, then χ_Z has property $\mathscr{F}$. Hence, for almost every $\mathbf{x} \in \mathbf{R}^n$, the set $\{\mathbf{y} : (\mathbf{x},\mathbf{y}) \in Z\}$ has $\mathbf{R}^m$-measure zero.*

Proof. Using (3.8), select a set H of type G_δ such that $Z \subset H$ and $|H| = 0$. If $H = \bigcap G_k$, we may assume that G_1 has finite measure, so that by (6.4),

$$\int\left[\int \chi_H(\mathbf{x},\mathbf{y})\, d\mathbf{y}\right] d\mathbf{x} = \int\int \chi_H(\mathbf{x},\mathbf{y})\, d\mathbf{x}\, d\mathbf{y} = 0.$$

Therefore, by (5.11), $|\{\mathbf{y} : (\mathbf{x},\mathbf{y}) \in H\}| = \int \chi_H(\mathbf{x},\mathbf{y})\, d\mathbf{y} = 0$ for almost every $\mathbf{x}$. If $\{\mathbf{y} : (\mathbf{x},\mathbf{y}) \in H\}$ has $\mathbf{R}^m$-measure zero, so does $\{\mathbf{y} : (\mathbf{x},\mathbf{y}) \in Z\}$ since $Z \subset H$. It follows that for almost every $\mathbf{x}$, $\chi_Z(\mathbf{x},\mathbf{y})$ is measurable in $\mathbf{y}$ and $\int \chi_Z(\mathbf{x},\mathbf{y})\, d\mathbf{y} = 0$. Hence, $\int[\int \chi_Z(\mathbf{x},\mathbf{y})\, d\mathbf{y}]\, d\mathbf{x} = 0$, which proves the lemma since $\int\int \chi_Z(\mathbf{x},\mathbf{y})\, d\mathbf{x}\, d\mathbf{y} = |Z| = 0$.

(6.6) Lemma *Let $E \subset \mathbf{R}^{n+m}$. If E is measurable with finite measure, then χ_E has property $\mathscr{F}$.*

Proof. Using (3.28), write $E = H - Z$, where H is of type G_δ and Z has measure zero. If $H = \bigcap G_k$, choose G_1 with finite measure [see the proof of (3.28)]. Since $\chi_E = \chi_H - \chi_Z$, the lemma follows from (6.4), (6.5), and (6.2).

Proof of Fubini's theorem. We must show that every $f \in L(d\mathbf{x}\, d\mathbf{y})$ has property $\mathscr{F}$. Since $f = f^+ - f^-$, we may assume by (6.2) that $f \geq 0$. Then, by (4.13), there are simple measurable $f_k \nearrow f$, $f_k \geq 0$. Each $f_k \in L(d\mathbf{x}\, d\mathbf{y})$, and by (6.3), it is enough to show that these have property $\mathscr{F}$. Hence, we may assume that f is simple and integrable, say $f = \sum_{j=1}^N v_j \chi_{E_j}$. Since each E_j must have finite measure, the result follows from (6.6) and (6.2).

If $f \in L(\mathbf{R}^{n+m})$, then by Fubini's theorem, $f(\mathbf{x},\mathbf{y})$ is a measurable function of $\mathbf{y}$ for almost every $\mathbf{x} \in \mathbf{R}^n$. We now show that the same conclusion holds if f is merely measurable.

(6.7) Theorem *Let $f(\mathbf{x},\mathbf{y})$ be a measurable function on $\mathbf{R}^{n+m}$. Then for almost every $\mathbf{x} \in \mathbf{R}^n$, $f(\mathbf{x},\mathbf{y})$ is a measurable function of $\mathbf{y} \in \mathbf{R}^m$.*

In particular, if E is a measurable subset of $\mathbf{R}^{n+m}$, then the set

$$E_{\mathbf{x}} = \{\mathbf{y} : (\mathbf{x},\mathbf{y}) \in E\}$$

is measurable in $\mathbf{R}^m$ for almost every $\mathbf{x} \in \mathbf{R}^n$.

Proof. Note that if f is the characteristic function χ_E of a measurable $E \subset \mathbf{R}^{n+m}$, then the two statements above are equivalent. To prove the result in this case, write $E = H \cup Z$, where H is of type F_σ in $\mathbf{R}^{n+m}$ and $|Z|_{n+m} = 0$. Then $E_{\mathbf{x}} = H_{\mathbf{x}} \cup Z_{\mathbf{x}}$, $H_{\mathbf{x}}$ is of type F_σ in $\mathbf{R}^m$, and for almost every $\mathbf{x} \in \mathbf{R}^n$, $|Z_{\mathbf{x}}|_m = 0$ by (6.5). Therefore, $E_{\mathbf{x}}$ is measurable for almost every $\mathbf{x}$.

If f is any measurable function on $\mathbf{R}^{n+m}$, consider the set $E(a) = \{(\mathbf{x},\mathbf{y}) : f(\mathbf{x},\mathbf{y}) > a\}$. Since $E(a)$ is measurable in $\mathbf{R}^{n+m}$, the set $E(a)_{\mathbf{x}} = \{\mathbf{y} : (\mathbf{x},\mathbf{y}) \in E(a)\}$ is measurable in $\mathbf{R}^m$ for almost every $\mathbf{x} \in \mathbf{R}^n$. The exceptional set of $\mathbf{R}^n$-measure zero depends on a. The union Z of these exceptional sets for all rational a still has $\mathbf{R}^n$-measure zero. If $\mathbf{x} \notin Z$, then $\{y : f(\mathbf{x},\mathbf{y}) > a\}$ is measurable for all rational a, and so for all a by (4.4). This completes the proof.

We will now extend Fubini's theorem to functions defined on subsets of $\mathbf{R}^{n+m}$.

(6.8) Theorem *Let $f(\mathbf{x},\mathbf{y})$ be a measurable function defined on a measurable subset E of $\mathbf{R}^{n+m}$, and let $E_{\mathbf{x}} = \{\mathbf{y} : (\mathbf{x},\mathbf{y}) \in E\}$.*

(i) *For almost every $\mathbf{x} \in \mathbf{R}^n$, $f(\mathbf{x},\mathbf{y})$ is a measurable function of $\mathbf{y}$ on $E_{\mathbf{x}}$.*

(ii) *If $f(\mathbf{x},\mathbf{y}) \in L(E)$, then for almost every $\mathbf{x} \in \mathbf{R}^n$, $f(\mathbf{x},\mathbf{y})$ is integrable on $E_{\mathbf{x}}$ with respect to $\mathbf{y}$; moreover, $\int_{E_{\mathbf{x}}} f(\mathbf{x},\mathbf{y})\, d\mathbf{y}$ is an integrable function of $\mathbf{x}$ and*

$$\iint_E f(\mathbf{x},\mathbf{y})\, d\mathbf{x}\, d\mathbf{y} = \int_{\mathbf{R}^n} \left[\int_{E_{\mathbf{x}}} f(\mathbf{x},\mathbf{y})\, d\mathbf{y} \right] d\mathbf{x}.$$

Proof. Let $\bar{f}$ be the function equal to f in E and to zero elsewhere in $\mathbf{R}^{n+m}$. Since f is measurable on E, $\bar{f}$ is measurable on $\mathbf{R}^{n+m}$. Therefore, by (6.7), $\bar{f}(\mathbf{x},\mathbf{y})$ is a measurable function of $\mathbf{y}$ for almost every $\mathbf{x} \in \mathbf{R}^n$. Since $E_{\mathbf{x}}$ is measurable for almost every $\mathbf{x} \in \mathbf{R}^n$, it follows that $f(\mathbf{x},\mathbf{y})$ is measurable on almost every $E_{\mathbf{x}}$. This proves (i).

If $f \in L(E)$, then $\bar{f} \in L(\mathbf{R}^{n+m})$ and

$$\iint_E f(\mathbf{x},\mathbf{y})\, d\mathbf{x}\, d\mathbf{y} = \iint_{\mathbf{R}^{n+m}} \bar{f}(\mathbf{x},\mathbf{y})\, d\mathbf{x}\, d\mathbf{y} = \int_{\mathbf{R}^n} \left[\int_{\mathbf{R}^m} \bar{f}(\mathbf{x},\mathbf{y})\, d\mathbf{y} \right] d\mathbf{x}.$$

Since $E_{\mathbf{x}}$ is measurable for almost every $\mathbf{x}$, we obtain by theorem (5.24) that

$\int_{\mathbf{R}^m} \bar{f}(\mathbf{x},\mathbf{y})\,d\mathbf{y} = \int_{E_\mathbf{x}} f(\mathbf{x},\mathbf{y})\,d\mathbf{y}$ for almost every $\mathbf{x} \in \mathbf{R}^n$. Part (ii) follows by combining equalities.

2. Tonelli's Theorem

By Fubini's theorem, the finiteness of a multiple integral implies that of the corresponding iterated integrals. The converse is not true, even if all the iterated integrals are equal, as shown by the following example.

(6.9) Example Let $n = m = 1$, let I be the unit square and $\{I_k\}$ be the infinite sequence of subsquares shown below:

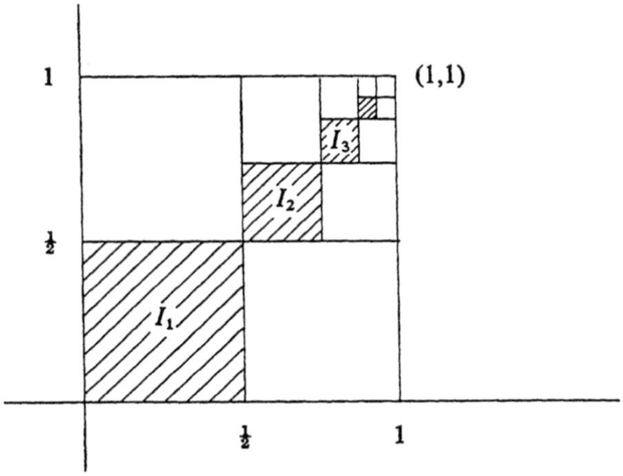

Subdivide each I_k into four equal subsquares by lines parallel to the x and y axes.

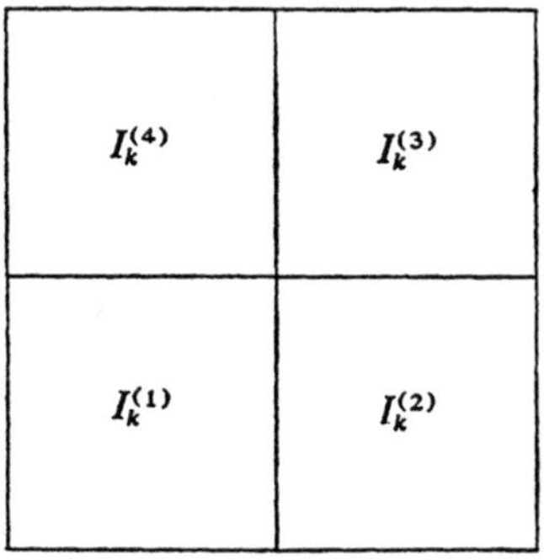

I_k

For each k, let $f = 1/|I_k|$ on the interiors of $I_k^{(1)}$ and $I_k^{(3)}$, and let $f = -1/|I_k|$ on the interiors of $I_k^{(2)}$ and $I_k^{(4)}$. Let $f = 0$ on the rest of I, that is, outside

$\bigcup I_k$ and on the boundaries of all the subsquares. Clearly, $\int_0^1 f(x,y)\,dy = 0$ for all x, and $\int_0^1 f(x,y)\,dx = 0$ for all y. Therefore,

$$\int_0^1 \left[\int_0^1 f(x,y)\,dy \right] dx = \int_0^1 \left[\int_0^1 f(x,y)\,dx \right] dy = 0.$$

However,

$$\iint_I |f(x,y)|\,dx\,dy = \sum_k \iint_{I_k} |f(x,y)|\,dx\,dy = \sum_k 1 = +\infty.$$

Hence, finiteness of the iterated integrals does not in general imply that of the multiple integral. However, for nonnegative f, we have the following basic result.

(6.10) Theorem (*Tonelli's Theorem*) *Let $f(\mathbf{x},\mathbf{y})$ be nonnegative and measurable on an interval $I = I_1 \times I_2$ of $\mathbf{R}^{n+m}$. Then, for almost every $\mathbf{x} \in I_1$, $f(\mathbf{x},\mathbf{y})$ is a measurable function of $\mathbf{y}$ on I_2. Moreover, as a function of $\mathbf{x}$, $\int_{I_2} f(\mathbf{x},\mathbf{y})\,d\mathbf{y}$ is measurable on I_1, and*

$$\iint_I f(\mathbf{x},\mathbf{y})\,d\mathbf{x}\,d\mathbf{y} = \int_{I_1} \left[\int_{I_2} f(\mathbf{x},\mathbf{y})\,d\mathbf{y} \right] d\mathbf{x}.$$

Proof. This is actually a corollary of Fubini's theorem. For $k = 1, 2, \dots$, let $f_k(\mathbf{x},\mathbf{y}) = 0$ if $|(\mathbf{x},\mathbf{y})| > k$ and $f_k(\mathbf{x},\mathbf{y}) = \min\{k, f(\mathbf{x},\mathbf{y})\}$ if $|(\mathbf{x},\mathbf{y})| \le k$. Then $f_k \ge 0$, $f_k \nearrow f$ on I and $f_k \in L(I)$ (f_k is bounded and vanishes outside a compact set). Hence, Fubini's theorem applies to each f_k. The statement concerning the measurability of $\int_{I_2} f(\mathbf{x},\mathbf{y})\,d\mathbf{y}$ then follows from its analogue for f_k; in fact, by the monotone convergence theorem, $\int_{I_2} f_k(\mathbf{x},\mathbf{y})\,d\mathbf{y} \nearrow \int_{I_2} f(\mathbf{x},\mathbf{y})\,d\mathbf{y}$. [The measurability of $f(\mathbf{x},\mathbf{y})$ as a function of $\mathbf{y}$ was proved in (6.8).] By the monotone convergence theorem again,

$$\text{(a)} \qquad \iint_I f_k(\mathbf{x},\mathbf{y})\,d\mathbf{x}\,d\mathbf{y} \to \iint_I f(\mathbf{x},\mathbf{y})\,d\mathbf{x}\,d\mathbf{y},$$

$$\text{(b)} \qquad \int_{I_1} \left[\int_{I_2} f_k(\mathbf{x},\mathbf{y})\,d\mathbf{y} \right] d\mathbf{x} \to \int_{I_1} \left[\int_{I_2} f(\mathbf{x},\mathbf{y})\,d\mathbf{y} \right] d\mathbf{x}.$$

Since $f_k \in L(I)$, the left-hand sides of (a) and (b) are equal. Therefore, so are the right-hand sides, and the theorem follows.

An extension of Tonelli's theorem to functions defined over arbitrary measurable sets E is straightforward.

Since the roles of $\mathbf{x}$ and $\mathbf{y}$ can be interchanged above, it follows that if f is nonnegative and measurable, then

$$\iint_I f(\mathbf{x},\mathbf{y})\,d\mathbf{x}\,d\mathbf{y} = \int_{I_1} \left[\int_{I_2} f(\mathbf{x},\mathbf{y})\,d\mathbf{y} \right] d\mathbf{x} = \int_{I_2} \left[\int_{I_1} f(\mathbf{x},\mathbf{y})\,d\mathbf{x} \right] d\mathbf{y}.$$

In particular, we obtain the important fact that *for $f \ge 0$, the finiteness of* *any one of Fubini's three integrals implies that of the other two.* Hence, for any measurable f, the finiteness of one of these integrals for $|f|$ implies that f is integrable and that all three Fubini integrals of f are equal.

An easy consequence of Tonelli's theorem is that the conclusion

$$\iint_I f(\mathbf{x},\mathbf{y})\,d\mathbf{x}\,d\mathbf{y} = \int_{I_1}\left[\int_{I_2} f(\mathbf{x},\mathbf{y})\,d\mathbf{y}\right]d\mathbf{x}$$

of Fubini's theorem holds for measurable f even if $\iint_I f = \pm\infty$ (that is, it holds if $\iint_I f$ merely *exists*). In fact, if $\iint_I f = +\infty$, we have $\iint_I f^+ = +\infty$ and $f^- \in L(I)$. By Tonelli's theorem,

$$\iint_I f^+ = \int_{I_1}\left(\int_{I_2} f^+\,d\mathbf{y}\right)d\mathbf{x}, \qquad \iint_I f^- = \int_{I_1}\left(\int_{I_2} f^-\,d\mathbf{y}\right)d\mathbf{x}.$$

Since $\iint_I f^-$ is finite, the desired formula follows by subtraction.

3. Applications of Fubini's Theorem

We shall derive several important results as corollaries of Fubini's and Tonelli's theorems. The first one is the necessity of the condition in theorem (5.1). Using the notation of Chapter 5, we will prove the following result.

(6.11) Theorem *Let f be a nonnegative function defined on a measurable set $E \subset \mathbf{R}^n$. If $R(f,E)$, the region under f over E, is a measurable subset of $\mathbf{R}^{n+1}$, then f is measurable.*

Proof. For $0 \le y < +\infty$, we have

$$\{\mathbf{x} \in E : f(\mathbf{x}) \ge y\} = \{\mathbf{x} : (\mathbf{x},y) \in R(f,E)\}.$$

Since $R(f,E)$ is measurable, it follows from (6.7) that $\{\mathbf{x} \in E : f(\mathbf{x}) \ge y\}$ is measurable (in $\mathbf{R}^n$) for almost all (linear measure) such y. In particular, $\{\mathbf{x} \in E : f(\mathbf{x}) \ge y\}$ is measurable for all y in a dense subset of $(0,\infty)$. If y is negative, then $\{\mathbf{x} \in E : f(\mathbf{x}) \ge y\} = E$, which is measurable. We conclude that f is measurable [cf. (4.4)].

As a second application of Fubini's theorem, we will prove a result about the convolution of two functions. More general results of this kind will be proved in Chapter 9. If f and g are measurable in $\mathbf{R}^n$, their *convolution* $(f * g)(\mathbf{x})$ is defined by

$$(f * g)(\mathbf{x}) = \int_{\mathbf{R}^n} f(\mathbf{x} - \mathbf{t})g(\mathbf{t})\,d\mathbf{t},$$

provided the integral exists.

We first claim that $f * g = g * f$, that is, that

$$\int_{\mathbf{R}^n} f(\mathbf{x} - \mathbf{t})g(\mathbf{t})\,d\mathbf{t} = \int_{\mathbf{R}^n} f(\mathbf{t})g(\mathbf{x} - \mathbf{t})\,d\mathbf{t}. \tag{6.12}$$

This is actually a special case of results dealing with changes of variable. In this simple case, however, it amounts to the statement that if $\mathbf{x} \in \mathbf{R}^n$, then

$$\int_{\mathbf{R}^n} r(\mathbf{t})\,dt = \int_{\mathbf{R}^n} r(\mathbf{x} - \mathbf{t})\,dt \tag{6.13}$$

when $r(\mathbf{t}) = f(\mathbf{x} - \mathbf{t})g(\mathbf{t})$. For fixed $\mathbf{x}$, $\mathbf{x} - \mathbf{t}$ ranges over $\mathbf{R}^n$ as $\mathbf{t}$ does. Therefore, for any measurable $r \geq 0$, (6.13) follows from the geometric interpretation of the integral. [See (5.1) and the definition of the integral of a nonnegative function.] For any measurable r, it follows by writing $r = r^+ - r^-$. (For the effect of a *linear* change of variables, see Exercise 20 of Chapter 5.)

The result we wish to prove for convolutions is the following.

(6.14) Theorem *If $f \in L(\mathbf{R}^n)$ and $g \in L(\mathbf{R}^n)$, then $(f * g)(\mathbf{x})$ exists for almost every $\mathbf{x} \in \mathbf{R}^n$. Moreover, $f * g \in L(\mathbf{R}^n)$ and*

$$\int_{\mathbf{R}^n} |f * g|\,d\mathbf{x} \leq \left(\int_{\mathbf{R}^n} |f|\,d\mathbf{x}\right)\left(\int_{\mathbf{R}^n} |g|\,d\mathbf{x}\right).$$

In order to prove this, we need a lemma.

(6.15) Lemma *If $f(\mathbf{x})$ is measurable in $\mathbf{R}^n$, then the function $F(\mathbf{x},\mathbf{t}) = f(\mathbf{x} - \mathbf{t})$ is measurable in $\mathbf{R}^n \times \mathbf{R}^n = \mathbf{R}^{2n}$.*

Proof. Let $F_1(\mathbf{x},\mathbf{t}) = f(\mathbf{x})$. Since f is measurable, it follows from (5.2) that $F_1(\mathbf{x},\mathbf{t})$ is measurable in $\mathbf{R}^{2n}$: in fact, the set $\{(\mathbf{x},\mathbf{t}) : F_1(\mathbf{x},\mathbf{t}) > a\}$, which equals $\{(\mathbf{x},\mathbf{t}) : f(\mathbf{x}) > a, \mathbf{t} \in \mathbf{R}^n\}$, is a cylinder set with measurable base $\{\mathbf{x} : f(\mathbf{x}) > a\}$ in $\mathbf{R}^n$. For $(\xi,\eta) \in \mathbf{R}^{2n}$, consider the transformation $\mathbf{x} = \xi - \eta$, $\mathbf{t} = \xi + \eta$. This is a nonsingular linear transformation of $\mathbf{R}^{2n}$, and therefore, by (3.33) (see Exercise 4 of Chapter 4), the function F_2 defined by $F_2(\xi,\eta) = F_1(\xi - \eta, \xi + \eta)$ is measurable in $\mathbf{R}^{2n}$. Since $F_2(\xi,\eta) = f(\xi - \eta)$, the lemma follows.

Proof of Theorem (6.14). Suppose first that both f and g are nonnegative on $\mathbf{R}^n$. By (6.15), $f(\mathbf{x} - \mathbf{t})g(\mathbf{t})$ is measurable on $\mathbf{R}^n \times \mathbf{R}^n$ since it is the product of two such functions. Hence, the integral

$$I = \iint f(\mathbf{x} - \mathbf{t})g(\mathbf{t})\,dt\,d\mathbf{x}$$

is well-defined. By Tonelli's theorem and (6.13),

$$\begin{aligned} I &= \int\left[\int f(\mathbf{x} - \mathbf{t})g(\mathbf{t})\,dt\right]d\mathbf{x} \\ &= \int g(\mathbf{t})\left[\int f(\mathbf{x} - \mathbf{t})\,d\mathbf{x}\right]dt = \left[\int f(\mathbf{x})\,d\mathbf{x}\right]\left[\int g(\mathbf{t})\,dt\right]. \end{aligned}$$

The first of these equations can be written $I = \int (f * g)(\mathbf{x})\,d\mathbf{x}$, so that

$$\int (f * g)(\mathbf{x})\,d\mathbf{x} = \left[\int f(\mathbf{x})\,d\mathbf{x}\right]\left[\int g(\mathbf{x})\,d\mathbf{x}\right].$$

This proves the theorem for $f \geq 0$ and $g \geq 0$. The general case can be reduced to the nonnegative case by observing that $|f * g| \leq |f| * |g|$. Hence,

$$\int |f * g|\, d\mathbf{x} \leq \int (|f| * |g|)\, d\mathbf{x} = \left(\int |f|\, d\mathbf{x}\right)\left(\int |g|\, d\mathbf{x}\right).$$

In the proof above, we showed the following useful fact.

(6.16) Corollary *If f and g are nonnegative and measurable on $\mathbf{R}^n$, then*

$$\int_{\mathbf{R}^n} (f * g)\, d\mathbf{x} = \left(\int_{\mathbf{R}^n} f\, d\mathbf{x}\right)\left(\int_{\mathbf{R}^n} g\, d\mathbf{x}\right).$$

Our last application of Fubini's theorem is an important theorem of Marcinkiewicz concerning the structure of closed sets. For simplicity, we will restrict our attention to the one-dimensional case. Given a closed subset F of $\mathbf{R}^1$ and a point x, let

$$\delta(x) = \delta(x,F) = \inf\{|x - y| : y \in F\}$$

denote the *distance of x from F*. Thus, $\delta(x) = 0$ if and only if $x \in F$. By (1.10), the complement of F is a union $\bigcup_k (a_k,b_k)$ of disjoint open intervals. At most two of these intervals can be infinite. The graph of $\delta(x)$ is thus an irregular sawtooth curve: over any finite interval (a_k,b_k), the graph is the sides of the isosceles triangle with base (a_k,b_k) and altitude $\frac{1}{2}(b_k - a_k)$; outside the terminal points of F the graph is linear. If we move from a point x to a point y, the distance from F cannot increase by more than $|x - y|$. Hence,

$$|\delta(x) - \delta(y)| \leq |x - y|;$$

that is, δ satisfies a Lipschitz condition.

We shall prove the following theorem. [See also Exercises 7–9, and theorem (9.19).]

(6.17) Theorem (*Marcinkiewicz*) *Let F be a closed subset of a bounded open interval (a,b), and let $\delta(x) = \delta(x;F)$ be the corresponding distance function. Then, given $\lambda > 0$, the integral*

$$M_\lambda(x) = M_\lambda(x;F) = \int_a^b \frac{\delta^\lambda(y)}{|x - y|^{1+\lambda}}\, dy$$

is finite a.e. in F. Moreover, $M_\lambda \in L(F)$ and

$$\int_F M_\lambda \leq 2\lambda^{-1}|G|,$$

where $G = (a,b) - F$.

Before the proof, we note that what makes the finiteness of $M_\lambda(x)$ remarkable is the singular behavior of $\delta^\lambda(y)/|x - y|^{1+\lambda}$ as $y \to x$. Since $\delta(y) \to$

$\delta(x)$ as $y \to x$, it follows that $M_\lambda(x) = +\infty$ if $x \notin F$ (see Exercise 9). If $x \in F$, then $\delta(x) = 0$, but the mere Lipschitz character of δ in the estimate

$$\frac{\delta^\lambda(y)}{|x-y|^{1+\lambda}} = \frac{[\delta(y) - \delta(x)]^\lambda}{|x-y|^{1+\lambda}} \le \frac{|x-y|^\lambda}{|x-y|^{1+\lambda}} = \frac{1}{|x-y|}$$

is not enough to imply that $M_\lambda(x)$ is finite since $\int_a^b dy/|x-y| = +\infty$. The key to the convergence of M_λ at a point x is the fact that $\delta(y)$ vanishes not only at x but also at every $y \in F$. Thus, roughly speaking, the finiteness of $M_\lambda(x)$ means that F is "very dense" near x.

Proof. Since $\delta = 0$ in F, integration in the integral defining M_λ can be restricted to the set $G = (a,b) - F$ without changing M_λ. Thus,

$$\int_F M_\lambda(x)\, dx = \int_G \delta^\lambda(y)\left(\int_F \frac{dx}{|x-y|^{1+\lambda}}\right) dy,$$

the change in the order of integration being justified by Tonelli's theorem. To estimate the inner integral, fix $y \in G$ and note that for any $x \in F$, we have $|x-y| \ge \delta(y) > 0$. Thus,

$$\int_F \frac{dx}{|x-y|^{1+\lambda}} \le \int_{|x-y|\ge\delta(y)} \frac{dx}{|x-y|^{1+\lambda}} = 2\int_{\delta(y)}^\infty \frac{dt}{t^{1+\lambda}} = 2\lambda^{-1}\delta(y)^{-\lambda}.$$

In particular,

$$\int_F M_\lambda(x)\, dx \le \int_G \delta^\lambda(y)[2\lambda^{-1}\delta(y)^{-\lambda}]\, dy = 2\lambda^{-1}|G| < +\infty.$$

Exercises

1. (a) Let E be a measurable subset of $\mathbf{R}^2$ such that for almost every $x \in \mathbf{R}^1$, $\{y : (x,y) \in E\}$ has $\mathbf{R}^1$-measure zero. Show that E has measure zero, and that for almost every $y \in \mathbf{R}^1$, $\{x : (x,y) \in E\}$ has measure zero.
   (b) Let $f(x,y)$ be nonnegative and measurable in $\mathbf{R}^2$. Suppose that for almost every $x \in \mathbf{R}^1$, $f(x,y)$ is finite for almost every y. Show that for almost every $y \in \mathbf{R}^1$, $f(x,y)$ is finite for almost every x.
2. If f and g are measurable in $\mathbf{R}^n$, show that the function $h(\mathbf{x},\mathbf{y}) = f(\mathbf{x})g(\mathbf{y})$ is measurable in $\mathbf{R}^n \times \mathbf{R}^n$. Deduce that if E_1 and E_2 are measurable subsets of $\mathbf{R}^n$, then their Cartesian product $E_1 \times E_2 = \{(\mathbf{x},\mathbf{y}) : \mathbf{x} \in E_1, \mathbf{y} \in E_2\}$ is measurable in $\mathbf{R}^n \times \mathbf{R}^n$, and $|E_1 \times E_2| = |E_1||E_2|$.
3. Let f be measurable on $(0,1)$. If $f(x) - f(y)$ is integrable over the square $0 \le x \le 1$, $0 \le y \le 1$, show that $f \in L(0,1)$.
4. Let f be measurable and periodic with period 1: $f(t+1) = f(t)$. Suppose that there is a finite c such that

$$\int_0^1 |f(a+t) - f(b+t)|\, dt \le c$$

for all a and b. Show that $f \in L(0,1)$. (Set $a = x$, $b = -x$, integrate with respect to x, and make the change of variables $\xi = x + t$, $\eta = -x + t$.)

5. (a) If f is nonnegative and measurable on E and $\omega(y) = |\{\mathbf{x} \in E : f(\mathbf{x}) > y\}|$, $y > 0$, use Tonelli's theorem to prove that $\int_E f = \int_0^\infty \omega(y)\,dy$. [By definition of the integral, we have $\int_E f = |R(f,E)| = \iint_{R(f,E)} d\mathbf{x}\,dy$. Use the observation in the proof of (6.11) that $\{\mathbf{x} \in E : f(\mathbf{x}) \geq y\} = \{\mathbf{x} : (\mathbf{x},y) \in R(f,E)\}$, and recall that $\omega(y) = |\{\mathbf{x} \in E : f(\mathbf{x}) \geq y\}|$ unless y is a point of discontinuity of ω.]

(b) Deduce from this special case the general formula

$$\int_E f^p = p \int_0^\infty y^{p-1}\omega(y)\,dy \qquad (f \geq 0,\ 0 < p < \infty).$$

6. For $f \in L(\mathbf{R}^1)$, define the *Fourier transform* $\hat{f}$ of f by

$$\hat{f}(x) = \int_{-\infty}^{+\infty} f(t)e^{-ixt}\,dt \qquad (x \in \mathbf{R}^1).$$

(For a complex-valued function $F = F_0 + iF_1$ whose real and imaginary parts F_0 and F_1 are integrable, we define $\int F = \int F_0 + i\int F_1$.) Show that if f and g belong to $L(\mathbf{R}^1)$, then

$$(f*g)^\wedge(x) = \hat{f}(x)\hat{g}(x).$$

7. Let F be a closed subset of $\mathbf{R}^1$ and let $\delta(x) = \delta(x,F)$ be the corresponding distance function. If $\lambda > 0$ and f is nonnegative and integrable over the complement of F, prove that the function

$$\int_{\mathbf{R}^1} \frac{\delta^\lambda(y)f(y)}{|x-y|^{1+\lambda}}\,dy$$

is integrable over F, and so is finite a.e. in F. [In case $f = \chi_{(a,b)}$, this reduces to (6.17).)]

8. Under the hypotheses of (6.17) and assuming that $b - a < 1$, prove that the function

$$M_0(x) = \int_a^b [\log 1/\delta(y)]^{-1}|x-y|^{-1}\,dy$$

is finite a.e. in F.

9. Show that $M_\lambda(x;F) = +\infty$ if $x \notin F$, $\lambda > 0$.

10. Let v_n be the volume of the unit ball in $\mathbf{R}^n$. Show by using Fubini's theorem that

$$v_n = 2v_{n-1}\int_0^1 (1-t^2)^{(n-1)/2}\,dt.$$

(We also observe that the integral can be expressed in terms of the Γ-function: $\Gamma(s) = \int_0^\infty e^{-t}t^{s-1}\,dt$, $s > 0$.]

11. Use Fubini's theorem to prove that

$$\int_{\mathbf{R}^n} e^{-|\mathbf{x}|^2}\,d\mathbf{x} = \pi^{n/2}.$$

[For $n = 1$, write $(\int_{-\infty}^{+\infty} e^{-x^2}\,dx)^2 = \int_{-\infty}^{+\infty}\int_{-\infty}^{+\infty} e^{-x^2-y^2}\,dx\,dy$ and use polar coordinates. For $n > 1$, use the formula $e^{-|\mathbf{x}|^2} = e^{-x_1^2}\cdots e^{-x_n^2}$ and Fubini's theorem to reduce to the case $n = 1$.]

Chapter 7

Differentiation

The main results in this chapter deal with problems of differentiability. A variety of topics is considered, but for the most part, the results are related to the analogue for Lebesgue integrals of the fundamental theorem of calculus.

1. The Indefinite Integral

If f is a Riemann integrable function on an interval $[a,b]$ in $\mathbf{R}^1$, then the familiar definition of its indefinite integral is

$$F(x) = \int_a^x f(y)\,dy, \qquad (a \leq x \leq b).$$

The fundamental theorem of calculus asserts that $F' = f$ if f is continuous. We will study an analogue of this result for Lebesgue integrable f and higher dimensions.

We must first find an appropriate definition of the indefinite integral. In two dimensions, for example, we might choose

$$F(x_1,x_2) = \int_{a_1}^{x_1} \int_{a_2}^{x_2} f(y_1,y_2)\,dy_1\,dy_2.$$

It turns out, however, to be better to abandon the notion that the indefinite integral be a function of point and adopt the idea that it be a function of set. Thus, given $f \in L(A)$, where A is a measurable subset of $\mathbf{R}^n$, we define the *indefinite integral of f* to be the function

$$F(E) = \int_E f,$$

where E is any measurable subset of A.

F is an example of a *set function*, by which we mean any real-valued function F defined on a σ-algebra Σ of measurable sets such that

(i) $F(E)$ is finite for every $E \in \Sigma$,
(ii) F is *countably additive*; i.e., if $E = \bigcup_k E_k$ is a union of disjoint $E_k \in \Sigma$, then

$$F(E) = \sum_k F(E_k).$$

By (5.5) and (5.24), the indefinite integral of an $f \in L(A)$ satisfies (i) and (ii) for the σ-algebra of measurable subsets of A. We shall systematically study set functions in Chapter 10.

We now discuss a continuity property of the indefinite integral. Recall (from p. 5) that the diameter of a set E is the number

$$\sup \{|\mathbf{x} - \mathbf{y}| : \mathbf{x},\mathbf{y} \in E\}.$$

A set function $F(E)$ is called *continuous* if $F(E)$ tends to zero as the diameter of E tends to zero; that is, $F(E)$ is continuous if, given $\varepsilon > 0$, there exists $\delta > 0$ such that $|F(E)| < \varepsilon$ whenever the diameter of E is less than δ. An example of a set function which is *not* continuous can be obtained by setting $F(E) = 1$ for any measurable E which contains the origin, and $F(E) = 0$ otherwise.

A set function F is called *absolutely continuous with respect to Lebesgue measure*, or simply *absolutely continuous*, if $F(E)$ tends to zero as the *measure* of E tends to zero. Thus, F is absolutely continuous if given $\varepsilon > 0$, there exists $\delta > 0$ such that $|F(E)| < \varepsilon$ whenever the measure of E is less than δ.

A set function which is absolutely continuous is clearly continuous. The converse, however, is false, as shown by the following example. Let A be the unit square in $\mathbf{R}^2$, let D be a diagonal of A, and consider the σ-algebra of measurable subsets E of A for which $E \cap D$ is linearly measurable. For such E, let $F(E)$ be the linear measure of $E \cap D$. Then F is a continuous set function. However, it is not absolutely continuous since there are sets E containing a fixed segment of D whose $\mathbf{R}^2$-measures are arbitrarily small.

(7.1) Theorem *If $f \in L(A)$, its indefinite integral is absolutely continuous.*

Proof. We may assume that $f \ge 0$ by considering f^+ and f^-. Fix k and write $f = g + h$, where $g = f$ whenever $f \le k$ and $g = k$ otherwise. Given $\varepsilon > 0$, we may choose k so large that $0 \le \int_A h < \frac{1}{2}\varepsilon$ and, *a fortiori*, $0 \le \int_E h < \frac{1}{2}\varepsilon$ for every measurable $E \subset A$. On the other hand, since $0 \le g \le k$, we have $0 \le \int_E g \le k|E| < \frac{1}{2}\varepsilon$ if $|E|$ is small enough. Thus

$$0 \le \int_E f = \int_E g + \int_E h < \tfrac{1}{2}\varepsilon + \tfrac{1}{2}\varepsilon = \varepsilon$$

if $|E|$ is small enough. This completes the proof.

We remark here that (7.1) has the following converse: If $F(E)$ is a set function which is absolutely continuous with respect to Lebesgue measure, then there exists an integrable f such that $F(E) = \int_E f$ for measurable E. A proof of this fact, known as the Radon-Nikodym theorem, is given in Chapter 10.

In the case of the real line, there is an alternate notion, also termed absolute continuity, which pertains to ordinary functions. This notion and its relation to the integral $\int_a^x f(y)\,dy$ are discussed in Section 5 of this chapter.

2. Lebesgue's Differentiation Theorem

We now come to a fundamental theorem of Lebesgue concerning differentiation of the indefinite integral. For $f \in L(\mathbf{R}^n)$, let F be the indefinite integral of f, and let Q denote an n-dimensional cube with edges parallel to the coordinate axes. Given $\mathbf{x}$, we consider those Q centered at $\mathbf{x}$, and ask whether the average

$$\frac{F(Q)}{|Q|} = \frac{1}{|Q|}\int_Q f(\mathbf{y})\,d\mathbf{y}$$

converges to $f(\mathbf{x})$ as Q contracts to $\mathbf{x}$. If this is the case, we write

$$\lim_{Q \searrow \mathbf{x}} \frac{F(Q)}{|Q|} = f(\mathbf{x}),$$

and say that *the indefinite integral of f is differentiable at* $\mathbf{x}$ *with derivative* $f(\mathbf{x})$. In case $n = 1$, the question is whether

$$\lim_{h\to 0} \frac{1}{2h}\int_{x-h}^{x+h} f(y)\,dy = f(x),$$

which we shall later see is essentially equivalent to

$$\lim_{h\to 0} \frac{1}{h}\int_{x}^{x+h} f(y)\,dy = f(x),$$

that is, to $d/dx \int_a^x f(y)\,dy = f(x)$.

Since f can be changed arbitrarily in a set of measure zero without affecting its indefinite integral, the best we can hope for is that F is differentiable to f almost everywhere. This is actually the case.

(7.2) Theorem (*Lebesgue's Differentiation Theorem*) *If $f \in L(\mathbf{R}^n)$, its indefinite integral is differentiable with derivative $f(\mathbf{x})$ at almost every $\mathbf{x} \in \mathbf{R}^n$.*

The proof of this basic result is difficult and requires several new ideas with wide applications. One of them is to consider the function

$$f^*(\mathbf{x}) = \sup \frac{1}{|Q|}\int_Q |f(\mathbf{y})|\,d\mathbf{y},$$

where the sup is taken over all Q with center $\mathbf{x}$. This function plays an important role in analysis.

Let us first observe that the theorem is easy to prove for continuous functions. In fact, if f is continuous at $\mathbf{x}$ and Q is a cube with center $\mathbf{x}$, then

$$\left|\frac{1}{|Q|}\int_Q f(\mathbf{y})\,d\mathbf{y} - f(\mathbf{x})\right| = \left|\frac{1}{|Q|}\int_Q [f(\mathbf{y}) - f(\mathbf{x})]\,d\mathbf{y}\right|$$
$$\leq \frac{1}{|Q|}\int_Q |f(\mathbf{y}) - f(\mathbf{x})|\,d\mathbf{y} \leq \sup_{\mathbf{y}\in Q}|f(\mathbf{y}) - f(\mathbf{x})|,$$

which tends to zero as Q shrinks to $\mathbf{x}$.

The strategy of the proof is to approximate a given $f \in L(\mathbf{R}^n)$ by continuous functions C_k. This approximation is stated in lemma (7.3), and is *global* in nature. Hence, it will be necessary to find a way to control the *local* behavior (i.e., the averages) of $f - C_k$ by this global estimate. This step is carried out in lemma (7.9), and consists of estimating the size of f^* in terms of $\int |f|$. Lemma (7.4) is a crucial covering lemma used to prove (7.9).

(7.3) Lemma *If $f \in L(\mathbf{R}^n)$, there exists a sequence $\{C_k\}$ of continuous functions with compact support such that*

$$\int_{\mathbf{R}^n} |f - C_k|\,d\mathbf{x} \to 0 \text{ as } k \to \infty.$$

Proof. If f is an integrable function for which the conclusion holds, we will say that f has property $\mathscr{A}$. We will prove the lemma by considering a series of special cases. To help in passing from one case to the next, we first show that

(1) A finite linear combination of functions with property $\mathscr{A}$ has property $\mathscr{A}$.
(2) If $\{f_k\}$ is a sequence of functions with property $\mathscr{A}$, and if $\int_{\mathbf{R}^n}|f - f_k| \to 0$, then f has property $\mathscr{A}$.

To prove (1), it is enough to show that any constant multiple, af, of a function with property $\mathscr{A}$ has property $\mathscr{A}$, and that the sum, $f_1 + f_2$, of two functions with property $\mathscr{A}$ has property $\mathscr{A}$. These facts follow easily from the relations

$$\int |af - aC| = |a| \int |f - C|,$$

$$\int |(f_1 + f_2) - (C_1 + C_2)| \leq \int |f_1 - C_1| + \int |f_2 - C_2|.$$

To prove (2), let $\{f_k\}$ and f satisfy the hypotheses of (2). First note that since f_k is integrable and $\int |f| \leq \int |f - f_k| + \int |f_k|$, it follows that f is integrable.

Next, given $\varepsilon > 0$, choose k_0 so that $\int |f - f_{k_0}| < \varepsilon/2$. Then choose a continuous C with compact support such that $\int |f_{k_0} - C| < \varepsilon/2$. Since

$$\int |f - C| \le \int |f - f_{k_0}| + \int |f_{k_0} - C| < \varepsilon/2 + \varepsilon/2 = \varepsilon,$$

we see that f has property $\mathscr{A}$.

To prove the lemma, let $f \in L(\mathbf{R}^n)$. Writing $f = f^+ - f^-$, we may assume by (1) that $f \ge 0$. Then, by (4.13), there exist nonnegative simple $f_k \nearrow f$. Thus, $f_k \in L(\mathbf{R}^n)$ and $\int |f - f_k| \to 0$, so that by (2), we may suppose that f is an integrable simple function. Hence, by (1), we may assume that $f = \chi_E$, $|E| < +\infty$. Given $\varepsilon > 0$, choose an open G such that $E \subset G$ and $|G - E| < \varepsilon$. Then

$$\int |\chi_G - \chi_E| = |G - E| < \varepsilon,$$

so we may assume that $f = \chi_G$ for an open G with $|G| < +\infty$. Using (1.11), write $G = \bigcup I_k$, where the I_k are disjoint, partly open intervals. If we let f_N be the characteristic function of $\bigcup_{k=1}^N I_k$, we obtain

$$\int |f - f_N| = \sum_{k=N+1}^{\infty} |I_k| \to 0$$

since $\sum_{k=1}^{\infty} |I_k| = |G| < +\infty$. By (2), it is thus enough to show that each f_N has property $\mathscr{A}$. But f_N is the sum of χ_{I_k}, $k = 1, \ldots, N$, so it suffices by (1) to show that the characteristic function of any interval I has property $\mathscr{A}$. This is practically self-evident: if I^* denotes an interval which contains the closure of I in its interior and which satisfies $|I^* - I| < \varepsilon$, then for any continuous C, $0 \le C \le 1$, which is 1 in I and 0 outside I^*, we have

$$\int |\chi_I - C| \le |I^* - I| < \varepsilon.$$

This completes the proof of (7.3).

The lemma which follows is a preliminary version of a covering lemma due to Vitali [theorem (7.17)] which has many applications.

(7.4) Lemma (*Simple Vitali Lemma*) *Let E be a subset of $\mathbf{R}^n$ with $|E|_e < +\infty$, and let K be a collection of cubes Q covering E. Then there exist a positive constant β, depending only on n, and a finite number of disjoint cubes $Q_1, \ldots, Q_N$ in K such that*

$$\sum_{j=1}^{N} |Q_j| \ge \beta |E|_e.$$

Proof. We will index the size of a cube $Q \in K$ by writing $Q = Q(t)$, where t is the edge length of Q. Let $K_1 = K$ and

$$t_1^* = \sup\{t : Q = Q(t) \in K_1\}.$$

If $t_1^* = +\infty$, then K_1 contains a sequence of cubes Q with $|Q| \to +\infty$. In this case, given $\beta > 0$, we simply choose one $Q \in K_1$ with $|Q| \geq \beta|E|_e$. If $t_1^* < +\infty$, the idea is still to pick a relatively large cube: choose $Q_1 = Q_1(t_1) \in K_1$ such that $t_1 > \frac{1}{2}t_1^*$. Now split $K_1 = K_2 \cup K_2'$, where K_2 consists of those cubes in K_1 which are disjoint from Q_1, and K_2' of those which intersect Q_1. Let Q_1^* denote the cube concentric with Q_1 whose edge length is $5t_1$. Thus, $|Q_1^*| = 5^n|Q_1|$, and since $2t_1 > t_1^*$, every cube in K_2' is contained in Q_1^*.

Starting with $j = 2$, continue this selection process for $j = 2, 3, \ldots$, by letting

$$t_j^* = \sup\{t : Q = Q(t) \in K_j\},$$

choosing a cube $Q_j = Q_j(t_j) \in K_j$ with $t_j > \frac{1}{2}t_j^*$, and splitting $K_j = K_{j+1} \cup K_{j+1}'$, where K_{j+1} consists of all those cubes of K_j which are disjoint from Q_j. If K_{j+1} is empty, the process ends. We have $t_j^* \geq t_{j+1}^*$; moreover, for each j, the $Q_1, \ldots, Q_j$ are disjoint from one another and from every cube in K_{j+1}, and every cube in K_{j+1}' is contained in the cube Q_j^* concentric with Q_j whose edge length is $5t_j$. Note that $|Q_j^*| = 5^n|Q_j|$.

Consider the sequence $t_1^* \geq t_2^* \geq \cdots$. If some K_{N+1} is empty (that is, if $t_j^* = 0$ for $j \geq N + 1$), then since

$$K_1 = K_2 \cup K_2' = \cdots = K_{N+1} \cup K_{N+1}' \cup \cdots \cup K_2',$$

and E is covered by the cubes in K_1, it follows that E is covered by the cubes in $K_{N+1}' \cup \cdots \cup K_2'$. Hence, $E \subset \bigcup_{j=1}^N Q_j^*$, so that

$$|E|_e \leq \sum_{j=1}^N |Q_j^*| = 5^n \sum_{j=1}^N |Q_j|.$$

This proves the lemma with $\beta = 5^{-n}$.

On the other hand, if no t_j^* is zero, then either there exists a $\delta > 0$ such that $t_j^* \geq \delta$ for all j, or $t_j^* \to 0$. In the first case, $t_j \geq \frac{1}{2}\delta$ for all j and, therefore, $\sum_{j=1}^N |Q_j| \to +\infty$ as $N \to \infty$. Given any $\beta > 0$, the lemma follows in this case by choosing N sufficiently large.

Finally, if $t_j^* \to 0$, we claim that every cube in K_1 is contained in $\bigcup_j Q_j^*$. Otherwise, there would be a cube $Q = Q(t)$ not intersecting any Q_j. Since this cube would belong to every K_j, t would satisfy $t \leq t_j^*$ for every j and, therefore, $t = 0$. This contradiction establishes the claim. Since E is covered by the cubes in K_1, it follows that

$$|E|_e \le \sum_j |Q_j^*| = 5^n \sum_j |Q_j|.$$

Hence, given β with $0 < \beta < 5^{-n}$, there exists an N such that $\sum_{j=1}^{N} |Q_j| \ge \beta |E|_e$. This completes the proof.

We stress that the lemma above does not presuppose the measurability of E, and that the proof can be shortened if E is measurable. In fact, if E is measurable, we can suppose it is closed and bounded [see, e.g., (3.22)]. Hence, assuming as we may that the cubes in K are open, it follows from the Heine-Borel theorem (1.12) that E can be covered by a finite number of cubes. For Q_1, we then choose the largest cube; similarly, in subsequent steps, we take Q_j to be the largest cube disjoint from $Q_1, \ldots, Q_{j-1}$. Thus, $E \subset \bigcup Q_j^*$, and the lemma follows.

Before stating the last lemma, we make a definition and a few remarks. If f is defined on $\mathbf{R}^n$ and integrable over every cube Q, let

(7.5)
$$f^*(\mathbf{x}) = \sup \frac{1}{|Q|} \int_Q |f(\mathbf{y})| \, d\mathbf{y},$$

where the supremum is taken over all Q with edges parallel to the coordinate axes and center $\mathbf{x}$. The function f^*, called the *Hardy-Littlewood maximal function of* f, is a gauge of the size of the averages of $|f|$ around $\mathbf{x}$. It clearly satisfies

(7.6)
$$\text{(i) } 0 \le f^*(\mathbf{x}) \le +\infty,$$
$$\text{(ii) } (f + g)^*(\mathbf{x}) \le f^*(\mathbf{x}) + g^*(\mathbf{x}),$$
$$\text{(iii) } (cf)^*(\mathbf{x}) = |c| f^*(\mathbf{x}).$$

If $f^*(\mathbf{x}_0) > \alpha$ for some $\mathbf{x}_0 \in \mathbf{R}^n$ and $\alpha > 0$, it follows from the absolute continuity of the indefinite integral that $f^*(\mathbf{x}) > \alpha$ for all $\mathbf{x}$ near $\mathbf{x}_0$. Hence, according to (4.14), f^* is lower semicontinuous in $\mathbf{R}^n$. In particular, it is measurable.

We now investigate the size of f^*. For any measurable E,

$$\chi_E^*(\mathbf{x}) = \sup \left\{ \frac{|E \cap Q|}{|Q|} : Q \text{ has center } \mathbf{x} \right\}.$$

If E is bounded and $Q^{\mathbf{x}}$ denotes the smallest cube with center $\mathbf{x}$ containing E, then

$$\frac{|E \cap Q^{\mathbf{x}}|}{|Q^{\mathbf{x}}|} = \frac{|E|}{|Q^{\mathbf{x}}|}.$$

It follows that there are positive constants c_1 and c_2 such that

(7.7)
$$c_1 \frac{|E|}{|\mathbf{x}|^n} \le \chi_E^*(\mathbf{x}) \le c_2 \frac{|E|}{|\mathbf{x}|^n} \quad \text{for large } |\mathbf{x}|.$$

In particular, if $|E| > 0$, χ_E^* is not integrable over $\mathbf{R}^n$. We leave it as an exercise to show that for any measurable f which is different from zero on a set of positive measure, there is a positive constant c such that

$$f^*(\mathbf{x}) \ge \frac{c}{|\mathbf{x}|^n} \quad \text{for} \quad |\mathbf{x}| \ge 1.$$

Therefore, *f^* is never integrable unless $f = 0$ a.e.*

To find a way to estimate the size of f^*, recall that by Tchebyshev's inequality

$$|\{\mathbf{x} \in \mathbf{R}^n : |f(\mathbf{x})| > \alpha\}| \le \frac{1}{\alpha}\int_{\mathbf{R}^n} |f(\mathbf{x})|\, d\mathbf{x}.$$

Hence, if $f \in L(\mathbf{R}^n)$, there is a constant c independent of α such that

$$|\{\mathbf{x} \in \mathbf{R}^n : |f(\mathbf{x})| > \alpha\}| \le \frac{c}{\alpha} \qquad (\alpha > 0). \tag{7.8}$$

Any measurable f, integrable or not, for which (7.8) is valid is said to belong to *weak* $L(\mathbf{R}^n)$. Thus, any function in $L(\mathbf{R}^n)$ is also in weak $L(\mathbf{R}^n)$. The function $|\mathbf{x}|^{-n}$ is an example of a function in weak $L(\mathbf{R}^n)$ which is not in $L(\mathbf{R}^n)$.

(7.9) **Lemma (*Hardy-Littlewood*)** *If $f \in L(\mathbf{R}^n)$, then f^* belongs to weak $L(\mathbf{R}^n)$. Moreover, there is a constant c independent of f and α such that*

$$|\{\mathbf{x} \in \mathbf{R}^n : f^*(\mathbf{x}) > \alpha\}| \le \frac{c}{\alpha}\int_{\mathbf{R}^n} |f|, \qquad \alpha > 0.$$

Proof. First suppose that in addition to being integrable, f has compact support. Then, as in (7.7), there is a constant c_1 depending on f such that $f^*(\mathbf{x}) \le c_1|\mathbf{x}|^{-n}$ for sufficiently large $|\mathbf{x}|$. In particular, $\{f^* > \alpha\}$ has finite measure for every $\alpha > 0$. Now fix $\alpha > 0$ and let

$$E = \{f^* > \alpha\}.$$

If $\mathbf{x} \in E$, then by the definitions of E and f^*, there is a cube $Q_{\mathbf{x}}$ with center $\mathbf{x}$ such that $|Q_{\mathbf{x}}|^{-1}\int_{Q_{\mathbf{x}}} |f| > \alpha$. Equivalently,

$$|Q_{\mathbf{x}}| < \frac{1}{\alpha}\int_{Q_{\mathbf{x}}} |f|.$$

The collection of such $Q_{\mathbf{x}}$ covers E, so by (7.4), there exist $\beta > 0$ and $\mathbf{x}_1, \dots, \mathbf{x}_N$ in E such that $Q_{\mathbf{x}_1}, \dots, Q_{\mathbf{x}_N}$ are disjoint and $|E| < \beta^{-1}\sum_{j=1}^N |Q_{\mathbf{x}_j}|$. Therefore,

$$|E| < \frac{1}{\beta}\sum_{j=1}^N \frac{1}{\alpha}\int_{Q_{\mathbf{x}_j}} |f| = \frac{1}{\beta\alpha}\int_{\bigcup_{j=1}^N Q_{\mathbf{x}_j}} |f| \le \frac{1}{\beta\alpha}\int_{\mathbf{R}^n} |f|.$$

This proves the result for such f, with $c = \beta^{-1}$.

Given any $f \in L(\mathbf{R}^n)$, we may assume without loss of generality that $f \geq 0$ since replacing f by $|f|$ does not alter f^*. Let $\{f_k\}$ be a sequence of integrable functions with compact support such that $0 \leq f_k \nearrow f$. Then there is a constant c independent of k and $\alpha > 0$ such that

$$|\{\mathbf{x} \in \mathbf{R}^n : f_k^*(\mathbf{x}) > \alpha\}| \leq \frac{c}{\alpha} \int_{\mathbf{R}^n} f_k \leq \frac{c}{\alpha} \int_{\mathbf{R}^n} f.$$

Since $f_k^* \nearrow f^*$, it follows from (3.26) that

$$|\{\mathbf{x} \in \mathbf{R}^n : f^*(\mathbf{x}) > \alpha\}| \leq \frac{c}{\alpha} \int_{\mathbf{R}^n} f,$$

which completes the proof.

Proof of Lebesgue's theorem. Given $f \in L(\mathbf{R}^n)$, there exists by (7.3) a sequence of continuous integrable C_k such that $\int_{\mathbf{R}^n} |f - C_k| \to 0$. Let $F(Q) = \int_Q f$ and $F_k(Q) = \int_Q C_k$. Then for any k,

$$\limsup_{Q \searrow \mathbf{x}} \left| \frac{F(Q)}{|Q|} - f(\mathbf{x}) \right| \leq \limsup_{Q \searrow \mathbf{x}} \left| \frac{F(Q)}{|Q|} - \frac{F_k(Q)}{|Q|} \right| + \limsup_{Q \searrow \mathbf{x}} \left| \frac{F_k(Q)}{|Q|} - C_k(\mathbf{x}) \right| + |C_k(\mathbf{x}) - f(\mathbf{x})|,$$

where the limsup is taken for cubes with center $\mathbf{x}$ which shrink to $\mathbf{x}$. Since C_k is continuous, the second term on the right is zero. Moreover,

$$\left| \frac{F(Q)}{|Q|} - \frac{F_k(Q)}{|Q|} \right| \leq \frac{1}{|Q|} \int_Q |f - C_k| \leq (f - C_k)^*(\mathbf{x}),$$

and therefore, the first term on the right is majorized by $(f - C_k)^*(\mathbf{x})$. Hence, for every k,

$$\text{(7.10)} \qquad \limsup_{Q \searrow \mathbf{x}} \left| \frac{F(Q)}{|Q|} - f(\mathbf{x}) \right| \leq (f - C_k)^*(\mathbf{x}) + |f(\mathbf{x}) - C_k(\mathbf{x})|.$$

Given $\varepsilon > 0$, let E_ε be the set on which the left side of (7.10) exceeds ε. By (7.10),

$$E_\varepsilon \subset \left\{\mathbf{x} : (f - C_k)^*(\mathbf{x}) > \frac{\varepsilon}{2}\right\} \cup \left\{\mathbf{x} : |f(\mathbf{x}) - C_k(\mathbf{x})| > \frac{\varepsilon}{2}\right\}.$$

Applying (7.9) to the first set on the right and Tchebyshev's inequality to the second, we obtain

$$|E_\varepsilon|_e \leq c\left(\frac{\varepsilon}{2}\right)^{-1} \int_{\mathbf{R}^n} |f - C_k| + \left(\frac{\varepsilon}{2}\right)^{-1} \int_{\mathbf{R}^n} |f - C_k|.$$

Since c is independent of k, it follows by letting $k \to \infty$ that $|E_\varepsilon|_e = 0$.

Let E be the set where the left side of (7.10) is positive. Then $E = \bigcup_k E_{\varepsilon_k}$ for any sequence $\varepsilon_k \to 0$, and therefore $|E| = 0$. This means that $\lim_{Q \searrow \mathbf{x}} F(Q)/|Q|$ exists and equals $f(\mathbf{x})$ for almost every $\mathbf{x}$, which completes the proof.

We now list several extensions and corollaries of Lebesgue's theorem.

(I) A measurable function f defined on $\mathbf{R}^n$ is said to be *locally integrable on* $\mathbf{R}^n$ if it is integrable over every bounded measurable subset of $\mathbf{R}^n$.

(7.11) Theorem *The conclusion of Lebesgue's theorem is valid if, instead of being integrable, f is locally integrable on $\mathbf{R}^n$.*

Proof. It is enough to show that the conclusion holds a.e. in every open ball. Fix a ball and replace f by zero outside it. This new function is integrable over $\mathbf{R}^n$, its integral is differentiable a.e., and since differentiability is a local property, the initial function f is differentiable a.e. in the ball. This completes the proof.

(II) For any measurable E, note that

$$\frac{1}{|Q|}\int_Q \chi_E = \frac{|E \cap Q|}{|Q|}.$$

By (7.11), the left-hand side tends to $\chi_E(\mathbf{x})$ a.e. as $Q \searrow \mathbf{x}$; that is,

$$\lim_{Q \searrow \mathbf{x}} \frac{|E \cap Q|}{|Q|} = \chi_E(\mathbf{x}) \qquad \text{a.e.} \tag{7.12}$$

A point $\mathbf{x}$ for which this limit is 1 is called a *point of density of E, and a* point for which it is zero is called a *point of dispersion of E*. Since

$$\frac{|Q \cap E|}{|Q|} + \frac{|Q \cap CE|}{|Q|} = \frac{|Q|}{|Q|} = 1,$$

every point of density of E is a point of dispersion of CE, and vice versa. Formula (7.12) can be restated as follows.

(7.13) Theorem *Let E be a measurable set. Then almost every point of E is a point of density of E.*

Thus, roughly speaking, a set "clusters" around almost all of its points.

(III) The formula $\lim_{Q \searrow \mathbf{x}} (1/|Q|) \int_Q f(\mathbf{y})\, d\mathbf{y} = f(\mathbf{x})$ can be written

$$\lim_{Q \searrow \mathbf{x}} \frac{1}{|Q|}\int_Q [f(\mathbf{y}) - f(\mathbf{x})]\, d\mathbf{y} = 0,$$

and is valid for almost every $\mathbf{x}$ if f is locally integrable. A point $\mathbf{x}$ at which the stronger statement

$$\lim_{Q \searrow \mathbf{x}} \frac{1}{|Q|}\int_Q |f(\mathbf{y}) - f(\mathbf{x})|\, d\mathbf{y} = 0 \tag{7.14}$$

is valid is called a *Lebesgue point of f*, and the collection of all such points is called the *Lebesgue set of f*.

(7.15) Theorem *Let f be locally integrable in* $\mathbf{R}^n$. *Then almost every point of* $\mathbf{R}^n$ *is a Lebesgue point of f; that is, there exists a set Z (depending on f) of measure zero such that (7.14) holds for* $\mathbf{x} \notin Z$.

Proof. Let $\{r_k\}$ be the rational numbers, and let Z_k be the set where the formula

$$\lim_{Q \searrow \mathbf{x}} \frac{1}{|Q|} \int_Q |f(\mathbf{y}) - r_k| \, d\mathbf{y} = |f(\mathbf{x}) - r_k|$$

is *not* valid. Since $|f(\mathbf{y}) - r_k|$ is locally integrable, we have $|Z_k| = 0$. Let $Z = \bigcup Z_k$; then $|Z| = 0$. For any Q, $\mathbf{x}$, and r_k,

$$\begin{aligned} \frac{1}{|Q|} \int_Q |f(\mathbf{y}) - f(\mathbf{x})| \, d\mathbf{y} &\leq \frac{1}{|Q|} \int_Q |f(\mathbf{y}) - r_k| \, d\mathbf{y} + \frac{1}{|Q|} \int_Q |f(\mathbf{x}) - r_k| \, d\mathbf{y} \\ &= \frac{1}{|Q|} \int_Q |f(\mathbf{y}) - r_k| \, d\mathbf{y} + |f(\mathbf{x}) - r_k|. \end{aligned}$$

Therefore, if $\mathbf{x} \notin Z$,

$$\limsup_{Q \searrow \mathbf{x}} \frac{1}{|Q|} \int_Q |f(\mathbf{y}) - f(\mathbf{x})| \, d\mathbf{y} \leq 2|f(\mathbf{x}) - r_k|$$

for every r_k. For an $\mathbf{x}$ at which $f(\mathbf{x})$ is finite (in particular, almost everywhere), we can choose r_k such that $|f(\mathbf{x}) - r_k|$ is arbitrarily small. This shows that the left side of the last formula is zero a.e., and completes the proof.

(IV) So far, the sets contracting to $\mathbf{x}$ have been cubes centered at $\mathbf{x}$ with edges parallel to the coordinate axes. Many other sets can be used. A family $\{S\}$ of measurable sets is said to *shrink regularly to* $\mathbf{x}$ provided

(i) The diameters of the sets S tend to zero.
(ii) If Q is the smallest cube with center $\mathbf{x}$ containing S, there is a constant k independent of S such that

$$|Q| \leq k|S|.$$

The sets S need not contain $\mathbf{x}$.

(7.16) Theorem *Let f be locally integrable in* $\mathbf{R}^n$. *Then at every point* $\mathbf{x}$ *of the Lebesgue set of f (in particular, almost everywhere),*

$$\frac{1}{|S|} \int_S |f(\mathbf{y}) - f(\mathbf{x})| \, d\mathbf{y} \to 0$$

for any family $\{S\}$ *which shrinks regularly to* $\mathbf{x}$. *Thus also*

$$\frac{1}{|S|}\int_S f(\mathbf{y})\,d\mathbf{y} \to f(\mathbf{x}) \qquad \text{a.e.}$$

Proof. If $S \subset Q$, we have

$$\int_S |f(\mathbf{y}) - f(\mathbf{x})|\,d\mathbf{y} \le \int_Q |f(\mathbf{y}) - f(\mathbf{x})|\,d\mathbf{y}.$$

Hence, if $\{S\}$ shrinks regularly to $\mathbf{x}$ and Q is the least cube with center $\mathbf{x}$ containing S, then

$$\begin{aligned}\frac{1}{|S|}\int_S |f(\mathbf{y}) - f(\mathbf{x})|\,d\mathbf{y} &\le \frac{|Q|}{|S|}\frac{1}{|Q|}\int_Q |f(\mathbf{y}) - f(\mathbf{x})|\,d\mathbf{y}\\ &\le k\frac{1}{|Q|}\int_Q |f(\mathbf{y}) - f(\mathbf{x})|\,d\mathbf{y}.\end{aligned}$$

If $\mathbf{x}$ is a Lebesgue point of f, the last expression tends to zero, and the theorem follows.

In particular, for functions of a single variable, we obtain

$$\lim_{h\to 0}\frac{1}{h}\int_x^{x+h} f(y)\,dy = f(x) \qquad \text{a.e.}$$

3. The Vitali Covering Lemma

The theorem which follows is a refinement of the simple Vitali lemma (7.4). We shall assume that each point of a set is covered not just by a single cube, but by a sequence of cubes with diameters tending to zero. In this case, it turns out that we can cover *almost all* points of the set by a sequence of disjoint cubes. The result will be essentially a corollary of (7.4).

A family K of cubes is said to cover a set E in the *Vitali sense* if for every $\mathbf{x} \in E$ and $\eta > 0$, there is a cube in K containing $\mathbf{x}$ whose diameter is less than η.

(7.17) Theorem (*Vitali Covering Lemma*) *Suppose that* E *is covered in the Vitali sense by a family* K *of cubes and that* $0 < |E|_e < +\infty$. *Then, given* $\varepsilon > 0$, *there is a sequence* $\{Q_j\}$ *of disjoint cubes in* K *such that*

$$\Big|E - \bigcup_j Q_j\Big| = 0, \qquad \sum_j |Q_j| < (1 + \varepsilon)|E|_e.$$

Proof. The second relation is automatically satisfied if we choose an open set G containing E with $|G| < (1 + \varepsilon)|E|_e$ and consider only those Q in K which lie in G.

By (7.4), there exist a constant β, $0 < \beta < 1$, depending only on the dimension, and disjoint $Q_1, \ldots, Q_{N_1}$ in K such that $\sum_{j=1}^{N_1} |Q_j| > \beta |E|_e$. Therefore,

$$\left| E - \bigcup_{j=1}^{N_1} Q_j \right|_e \leq \left| G - \bigcup_{j=1}^{N_1} Q_j \right| = |G| - \sum_{j=1}^{N_1} |Q_j| < |E|_e(1 + \varepsilon - \beta).$$

Hence, by considering from the start only those ε with $0 < \varepsilon < \beta/2$, we have

$$\left| E - \bigcup_{j=1}^{N_1} Q_j \right|_e < |E|_e\left(1 - \frac{\beta}{2}\right).$$

Thus, the part of E not covered by the cubes obtained from the simple Vitali lemma has outer measure less than $|E|_e(1 - \beta/2)$. We now repeat the process for the set $E_1 = E - \bigcup_{j=1}^{N_1} Q_j$, which is still covered in the Vitali sense by those cubes in K which are disjoint from $Q_1, \ldots, Q_{N_1}$. We obtain $Q_{N_1+1}, \ldots, Q_{N_2}$, disjoint from each other and from $Q_1, \ldots, Q_{N_1}$, such that

$$\left| E - \bigcup_{j=1}^{N_2} Q_j \right|_e = \left| E_1 - \bigcup_{j=N_1+1}^{N_2} Q_j \right|_e < |E_1|_e\left(1 - \frac{\beta}{2}\right) < |E|_e\left(1 - \frac{\beta}{2}\right)^2.$$

Continuing in this way, we obtain at the mth stage disjoint $Q_1, \ldots, Q_{N_m}$ in K such that

$$\left| E - \bigcup_{j=1}^{N_m} Q_j \right|_e < |E|_e\left(1 - \frac{\beta}{2}\right)^m.$$

Since $(1 - \beta/2)^m \to 0$ as $m \to \infty$, there is a sequence of disjoint cubes in K with the desired properties.

(7.18) Corollary *Suppose that E is covered in the Vitali sense by a family K of cubes and $0 < |E|_e < +\infty$. Then, given $\varepsilon > 0$, there is a finite collection $Q_1, \ldots, Q_N$ of disjoint cubes in K such that*

$$\left| E - \bigcup_{j=1}^{N} Q_j \right|_e < \varepsilon, \qquad \sum_{j=1}^{N} |Q_j| < (1 + \varepsilon)|E|_e.$$

This is part of the proof of Vitali's lemma.

Note by Carathéodory's theorem that

$$|E|_e = \left| E - \bigcup_{j=1}^{N} Q_j \right|_e + \left| E \cap \bigcup_{j=1}^{N} Q_j \right|_e,$$

so that for E and $\{Q_j\}_{j=1}^N$ as in (7.18), we have

(7.19)
$$|E|_e - \varepsilon < \left| E \cap \bigcup_{j=1}^{N} Q_j \right|_e.$$

In particular,

(7.20)
$$|E|_e - \varepsilon < \sum_{j=1}^{N} |Q_j|.$$

4. Differentiation of Monotone Functions

As an application of Vitali's covering lemma, we will prove a basic result concerning the differentiability of monotone functions on $\mathbf{R}^1$. If $f(x)$, $x \in \mathbf{R}^1$, is a real-valued function defined and finite in a neighborhood of x_0, consider the four *Dini numbers* (or *derivates*)

$$D_1 f(x_0) = \limsup_{h \to 0+} \frac{f(x_0 + h) - f(x_0)}{h},$$

$$D_2 f(x_0) = \liminf_{h \to 0+} \frac{f(x_0 + h) - f(x_0)}{h},$$

$$D_3 f(x_0) = \limsup_{h \to 0-} \frac{f(x_0 + h) - f(x_0)}{h},$$

$$D_4 f(x_0) = \liminf_{h \to 0-} \frac{f(x_0 + h) - f(x_0)}{h}.$$

Clearly, $D_2 f \leqslant D_1 f$ and $D_4 f \leqslant D_3 f$. If all four Dini numbers are equal, we say that f *has a derivative at* x_0, and call the common value the *derivative* $f'(x_0)$ at x_0.

(7.21) Theorem *Let $f(x)$, $x \in \mathbf{R}^1$, be monotone increasing and finite on an interval (a,b). Then f has a measurable, nonnegative derivative f' almost everywhere in (a,b). Moreover,*

$$0 \le \int_a^b f' \le f(b-) - f(a+). \tag{7.22}$$

Proof. We may assume that (a,b) is finite; the general case follows from this by passage to the limit. We will show that the set $\{x \in (a,b) : D_1 f(x) > D_4 f(x)\}$ has measure zero. A similar argument will apply to any two Dini numbers of f. It is enough to show that each set

$$A_{r,s} = \{x \in (a,b) : D_1 f(x) > r > s > D_4 f(x)\}$$

has measure zero since the original set is the union of these over rational r and s. We may assume r, $s > 0$ since f is increasing.

Fix r and s, $r > s > 0$, write $A = A_{r,s}$, and suppose that $|A|_e > 0$. If $x \in A$, the fact that $D_4 f(x) < s$ implies the existence of arbitrarily small $h > 0$ such that

$$\frac{f(x - h) - f(x)}{-h} < s.$$

By (7.18) and (7.19), given $\varepsilon > 0$, there exist disjoint intervals $[x_j - h_j, x_j]$, $j = 1, \ldots, N$, such that

(i) $f(x_j) - f(x_j - h_j) < s h_j$, $j = 1, \ldots, N$.
(ii) $|A \cap \bigcup_{j=1}^N [x_j - h_j, x_j]|_e > |A|_e - \varepsilon$.

(iii) $\sum_{j=1}^{N} h_j < (1 + \varepsilon)|A|_e$.

Combining (i) and (iii), we obtain

(iv) $\sum_{j=1}^{N} [f(x_j) - f(x_j - h_j)] < s(1 + \varepsilon)|A|_e$.

Let $B = A \cap \bigcup_{j=1}^{N} [x_j - h_j, x_j]$. For every $y \in B$ which is not an endpoint of some $[x_j - h_j, x_j]$, the fact that $D_1 f(y) > r$ implies that there exist arbitrarily small $k > 0$ such that $[y, y + k]$ lies in some $[x_j - h_j, x_j]$ and

$$\frac{f(y + k) - f(y)}{k} > r.$$

Hence, by (7.18) and (7.20), there exist disjoint $[y_i, y_i + k_i]$, $i = 1, \ldots, M$, such that

(v) Each $[y_i, y_i + k_i]$ lies in some $[x_j - h_j, x_j]$.
(vi) $f(y_i + k_i) - f(y_i) > rk_i$, $i = 1, \ldots, M$.
(vii) $\sum_{i=1}^{M} k_i > |B|_e - \varepsilon > |A|_e - 2\varepsilon$ [by (ii)].

Therefore,

(viii) $\sum_{i=1}^{M} [f(y_i + k_i) - f(y_i)] > r(|A|_e - 2\varepsilon)$.

Since f is increasing, it follows from (v) that

$$\sum_{i=1}^{M} [f(y_i + k_i) - f(y_i)] \le \sum_{j=1}^{N} [f(x_j) - f(x_j - h_j)].$$

Combining this inequality with (iv) and (viii), we obtain $r(|A|_e - 2\varepsilon) < s(1 + \varepsilon)|A|_e$. Since $\varepsilon > 0$ is arbitrary, this gives $r \le s$, which is a contradiction. Hence, $|A|_e = 0$.

Since an analogous argument applies to any two Dini numbers, it follows that $f'(x)$ exists for almost every x in (a,b). Extend the definition of f by setting $f(x) = f(b-)$ for $x \ge b$, and let

$$f_k(x) = \frac{f(x + h) - f(x)}{h} \qquad \left(h = \frac{1}{k}\right)$$

for $x \in (a,b)$ and $k = 1, 2, \ldots$. Each f_k is nonnegative and measurable, and $f_k \to f'$ a.e. in (a,b). Hence, f' is measurable on (a,b), and by Fatou's lemma,

$$\int_a^b f' \le \liminf_{k \to \infty} \int_a^b f_k.$$

If $f(b-)$ is finite (otherwise, the result is obvious), we have

$$\int_a^b f_k = \frac{1}{h}\int_{a+h}^{b+h} f - \frac{1}{h}\int_a^b f$$
$$= \frac{1}{h}\int_b^{b+h} f - \frac{1}{h}\int_a^{a+h} f = f(b-) - \frac{1}{h}\int_a^{a+h} f.$$

Since $f(a+) \le (1/h) \int_a^{a+h} f \le f(a+h)$, we obtain

$$\int_a^b f' \le \liminf_{k\to\infty} \int_a^b f_k = \lim_{k\to\infty} \int_a^b f_k = f(b-) - f(a+),$$

which proves the theorem.

The inequality in formula (7.22) cannot in general be replaced by equality, even if f is continuous on $[a,b]$. To see this, let f be the Cantor-Lebesgue function on $[0,1]$ (p. 35). Then f is continuous and monotone increasing on $[0,1]$, $f(0) = 0$, $f(1) = 1$, and since f is constant on every interval removed in constructing the Cantor set, $f' = 0$ a.e. Thus, $\int_0^1 f' = 0$, while $f(1) - f(0) = 1$.

We shall return in theorem (7.29) to the question of equality in (7.22).

By (2.7), any function of bounded variation can be written as the difference of two bounded monotone increasing functions. Hence, we obtain the following result. (See also Exercise 9.)

(7.23) Corollary *If f is of bounded variation on $[a,b]$, then f' exists a.e. in $[a,b]$, and $f' \in L[a,b]$.*

If f is of bounded variation on $[a,b]$ and $V(x)$ denotes its (total) variation on $[a,x]$, $a \le x \le b$, then V is monotone increasing by (2.2). The following result gives an important relation between V' and f'.

(7.24) Theorem *If f is of bounded variation on $[a,b]$ and $V(x)$ is the variation of f on $[a,x]$, $a \le x \le b$, then*

$$V'(x) = |f'(x)| \text{ for a.e. } x \in [a,b].$$

We will prove this with the help of the next lemma, which is of independent interest.

(7.25) Lemma (*Fubini*) *Let $\{f_k\}$ be a sequence of monotone increasing functions on $[a,b]$. If the series $s(x) = \sum f_k(x)$ converges on $[a,b]$, then*

$$s'(x) = \sum f_k'(x) \text{ a.e. in } [a,b].$$

Proof. Let $s_m = \sum_{k=1}^m f_k$, $r_m = \sum_{k=m+1}^\infty f_k$. Then s_m and r_m are monotone increasing functions and $s = s_m + r_m$. With the exception of a set Z_m, $|Z_m| = 0$, these three terms together with $f_1, \ldots, f_m$ are differentiable and $s' = s_m' + r_m'$. In particular, $s' \ge s_m' = \sum_{k=1}^m f_k'$ except in Z_m. It follows that

$$\sum_{k=1}^\infty f_k' \le s'$$

except in $Z = \bigcup_{m=1}^\infty Z_m$, $|Z| = 0$.

To prove that in the last inequality we actually have equality a.e., it is enough to show that $r_m' \to 0$ a.e. for m running through a sequence of values $m_1 < m_2 < \cdots$. Select $\{m_j\}$ increasing so rapidly that $\sum r_{m_j}(x)$ converges at both $x = a$ and $x = b$. This implies the convergence of $\sum \{r_{m_j}(b) - r_{m_j}(a)\}$ and also, in view of the monotonicity of r_{m_j}, of $\sum \{r_{m_j}(b-) - r_{m_j}(a+)\}$. By (7.22), we have

$$0 \le \int_a^b \sum r'_{m_j} = \sum \int_a^b r'_{m_j} \le \sum \{r_{m_j}(b-) - r_{m_j}(a+)\}.$$

Thus, $\sum r'_{m_j}$ is integrable over (a,b), and therefore, it is finite a.e. in (a,b). Thus, $r'_{m_j} \to 0$ a.e., and the proof is complete.

Proof of Theorem (7.24). Let f be of bounded variation on $[a,b]$, and let $V(x) = V[a,x]$ be its variation on $[a,x]$, $a \le x \le b$. Then $V(a) = 0$ and $V(b)$ is the variation of f on $[a,b]$. Select a sequence $\{\Gamma_k : \Gamma_k = \{x_j^k\}\}$ of partitions of $[a,b]$ such that $0 \le V(b) - S_{\Gamma_k} < 2^{-k}$, where

$$S_{\Gamma_k} = \sum_j |f(x_j^k) - f(x_{j-1}^k)|.$$

For each k, define a function f_k on $[a,b]$ as follows: if $x \in [x_{j-1}^k, x_j^k]$, let

$$f_k(x) = \begin{cases} f(x) + c_j^k & \text{if } f(x_j^k) \ge f(x_{j-1}^k), \\ -f(x) + c_j^k & \text{if } f(x_j^k) < f(x_{j-1}^k), \end{cases}$$

where the c_j^k are constants chosen so that $f_k(a) = 0$ and f_k is well-defined (i.e., single-valued) at x_j for every j. Then, for all k and j,

$$f_k(x_j^k) - f_k(x_{j-1}^k) = |f(x_j^k) - f(x_{j-1}^k)|,$$

so that

$$S_{\Gamma_k} = \sum_j [f_k(x_j^k) - f_k(x_{j-1}^k)] = f_k(b).$$

Hence, for any k, we obtain $0 \le V(b) - f_k(b) < 2^{-k}$.

We will show that each $V(x) - f_k(x)$ is an increasing function of x. This amounts to showing that if $x < y$, then $f_k(y) - f_k(x) \le V(y) - V(x)$. If x and y both belong to the same partitioning interval of Γ_k, then $f_k(y) - f_k(x) \le |f(y) - f(x)|$, and therefore,

$$f_k(y) - f_k(x) \le V[x,y] = V(y) - V(x).$$

In the general case, if $x_l^k, x_{l+1}^k, \ldots, x_m^k$ are the points of Γ_k between x and y, the result follows by adding the inequalities for the intervals $[x,x_l^k]$, $[x_l^k,x_{l+1}^k], \ldots, [x_m^k,y]$. Since $V(a) = f_k(a) = 0$ and $V(b) - f_k(b) < 2^{-k}$, it follows that $0 \le V(x) - f_k(x) < 2^{-k}$ for all $x \in [a,b]$. Hence,

$$\sum [V(x) - f_k(x)] < \sum 2^{-k} < +\infty$$

for $x \in [a,b]$, so that by (7.25) the series $\sum [V'(x) - f_k'(x)]$ converges a.e. in $[a,b]$. Hence, $f_k' \to V'$ a.e. However, $|f_k'| = |f'|$ a.e., so that $|f'| = |V'|$ a.e.

The theorem now follows from the fact that V' is nonnegative wherever it exists.

5. Absolutely Continuous and Singular Functions

We now turn to the question of equality in formula (7.22). As we know, the Cantor-Lebesgue function is an example of an increasing continuous f whose derivative is integrable on $[0,1]$, but for which $\int_0^1 f' \neq f(1) - f(0)$. A finite function f on a finite interval $[a,b]$ is said to be *absolutely continuous* on $[a,b]$ if given $\varepsilon > 0$, there exists $\delta > 0$ such that for any collection $\{[a_i,b_i]\}$ (finite or not) of nonoverlapping subintervals of $[a,b]$,

$$\sum |f(b_i) - f(a_i)| < \varepsilon \text{ if } \sum (b_i - a_i) < \delta.$$

For example, if g is integrable on $[a,b]$ and $G(x) = \int_a^x g$ denotes its indefinite integral, then

$$\sum |G(b_i) - G(a_i)| \leq \int_{\bigcup[a_i,b_i]} |g|$$

for any nonoverlapping $[a_i,b_i]$. By (7.1), $\int_E |g|$ is an absolutely continuous *set* function, and therefore, G is an absolutely continuous function. Thus, an indefinite integral is an absolutely continuous function. One of the main results proved below is the converse: Every absolutely continuous function is an indefinite integral.

Another example of an absolutely continuous function is any f which satisfies a Lipschitz condition:

(7.26) $$|f(x) - f(y)| \leq C|x - y|$$

for $x,y \in [a,b]$.

On the other hand, the Cantor-Lebesgue function f is an example of a continuous function which is not absolutely continuous, since if $C_k = \bigcup_j [a_j^k,b_j^k]$ denotes the intervals remaining at the kth stage of construction of the Cantor set, then $|C_k| \to 0$, while

$$\sum_j [f(b_j^k) - f(a_j^k)] = 1$$

for every k.

A function which is absolutely continuous is clearly continuous. We will also show that if f is absolutely continuous on $[a,b]$, then f' exists a.e. in $[a,b]$. This is an immediate corollary of (7.23) and the following theorem.

(7.27) Theorem *If f is absolutely continuous on $[a,b]$, then it is of bounded variation on $[a,b]$.*

Proof. Choose δ so that $\sum |f(b_i) - f(a_i)| \leq 1$ for any collection of nonoverlapping intervals with $\sum_i (b_i - a_i) \le \delta$. Then the variation of f over any subinterval of $[a,b]$ with length less than δ is at most 1. Hence, if we split $[a,b]$ into N intervals each with length less than δ, then $V[a,b] \le N$.

The assumption above that $[a,b]$ is finite is necessary as shown by the example $f(x) = x$.

A function f for which f' is zero a.e. in $[a,b]$ is said to be *singular* on $[a,b]$. The Cantor-Lebesgue function is an example of a nonconstant, singular function on $[0,1]$.

(7.28) Theorem *If f is both absolutely continuous and singular on $[a,b]$, then it is constant on $[a,b]$.*

Proof. It is enough to show that $f(a) = f(b)$ since this result applied to any subinterval proves that f is constant. Let E be the subset of (a,b), where $f' = 0$, so that $|E| = b - a$. Given $\varepsilon > 0$ and $x \in E$, we have $[x, x + h] \subset (a,b)$ and $|f(x + h) - f(x)| < \varepsilon h$ for all sufficiently small $h > 0$. Let δ be the number corresponding to ε in the definition of the absolute continuity of f. By (7.18) and (7.20), there exist disjoint $Q_j = [x_j, x_j + h_j]$, $j = 1, \ldots, N$, in (a,b) such that

(i) $|f(x_j + h_j) - f(x_j)| < \varepsilon h_j$,
(ii) $\sum_{j=1}^N |Q_j| > (b - a) - \delta$.

By (i),

$$\sum_{j=1}^{N} |f(x_j + h_j) - f(x_j)| < \varepsilon \sum_{j=1}^{N} |Q_j| \le \varepsilon(b - a).$$

Moreover, since by (ii) the total length of the complementary intervals is less than δ, the sum of the absolute values of the increments of f over them is less than ε. Thus, the sum of the absolute values of the increments of f over the Q_j and the complementary intervals is less than $\varepsilon(b - a) + \varepsilon$. Hence, $|f(b) - f(a)| < \varepsilon(b - a) + \varepsilon$, so that $f(b) = f(a)$. This completes the proof.

(7.29) Theorem *A function f is absolutely continuous on $[a,b]$ if and only if f' exists a.e. in $[a,b]$, f' is integrable on $[a,b]$ and*

$$f(x) - f(a) = \int_a^x f' \qquad (a \le x \le b).$$

Proof. We have already observed (see p. 115) that any function of the form $G(x) = \int_a^x g$, $g \in L(a,b)$, is absolutely continuous on $[a,b]$. Hence, the sufficiency of the condition follows.

Conversely, if f is absolutely continuous, let $F(x) = \int_a^x f'$. F is well-defined by virtue of (7.27) and (7.23). Moreover, by (7.16), $F' = f'$ a.e. in $[a,b]$. It follows that $F(x) - f(x)$ is both absolutely continuous and singular on $[a,b]$. Hence, by (7.28), we obtain $F(x) - f(x) = F(a) - f(a)$ for $x \in [a,b]$. Since $F(a) = 0$, the proof is complete.

(7.30) Theorem *If f is of bounded variation on $[a,b]$, then f can be written $f = g + h$, where g is absolutely continuous on $[a,b]$ and h is singular on $[a,b]$. Moreover, g and h are unique up to additive constants.*

Proof. Let $g(x) = \int_a^x f'$ and $h = f - g$. Then $h' = f' - g' = f' - f' = 0$ a.e. in $[a,b]$, so that h is singular, and the formula $f = g + h$ gives the desired decomposition. If $f = g_1 + h_1$ is another such decomposition, then $g - g_1 = h_1 - h$. Since $g - g_1$ is absolutely continuous and $h_1 - h$ is singular, it follows from (7.28) that $g - g_1 = h_1 - h =$ constant, which completes the proof.

The next theorem, which is an extension of (2.10), gives formulas for the variations of an absolutely continuous function.

(7.31) Theorem *Let f be absolutely continuous on $[a,b]$, and let $V(x)$, $P(x)$, and $N(x)$ denote its total, positive and negative variations on $[a,x]$, $a \le x \le b$. Then V, P, and N are absolutely continuous on $[a,b]$, and*

$$V(x) = \int_a^x |f'|,\ P(x) = \int_a^x \{f'\}^+,\text{ and } N(x) = \int_a^x \{f'\}^-.$$

Proof. We will first show that V is absolutely continuous. If $[\alpha,\beta]$ is a subinterval of $[a,b]$ and $\Gamma = \{x_k\}$ is a partition of $[\alpha,\beta]$, then

$$V(\beta) - V(\alpha) = V[\alpha,\beta] = \sup_\Gamma \sum |f(x_k) - f(x_{k-1})|$$
$$= \sup_\Gamma \sum \left| \int_{x_{k-1}}^{x_k} f' \right| \le \int_\alpha^\beta |f'|.$$

Hence, if $\{[\alpha_i,\beta_i]\}$ is a collection of nonoverlapping subintervals of $[a,b]$, then

$$\sum [V(\beta_i) - V(\alpha_i)] \le \int_{\bigcup[\alpha_i,\beta_i]} |f'|.$$

From this inequality and (7.1), it follows that V is absolutely continuous on $[a,b]$. Therefore, by (7.29) and the fact that $V(a) = 0$, we have $V(x) = \int_a^x V'$. Since $V' = |f'|$ a.e. by (7.24), we obtain $V(x) = \int_a^x |f'|$.

The fact that P and N are absolutely continuous and the formulas for P and N now follow from the relations $V(x) = \int_a^x |f'|$, $f(x) = f(a) + \int_a^x f'$, $P(x) = \frac{1}{2}[V(x) + f(x) - f(a)]$, and $N(x) = \frac{1}{2}[V(x) - f(x) + f(a)]$. [See (2.6).] This completes the proof.

In Chapter 2, p. 23, we proved that if g is continuous on $[a,b]$ and f is continuously differentiable on $[a,b]$, then

$$\int_a^b g\, df = \int_a^b gf'\, dx.$$

This is a special case of the first part of the following theorem.

(7.32) Theorem (*Integration by Parts*)

(i) *If g is continuous on $[a,b]$ and f is absolutely continuous on $[a,b]$, then*

$$\int_a^b g\,df = \int_a^b gf'\,dx.$$

(ii) *If both f and g are absolutely continuous on $[a,b]$, then*

$$\int_a^b gf'\,dx = g(b)f(b) - g(a)f(a) - \int_a^b g'f\,dx.$$

Proof. To prove (i), first note that the integrals in the conclusion exist and are finite by (2.24) and (5.30). Let $\Gamma = \{x_i\}$ be a partition of $[a,b]$ with norm $|\Gamma|$. Then

$$\int_a^b gf'\,dx = \sum \int_{x_{i-1}}^{x_i} gf'\,dx = \sum g(x_{i-1}) \int_{x_{i-1}}^{x_i} f'(x)\,dx + \sum \int_{x_{i-1}}^{x_i} [g(x) - g(x_{i-1})]\,f'(x)\,dx.$$

The first term on the right equals $\sum g(x_{i-1})[f(x_i) - f(x_{i-1})]$, which converges to $\int_a^b g\,df$ as $|\Gamma| \to 0$. The second term on the right is majorized in absolute value by

$$[\sup_{|x-y|\le|\Gamma|} |g(x) - g(y)|] \sum \int_{x_{i-1}}^{x_i} |f'|\,dx = [\sup_{|x-y|\le|\Gamma|} |g(x) - g(y)|] \int_a^b |f'|\,dx.$$

Since g is uniformly continuous on $[a,b]$, the last expression tends to 0 as $|\Gamma| \to 0$. This proves (i).

We can easily deduce (ii) from (i) by using (2.21). In fact, if f and g are absolutely continuous, then

$$\int_a^b gf'\,dx = \int_a^b g\,df = g(b)f(b) - g(a)f(a) - \int_a^b f\,dg = g(b)f(b) - g(a)f(a) - \int_a^b fg'\,dx.$$

This proves (ii).

For a generalization of part (i) above, see (9.14) in Chapter 9.

6. Convex Functions

Let ϕ be defined and finite on an interval (a,b). Then ϕ is said to be *convex* in (a,b) if for every $[x_1,x_2]$ in (a,b), the graph of ϕ on $[x_1,x_2]$ lies on or below the line segment connecting the points $(x_1,\phi(x_1))$ and $(x_2,\phi(x_2))$ of the graph of ϕ.

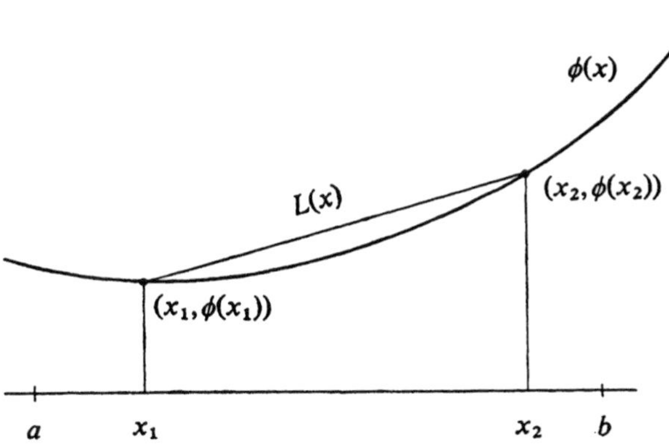

Let $y = L(x)$ be the equation of this line segment. Then since every $x \in [x_1,x_2]$ can be written $x = \theta x_1 + (1 - \theta)x_2$ for appropriate θ, $0 \le \theta \le 1$, the condition for convexity is

$$\phi(\theta x_1 + (1 - \theta)x_2) \le L(\theta x_1 + (1 - \theta)x_2)$$

for $[x_1,x_2] \subset (a,b)$ and $0 \le \theta \le 1$. Since

$$L(\theta x_1 + (1 - \theta)x_2) = \theta L(x_1) + (1 - \theta)L(x_2) = \theta\phi(x_1) + (1 - \theta)\phi(x_2),$$

it follows that ϕ is convex in (a,b) if and only if

(7.33) $$\phi(\theta x_1 + (1 - \theta)x_2) \le \theta\phi(x_1) + (1 - \theta)\phi(x_2)$$

for $a < x_1 < x_2 < b$, $0 \le \theta \le 1$. Equivalently, ϕ is convex in (a,b) if and only if

(7.34) $$\phi\left(\frac{p_1x_1 + p_2x_2}{p_1 + p_2}\right) \le \frac{p_1\phi(x_1) + p_2\phi(x_2)}{p_1 + p_2}$$

for $a < x_1 < x_2 < b$, $p_1 \ge 0$, $p_2 \ge 0$, $p_1 + p_2 > 0$.

(7.35) Theorem (*Jensen's Inequality*) *Let ϕ be convex in (a,b). Let $\{x_j\}_{j=1}^N$ be points of (a,b) and $\{p_j\}_{j=1}^N$ satisfy $p_j \ge 0$ and $\sum p_j > 0$. Then*

$$\phi\left(\frac{\sum p_jx_j}{\sum p_j}\right) \le \frac{\sum p_j\phi(x_j)}{\sum p_j}.$$

The proof follows by repeated application of (7.34).

(7.36) Theorem

(i) *If ϕ_1 and ϕ_2 are convex in (a,b), then $\phi_1 + \phi_2$ is convex in (a,b).*
(ii) *If ϕ is convex in (a,b) and c is a positive constant, then $c\phi$ is convex in (a,b).*
(iii) *If ϕ_k, $k = 1, 2, \ldots$, are convex in (a,b) and $\phi_k \to \phi$ in (a,b), then ϕ is convex in (a,b).*

The proof is left as an exercise.

(7.37) Theorem *If ϕ' exists and is monotone increasing in (a,b), then ϕ is convex in (a,b). In particular, if ϕ'' exists and is nonnegative in (a,b), then ϕ is convex in (a,b).*

Proof. We shall use the following inequality: If $b_1,b_2 > 0$, then

(7.38)
$$\min\left\{\frac{a_1}{b_1},\frac{a_2}{b_2}\right\} \le \frac{a_1 + a_2}{b_1 + b_2} \le \max\left\{\frac{a_1}{b_1},\frac{a_2}{b_2}\right\}.$$

To prove the first inequality in (7.38), let $m = \min\{a_1/b_1,a_2/b_2\}$. Then $mb_1 \le a_1$ and $mb_2 \le a_2$, so that $m(b_1 + b_2) \le a_1 + a_2$. This is the desired result. The second inequality is proved similarly.

The slope of the line connecting two points $(\alpha,\phi(\alpha))$ and $(\beta,\phi(\beta))$ is $[\phi(\beta) - \phi(\alpha)]/(\beta - \alpha)$. Hence, to show that ϕ is convex, we must show that if $a < x_1 < x < x_2 < b$, then

$$\frac{\phi(x_2) - \phi(x_1)}{x_2 - x_1} \ge \frac{\phi(x) - \phi(x_1)}{x - x_1}.$$

Write

$$\frac{\phi(x_2) - \phi(x_1)}{x_2 - x_1} = \frac{[\phi(x_2) - \phi(x)] + [\phi(x) - \phi(x_1)]}{[x_2 - x] + [x - x_1]}.$$

Since ϕ' exists in (a,b), the mean-value theorem implies that there are $\xi_1 \in (x_1,x)$ and $\xi_2 \in (x,x_2)$ such that

$$\frac{\phi(x) - \phi(x_1)}{x - x_1} = \phi'(\xi_1), \qquad \frac{\phi(x_2) - \phi(x)}{x_2 - x} = \phi'(\xi_2).$$

Since ϕ' is increasing, $\phi'(\xi_1) \le \phi'(\xi_2)$, and it follows from the first inequality in (7.38) that

$$\frac{\phi(x_2) - \phi(x_1)}{x_2 - x_1} \ge \frac{\phi(x) - \phi(x_1)}{x - x_1}.$$

This completes the proof.

As a corollary of (7.37), we see that

(7.39)
(i) x^p is convex in $(0,\infty)$ if $p \ge 1$.
(ii) e^{ax} is convex in $(-\infty,+\infty)$.
(iii) $\log 1/x = -\log x$ is convex in $(0,\infty)$.

(7.40) Theorem *If ϕ is convex in (a,b), then ϕ is continuous in (a,b). Moreover, ϕ' exists except at most in a countable set and is monotone increasing.*

Proof. Since ϕ is convex, the slope $[\phi(x + h) - \phi(x)]/h$, $h > 0$, decreases with h. Hence, the derivative on the right,

$$D^+\phi(x) = \lim_{h\to 0+} \frac{[\phi(x + h) - \phi(x)]}{h},$$

exists and is distinct from $+\infty$ in (a,b). Similarly, the derivative on the left,

$$D^-\phi(x) = \lim_{h\to 0+} \frac{[\phi(x) - \phi(x - h)]}{h},$$

exists and is distinct from $-\infty$ in (a,b). Since $[\phi(x) - \phi(x - h)]/h \le [\phi(x + h) - \phi(x)]/h$, $h > 0$, we obtain

(7.41) $$-\infty < D^-\phi(x) \le D^+\phi(x) < +\infty.$$

This shows in particular that ϕ is continuous in (a,b).

We next claim that

(7.42) $$D^+\phi(y) \le D^-\phi(x) \text{ if } a < y < x < b.$$

In fact, if $y < x$, then as seen from the discussion above, we have

$$D^+\phi(y) \le \frac{\phi(x) - \phi(y)}{x - y} \le D^-\phi(x).$$

This proves the claim. We therefore obtain $D^+\phi(y) \le D^-\phi(x) \le D^+\phi(x)$ if $y < x$, which shows that $D^+\phi$ is monotone increasing. Similarly, $D^-\phi$ is monotone increasing.

To complete the proof of the theorem, note that $D^+\phi$ can have at most a countable number of discontinuities since it is monotone and finite on (a,b). If x is a point of continuity of $D^+\phi$, then letting $y \to x-$ in the last inequalities, we obtain $D^+\phi(x) = D^-\phi(x)$. Therefore, ϕ' exists at every point of continuity of $D^+\phi$, and the theorem follows.

(7.43) Theorem *If ϕ is convex in (a,b), then it satisfies a Lipschitz condition on every closed subinterval of (a,b). In particular, if $a < x_1 < x_2 < b$, we have*

$$\phi(x_2) - \phi(x_1) = \int_{x_1}^{x_2} \phi'.$$

Proof. Let $[x_1,x_2]$ be a closed subinterval of (a,b), and let $x_1 \le y < x \le x_2$. Then as before

$$D^+\phi(y) \le \frac{\phi(x) - \phi(y)}{x - y} \le D^-\phi(x),$$

so that, since $D^+\phi$ and $D^-\phi$ are monotone increasing,

$$D^+\phi(x_1) \le \frac{\phi(x) - \phi(y)}{x - y} \le D^-\phi(x_2).$$

Hence, $|\phi(x) - \phi(y)| \le C|x - y|$, where C is the larger of $|D^+\phi(x_1)|$ and $|D^-\phi(x_2)|$. This shows that ϕ satisfies a Lipschitz condition on $[x_1,x_2]$, and the rest of the theorem follows.

The next result is a useful version of Jensen's inequality for integrals. We shall need the notion of a *supporting line*: If ϕ is convex on (a,b) and $x_0 \in (a,b)$, a supporting line at x_0 is a line through $(x_0,\phi(x_0))$ which lies on or below the graph of ϕ on (a,b). It follows from the discussion preceding (7.41) that any line through $(x_0,\phi(x_0))$ whose slope m satisfies $D^-\phi(x_0) \le m \le D^+\phi(x_0)$ is a supporting line at x_0.

(7.44) Theorem (*Jensen's Integral Inequality*) *Let f and p be measurable functions finite a.e. on a set $A \subset \mathbf{R}^n$. Suppose that fp and p are integrable on A, $p \geqslant 0$, and $\int_A p > 0$. If ϕ is convex in an interval containing the range of f, then*

$$\phi\left(\frac{\int_A fp}{\int_A p}\right) \le \frac{\int_A \phi(f)p}{\int_A p}.$$

Proof. By hypothesis, f is finite a.e. in A. Choose (a,b), $-\infty \le a < b \le +\infty$, so that ϕ is convex in (a,b), and so that $a < f(\mathbf{x}) < b$ for every $\mathbf{x}$ at which $f(\mathbf{x})$ is finite. The number γ defined by

$$\gamma = \frac{\int_A fp}{\int_A p}$$

is finite and satisfies $a < \gamma < b$. If m is the slope of a supporting line at γ and $a < t < b$, then $\phi(\gamma) + m(t - \gamma) \le \phi(t)$. Hence, for almost every $\mathbf{x} \in A$,

$$\phi(\gamma) + m[f(\mathbf{x}) - \gamma] \le \phi(f(\mathbf{x})).$$

Multiplying both sides of this inequality by $p(\mathbf{x})$ and integrating the result with respect to $\mathbf{x}$, we obtain

$$\phi(\gamma)\int_A p + m\left(\int_A fp - \gamma\int_A p\right) \le \int_A \phi(f)p.$$

Here the existence of $\int_A \phi(f)p$ follows from the integrability of p and fp. [The continuity of ϕ implies that $\phi(f)$ is measurable.] Since $\int_A fp - \gamma\int_A p = 0$, the last inequality reduces to

$$\phi(\gamma)\int_A p \le \int_A \phi(f)p,$$

which is the desired result.

Exercises

1. Let f be measurable in $\mathbf{R}^n$ and different from zero in some set of positive measure. Show that there is a positive constant c such that $f^*(\mathbf{x}) \ge c|\mathbf{x}|^{-n}$ for $|\mathbf{x}| \ge 1$.
2. Let $\phi(\mathbf{x})$, $\mathbf{x} \in \mathbf{R}^n$, be a bounded measurable function such that $\phi(\mathbf{x}) = 0$ for $|\mathbf{x}| \ge 1$ and $\int \phi = 1$. For $\varepsilon > 0$, let $\phi_\varepsilon(\mathbf{x}) = \varepsilon^{-n}\phi(\mathbf{x}/\varepsilon)$. ($\phi_\varepsilon$ is called an *approximation to the identity*.) If $f \in L(\mathbf{R}^n)$, show that
$$\lim_{\varepsilon\to 0}(f*\phi_\varepsilon)(\mathbf{x}) = f(\mathbf{x})$$
in the Lebesgue set of f. [Note that $\int \phi_\varepsilon = 1$, $\varepsilon > 0$, so that
$$(f*\phi_\varepsilon)(\mathbf{x}) - f(\mathbf{x}) = \int [f(\mathbf{x}-\mathbf{y}) - f(\mathbf{x})]\phi_\varepsilon(\mathbf{y})\,d\mathbf{y}.$$
Use (7.16).]
3. Show that the conclusion of (7.4) remains true for the case of two dimensions if instead of being squares, the sets Q covering E are rectangles with x-dimension equal to h and y-dimension equal to h^2. (Of course, h varies with Q.)
   Show that the same conclusion is valid if the y-dimension is any increasing function of h. Generalize this to higher dimensions.
4. If E_1 and E_2 are measurable subsets of $\mathbf{R}^1$ with $|E_1| > 0$ and $|E_2| > 0$, prove that the set $\{x : x = x_1 - x_2,\ x_1 \in E_1,\ x_2 \in E_2\}$ contains an interval. [cf. (3.37).]
5. Let f be of bounded variation on $[a,b]$. If $f = g + h$, where g is absolutely continuous and h is singular, show that
$$\int_a^b \phi\, df = \int_a^b \phi f'\,dx + \int_a^b \phi\, dh,$$
for any continuous ϕ.
6. Show that if $\alpha > 0$, x^α is absolutely continuous on every bounded subinterval of $[0,\infty)$.
7. Prove that f is absolutely continuous on $[a,b]$ if and only if given $\varepsilon > 0$, there exists $\delta > 0$ such that $|\sum [f(b_i) - f(a_i)]| < \varepsilon$ for any *finite* collection $\{[a_i,b_i]\}$ of nonoverlapping subintervals of $[a,b]$ with $\sum (b_i - a_i) < \delta$.
8. Prove the following converse of (7.31): If f is of bounded variation on $[a,b]$, and if the function $V(x) = V[a,x]$ is absolutely continuous on $[a,b]$, then f is absolutely continuous on $[a,b]$.
9. If f is of bounded variation on $[a,b]$, show that
$$\int_a^b |f'| \le V[a,b].$$
Show that if equality holds in this inequality, then f is absolutely continuous

on $[a,b]$. (For the second part, use (2.2)(ii) and (7.24) to show that $V(x)$ is absolutely continuous, and then use the result of Exercise 8.)

10. Show that if f is absolutely continuous on $[a,b]$ and Z is a subset of $[a,b]$ of measure zero, then the image set defined by $f(Z) = \{w : w = f(z), z \in Z\}$ also has measure zero. Deduce that the image under f of any measurable subset of $[a,b]$ is measurable. [Compare (3.33).] [HINT: use the fact that the image of an interval $[a_i,b_i]$ is an interval of length at most $V(b_i) - V(a_i)$.]

11. Prove the following result concerning *changes of variable*. Let $g(t)$ be monotone increasing and absolutely continuous on $[\alpha,\beta]$, and let f be bounded and measurable on $[a,b]$, $a = g(\alpha)$, $b = g(\beta)$. Then $f(g(t))$ is measurable on $[\alpha,\beta]$ and

$$\int_a^b f(x)\,dx = \int_\alpha^\beta f(g(t))g'(t)\,dt.$$

(Consider the cases when f is the characteristic function of an interval, an open set, etc.)

12. Use Jensen's inequality to prove that for $a,b \geq 0$, $p,q > 1$, $(1/p) + (1/q) = 1$, we have

$$ab \leq \frac{a^p}{p} + \frac{b^q}{q}.$$

More generally, show that

$$a_1 \cdots a_N \leq \sum_{j=1}^N \frac{a_j{}^{p_j}}{p_j}$$

where $a_j \geq 0$, $p_j > 1$, $\sum_{j=1}^N (1/p_j) = 1$. (Write $a_j = e^{x_j/p_j}$ and use the convexity of e^x.)

13. Prove theorem (7.36).

14. Prove that ϕ is convex on (a,b) if and only if it is continuous and

$$\phi\left(\frac{x_1 + x_2}{2}\right) \leq \frac{\phi(x_1) + \phi(x_2)}{2}$$

for $x_1,x_2 \in (a,b)$.

15. Theorem (7.43) shows that a convex function is the indefinite integral of a monotone increasing function. Prove the converse: If $\phi(x) = \int_a^x f(t)\,dt + \phi(a)$ in (a,b) and f is monotone increasing, then ϕ is convex in (a,b). (Use Exercise 14.)

16. Show that the formula

$$\int_{-\infty}^{+\infty} fg' = -\int_{-\infty}^{+\infty} f'g$$

for integration by parts may not hold if f is of bounded variation on $(-\infty,+\infty)$ and g is infinitely differentiable with compact support. (Let f be the Cantor-Lebesgue function on $[0,1]$, and let $f = 0$ elsewhere.)

17. A sequence $\{\phi_k\}$ of set functions is said to be *uniformly absolutely continuous* if given $\varepsilon > 0$, there exists $\delta > 0$ such that if E satisfies $|E| < \delta$, then $|\phi_k(E)| < \varepsilon$ for all k. If $\{f_k\}$ is a sequence of integrable functions on $(0,1)$ which converges pointwise a.e. to an integrable f, show that $\int_0^1 |f - f_k| \to 0$ if and only if the indefinite integrals of the f_k are uniformly absolutely continuous.

Chapter 8

L^p Classes

1. Definition of L^p

If E is a measurable subset of $\mathbf{R}^n$ and p satisfies $0 < p < \infty$, then $L^p(E)$ denotes the collection of measurable f for which $\int_E |f|^p$ is finite, that is,

$$L^p(E) = \left\{f: \int_E |f|^p < +\infty\right\}, \quad 0 < p < \infty.$$

Here, f may be complex-valued. (See Exercise 3 of Chapter 4 for the definition of measurability of vector-valued functions.) In this case, if $f = f_1 + if_2$ for measurable real-valued f_1 and f_2, we have $|f|^2 = f_1^2 + f_2^2$, so that

$$|f_1|,|f_2| \le |f| \le |f_1| + |f_2|.$$

It follows that $f \in L^p(E)$ if and only if both $f_1, f_2 \in L^p(E)$. (See Exercise 1.)

We shall write

$$\|f\|_{p,E} = \left(\int_E |f|^p\right)^{1/p} \qquad (0 < p < \infty);$$

thus, $L^p(E)$ is the class of measurable f for which $\|f\|_{p,E}$ is finite. Whenever it is clear from context what E is, we will write L^p for $L^p(E)$ and $\|f\|_p$ for $\|f\|_{p,E}$. Note that $L^1 = L$.

In order to define $L^\infty(E)$, let f be real-valued and measurable on a set E of positive measure. Define the *essential supremum* of f on E as follows: If $|\{\mathbf{x} \in E : f(\mathbf{x}) > \alpha\}| > 0$ for all real α, let $\text{ess}_E\sup f = +\infty$; otherwise, let

$$\operatorname*{ess\,sup}_E f = \inf\{\alpha : |\{\mathbf{x} \in E : f(\mathbf{x}) > \alpha\}| = 0\}.$$

Since the distribution function $\omega(\alpha) = |\{\mathbf{x} \in E : f(\mathbf{x}) > \alpha\}|$ is continuous from the right [see (5.39)], it follows that $\omega(\text{ess}_E \sup f) = 0$ if $\text{ess}_E\sup f$ is finite. Therefore, $\text{ess}_E\sup f$ is the smallest number M, $-\infty \le M \le +\infty$, such that $f(x) \le M$ except for a subset of E of measure zero.

A real- or complex-valued measurable f is said to be *essentially bounded*, or simply *bounded*, on E if $\text{ess}_E\text{sup}\,|f|$ is finite. The class of all functions that are essentially bounded on E is denoted by $L^\infty(E)$. Clearly, f belongs to $L^\infty(E)$ if and only if its real and imaginary parts do. We shall write

$$\|f\|_\infty = \|f\|_{\infty,E} = \operatorname*{ess\,sup}_E |f|.$$

Thus, $\|f\|_\infty$ is the smallest M such that $|f| \le M$ a.e. in E, and

$$L^\infty = L^\infty(E) = \{f : \|f\|_\infty < +\infty\}.$$

The following theorem gives some motivation for this notation.

(8.1) Theorem *If $|E| < +\infty$, then $\|f\|_\infty = \lim_{p\to\infty} \|f\|_p$.*

Proof. Let $M = \|f\|_\infty$. If $M' < M$, then the set $A = \{\mathbf{x} \in E : |f(\mathbf{x})| > M'\}$ has positive measure. Moreover, $\|f\|_p \ge (\int_A |f|^p)^{1/p} \ge M'|A|^{1/p}$. Since $|A|^{1/p} \to 1$ as $p \to \infty$, it follows that $\liminf_{p\to\infty} \|f\|_p \ge M'$, so that $\liminf_{p\to\infty} \|f\|_p \ge M$. However, we also have $\|f\|_p \le (\int_E M^p)^{1/p} = M|E|^{1/p}$. Therefore, $\limsup_{p\to\infty} \|f\|_p \le M$, which completes the proof.

Remark: This result may fail if $|E| = +\infty$. Consider, for example, the constant function $f(x) = c$, $c \ne 0$, in $(0,\infty)$. Clearly, $f \in L^\infty$ but $f \notin L^p$ for $0 < p < \infty$.

We will now study some basic properties of the L^p classes.

(8.2) Theorem *If $0 < p_1 < p_2 \le \infty$ and $|E| < +\infty$, then $L^{p_2} \subset L^{p_1}$.*

Proof. Write $E = E_1 \cup E_2$, E_1 being the set where $|f| \le 1$, and E_2 the set where $|f| > 1$. Then

$$\int_E |f|^p = \int_{E_1} |f|^p + \int_{E_2} |f|^p, \qquad 0 < p < \infty.$$

The first term on the right is majorized by $|E_1|$; the second increases with p since its integrand exceeds 1. It follows that if $f \in L^{p_2}$, $p_2 < \infty$, then $f \in L^{p_1}$, $p_1 < p_2$. If $p_2 = \infty$, then f is a bounded function on a set of finite measure, and so belongs to L^{p_1}.

Remarks:

(i) The hypothesis above that E have finite measure cannot be omitted: for example, x^{-1/p_1} belongs to $L^{p_2}(1,\infty)$ if $p_2 > p_1$, but does not belong to $L^{p_1}(1,\infty)$. Again, any nonzero constant is in L^∞, but is not in $L^{p_1}(E)$ if $|E| = +\infty$ and $p_1 < \infty$.

(ii) A function may belong to all L^{p_1} with $p_1 < p_2$ and yet not belong to L^{p_2}. In fact, if $p_2 < \infty$, x^{-1/p_2} belongs to $L^{p_1}(0,1)$, $p_1 < p_2$, but does not belong to $L^{p_2}(0,1)$; $\log 1/x$ is in $L^{p_1}(0,1)$ for $p_1 < \infty$, but is not in $L^\infty(0,1)$.

(iii) We leave it to the reader to show that any function which is bounded on

E ($|E| < +\infty$ or not) and which belongs to L^{p_1} also belongs to L^{p_2}, $p_2 > p_1$.

The next theorem states that the L^p classes are vector (i.e., linear) spaces. Its proof is left as an exercise.

(8.3) Theorem *If $f,g \in L^p(E)$, $p > 0$, then $f + g \in L^p(E)$ and $cf \in L^p(E)$ for any constant c.*

2. Hölder's Inequality; Minkowski's Inequality

In order to discuss the integrability of the product of two functions, we will use the following basic result.

(8.4) Theorem (*Young's Inequality*) *Let $y = \phi(x)$ be continuous, real-valued, and strictly increasing for $x \geq 0$, and let $\phi(0) = 0$. If $x = \psi(y)$ is the inverse of ϕ, then for $a,b > 0$,*

$$ab \leq \int_0^a \phi(x)\,dx + \int_0^b \psi(y)\,dy.$$

Equality holds if and only if $b = \phi(a)$.

Proof. A geometric proof is immediate if we interpret each term as an area and remember that the graph of ϕ also serves as that of ψ if we interchange the x and y axes. Equality holds if and only if the point (a,b) lies on the graph of ϕ.

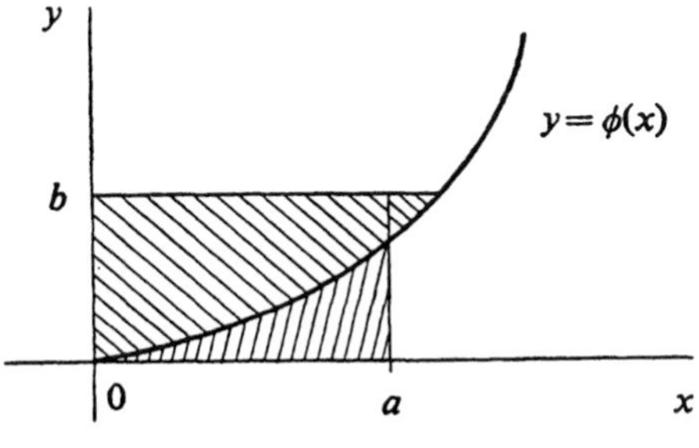

If $\phi(x) = x^\alpha$, $\alpha > 0$, then $\psi(y) = y^{1/\alpha}$, and Young's inequality becomes $ab \leq a^{1+\alpha}/(1 + \alpha) + b^{1+1/\alpha}/(1 + 1/\alpha)$. Setting $p = \alpha + 1$ and $p' = 1 + 1/\alpha$, we obtain

$$ab \leq \frac{a^p}{p} + \frac{b^{p'}}{p'} \text{ if } a,b \geq 0,\ 1 < p < \infty, \text{ and } \frac{1}{p} + \frac{1}{p'} = 1. \tag{8.5}$$

Two numbers p and p' which satisfy $1/p + 1/p' = 1$, $p,p' > 1$, are called *conjugate exponents*. Note that $p' = p/(p - 1)$, and that 2 is self-conjugate. We will adopt the conventions that $p' = \infty$ if $p = 1$, and $p' = 1$ if $p = \infty$.

(8.6) Theorem (***Hölder's Inequality***) *If* $1 \le p \le \infty$ *and* $1/p + 1/p' = 1$, *then* $\|fg\|_1 \le \|f\|_p \|g\|_{p'}$; *that is,*

$$\int_E |fg| \le \left(\int_E |f|^p\right)^{1/p} \left(\int_E |g|^{p'}\right)^{1/p'} \qquad (1 < p < \infty);$$

$$\int_E |fg| \le (\operatorname*{ess\,sup}_E |f|) \int_E |g|.$$

Proof. The last inequality, which corresponds to the case $p = \infty$, is obvious. Let us suppose then that $1 < p < \infty$. In case $\|f\|_p = \|g\|_{p'} = 1$, (8.5) implies that

$$\int_E |fg| \le \int_E \left(\frac{|f|^p}{p} + \frac{|g|^{p'}}{p'}\right) = \frac{\|f\|_p^p}{p} + \frac{\|g\|_{p'}^{p'}}{p'}$$
$$= \frac{1}{p} + \frac{1}{p'} = 1 = \|f\|_p \|g\|_{p'}.$$

For the general case, we may assume that neither $\|f\|_p$ nor $\|g\|_{p'}$ is zero; otherwise fg is zero a.e. in E, and the result is immediate. We may also assume that neither $\|f\|_p$ nor $\|g\|_{p'}$ is infinite. If we set $f_1 = f/\|f\|_p$ and $g_1 = g/\|g\|_{p'}$, then $\|f_1\|_p = \|g_1\|_{p'} = 1$. Therefore, by the case above, we have $\int_E |f_1 g_1| \le 1$; i.e., $\int_E |fg| \le \|f\|_p \|g\|_{p'}$, as desired.

The case $p = p' = 2$ is a classical inequality.

(8.7) Corollary (***Schwarz's Inequality***)

$$\int_E |fg| \le \left(\int_E |f|^2\right)^{1/2} \left(\int_E |g|^2\right)^{1/2}.$$

The theorem which follows is usually referred to as the "converse of Hölder's inequality." (See also Exercise 15 in Chapter 10.)

(8.8) Theorem *Let f be real-valued and measurable on E, let $1 \le p \le \infty$ and $1/p + 1/p' = 1$. Then*

$$\|f\|_p = \sup \int_E fg, \tag{8.9}$$

where the supremum is taken over all real-valued g such that $\|g\|_{p'} \le 1$ and $\int_E fg$ exists.

Proof. That the left-hand side of (8.9) majorizes the right-hand side follows from Hölder's inequality. To show the opposite inequality, let us consider first the case of $f \ge 0$, $1 < p < \infty$.

If $\|f\|_p = 0$, then $f = 0$ a.e. in E, and the result is obvious. If $0 < \|f\|_p <$

$+\infty$, we may further assume that $\|f\|_p = 1$ by dividing both sides of (8.9) by $\|f\|_p$. Now let $g = f^{p/p'}$. It is easy to verify that $\|g\|_{p'} = 1$ and $\int_E fg = 1$, which completes the proof in this case.

If $\|f\|_p = +\infty$, define functions f_k on E by setting

$$f_k(\mathbf{x}) = 0 \text{ if } |\mathbf{x}| > k, \qquad f_k(\mathbf{x}) = \min\,[f(\mathbf{x}),k] \text{ if } |\mathbf{x}| \le k.$$

Then each f_k belongs to L^p and $\|f_k\|_p \to \|f\|_p = +\infty$. By the case already considered, we have $\|f_k\|_p = \int_E f_k g_k$ for some $g_k \ge 0$ with $\|g_k\|_{p'} = 1$. Since $f \ge f_k$, it follows that

$$\int_E fg_k \ge \int_E f_k g_k \to +\infty.$$

This shows that

$$\sup_{\|g\|_{p'}=1} \int_E fg = +\infty = \|f\|_p.$$

To dispose of the restriction $f \ge 0$, apply the result above to $|f|$. Thus, there exist g_k with $\|g_k\|_{p'} = 1$ such that

$$\|f\|_p = \lim \int_E |f| g_k = \lim \int_E f\tilde{g}_k,$$

where $\tilde{g}_k = g_k\,(\operatorname{sign} f)$. (By sign x, we mean the function equal to $+1$ for $x > 0$ and to -1 for $x < 0$.) Since $\|\tilde{g}_k\|_{p'} = 1$, the result follows.

The cases $p = 1$ and ∞ are left as exercises.

We have already observed that the sum of two L^p functions is again in L^p. The next theorem gives a more specific result when $1 \le p \le \infty$.

(8.10) Theorem (*Minkowski's Inequality*) *If* $1 \le p \le \infty$, *then* $\|f + g\|_p \le \|f\|_p + \|g\|_p$; *that is,*

$$\left(\int_E |f + g|^p\right)^{1/p} \le \left(\int_E |f|^p\right)^{1/p} + \left(\int_E |g|^p\right)^{1/p} \qquad (1 \le p < \infty),$$

$$\operatorname*{ess\,sup}_E |f + g| \le \operatorname*{ess\,sup}_E |f| + \operatorname*{ess\,sup}_E |g|.$$

Proof. If $p = 1$, the result is obvious. If $p = \infty$, we have $|f| \le \|f\|_\infty$ a.e. in E and $|g| \le \|g\|_\infty$ a.e. in E. Therefore, $|f + g| \le \|f\|_\infty + \|g\|_\infty$ a.e. in E, so that $\|f + g\|_\infty \le \|f\|_\infty + \|g\|_\infty$.

For $1 < p < \infty$,

$$\begin{aligned}\|f + g\|_p^p = \int_E |f + g|^p &= \int_E |f + g|^{p-1}|f + g| \\ &\le \int_E |f + g|^{p-1}|f| + \int_E |f + g|^{p-1}|g|.\end{aligned}$$

In the last integral, apply Hölder's inequality to $|f+g|^{p-1}$ and $|g|$ with exponents $p' = p/(p-1)$ and p, respectively. This gives

$$\int_E |f+g|^{p-1}|g| \le \|f+g\|_p^{p-1}\|g\|_p.$$

Since a similar result holds for $\int_E |f+g|^{p-1}|f|$, we obtain $\|f+g\|_p^p \le \|f+g\|_p^{p-1}(\|f\|_p + \|g\|_p)$, and the theorem follows by dividing both sides by $\|f+g\|_p^{p-1}$. [Note that if $\|f+g\|_p = 0$, there is nothing to prove; if $\|f+g\|_p = +\infty$, then either $\|f\|_p = +\infty$ or $\|g\|_p = +\infty$ by (8.3), and the result is obvious again.]

Remark: Minkowski's inequality fails for $0 < p < 1$. To see this, take $E = (0,1)$, $f = \chi_{(0,\frac{1}{2})}$ and $g = \chi_{(\frac{1}{2},1)}$. Then $\|f+g\|_p = 1$, while $\|f\|_p + \|g\|_p = 2^{-1/p} + 2^{-1/p} = 2^{1-1/p} < 1$. [See also (8.17).]

3. Classes l^p

Let $a = \{a_k\}$ be a sequence of real or complex numbers, and let

$$\|a\|_p = \Big(\sum_k |a_k|^p\Big)^{1/p},\ 0 < p < \infty; \qquad \|a\|_\infty = \sup_k |a_k|.$$

Then a is said to belong to l^p, $0 < p < \infty$, if $\|a\|_p < +\infty$, and to belong to l^∞ if $\|a\|_\infty < +\infty$.

Let us show that if $0 < p_1 < p_2 \le \infty$, then $l^{p_1} \subset l^{p_2}$. [The opposite inclusion holds for $L^p(E)$, $|E| < +\infty$, by (8.2).] For $p_2 = \infty$, this is clear, and for $p_2 < \infty$, it follows from the fact that if $|a_k| \le 1$, then $|a_k|^{p_2} \le |a_k|^{p_1}$. An example of a sequence which is in l^{p_2} for a given $p_2 < \infty$ but which is not in l^{p_1} for $p_1 < p_2$ is $\{(1/k\log^2 k)^{1/p_2} : k \ge 2\}$. Any constant sequence $\{a_k\}$, $a_k = c \ne 0$, belongs to l^∞ but not to l^p for $p < \infty$. The same is true for $\{1/\log k : k \ge 2\}$, whose terms even tend to zero.

(8.11) Theorem *If $a = \{a_k\}$ belongs to l^p for some $p < \infty$, then* $\lim_{p\to\infty} \|a\|_p = \|a\|_\infty$.

Proof. If $a \in l^{p_0}$, then $a \in l^p$ for $p_0 \le p \le \infty$. Since $|a_k| \to 0$, there is a largest $|a_k|$, say $|a_{k_0}|$. Thus, $\|a\|_\infty = |a_{k_0}|$. Write $\sum |a_k|^p = |a_{k_0}|^p \sum |a_k/a_{k_0}|^p$. Since $|a_k/a_{k_0}| \le 1$, we see that $\sum |a_k/a_{k_0}|^p$ decreases (and so is bounded) as $p \to \infty$. Hence, there is a constant c such that $|a_{k_0}|^p \le \|a\|_p^p \le c|a_{k_0}|^p$. Since $c^{1/p} \to 1$ as $p \to \infty$, the theorem follows.

The next two results are analogues for series of Hölder's and Minkowski's inequalities. Their proofs are left as exercises. If $a = \{a_k\}$ and $b = \{b_k\}$, we use the notation

$$ab = \{a_k b_k\}, \qquad a + b = \{a_k + b_k\}, \text{ etc.}$$

(8.12) Theorem (*Hölder's Inequality*) *Suppose that* $1 \le p \le \infty$, $1/p + 1/p'$

$= 1$, $a = \{a_k\}$, $b = \{b_k\}$, *and* $ab = \{a_k b_k\}$. *Then* $\|ab\|_1 \le \|a\|_p \|b\|_{p'}$; *that is,*

$$\sum |a_k b_k| \le \left(\sum |a_k|^p\right)^{1/p} \left(\sum |b_k|^{p'}\right)^{1/p'} \qquad (1 < p < \infty);$$
$$\sum |a_k b_k| \le (\sup |a_k|) \sum |b_k|.$$

(8.13) Theorem (***Minkowski's Inequality***) *Suppose that* $1 \le p \le \infty$, $a = \{a_k\}, b = \{b_k\}$, *and* $a + b = \{a_k + b_k\}$. *Then* $\|a + b\|_p \le \|a\|_p + \|b\|_p$; *that is,*

$$\left(\sum |a_k + b_k|^p\right)^{1/p} \le \left(\sum |a_k|^p\right)^{1/p} + \left(\sum |b_k|^p\right)^{1/p} \qquad (1 \le p < \infty);$$
$$\sup |a_k + b_k| \le \sup |a_k| + \sup |b_k|.$$

Even though Minkowski's inequality fails when $p < 1$ (see Exercise 3), l^p is still a vector space for $0 < p < 1$; that is, $a + b \in l^p$ and $\alpha a = \{\alpha a_k\} \in l^p$ if $a,b \in l^p$ and α is any constant.

4. Banach and Metric Space Properties

We now define a notion which incorporates the main properties of L^p and l^p when $p \ge 1$. A set X is called a *Banach space over the complex numbers* if it satisfies the following conditions:

(B_1) X is a *linear space* over the complex numbers **C**; that is, if $x,y \in X$ and $\alpha \in \mathbf{C}$, then $x + y \in X$ and $\alpha x \in X$.

(B_2) X is a *normed space*; that is, for every $x \in X$ there is a non-negative number $\|x\|$ such that

(a) $\|x\| = 0$ if and only if x is the zero element of X.
(b) $\|\alpha x\| = |\alpha| \|x\|$ for $\alpha \in \mathbf{C}$; $x \in X$.
(c) $\|x + y\| \le \|x\| + \|y\|$.

If these conditions are fulfilled, $\|x\|$ is called the *norm* of x.

(B_3) X is *complete* with respect to its norm; that is, every Cauchy sequence in X converges in X, or if $\|x_k - x_m\| \to 0$ as $k,m \to \infty$, then there is an $x \in X$ such that $\|x_k - x\| \to 0$.

A set X which satisfies (B_1) and (B_2), but not necessarily (B_3), is called a *normed linear space* over the complex numbers. A sequence $\{x_k\}$ such that $\|x_k - x\| \to 0$ as $k \to \infty$ is said to *converge in norm* to x.

Restricting the scalars α in (B_1) and (B_2) to be real numbers, we obtain definitions for a Banach space over the real numbers and for a normed linear space over the real numbers. Unless specifically stated to the contrary, we will take the scalar field to be the complex numbers.

If X is a Banach space, define $d(x,y) = \|x - y\|$ to be the *distance between x and y*. Then,

(M_1) $d(x,y) \geq 0$; $d(x,y) = 0$ if and only if $x = y$.
(M_2) $d(x,y) = d(y,x)$.
(M_3) $d(x,y) \leq d(x,z) + d(z,y)$ (triangle inequality).

Any set which has a distance function $d(x,y)$ satisfying (M_1), (M_2), and (M_3) is called a *metric space* with *metric d*. Therefore, a Banach space is a metric space whose metric is the norm. Moreover, by (B_3), a Banach space X is a complete metric space; that is, if $d(x_k,x_m) \to 0$ as $k,m \to \infty$, then there is an $x \in X$ such that $d(x_k,x) \to 0$.

(8.14) Theorem *For* $1 \leq p \leq \infty$, $L^p(E)$ *is a Banach space with norm* $\|f\| = \|f\|_{p,E}$.

Proof. Parts (B_1) and (B_2) in the definition of a Banach space are clearly fulfilled by $L^p(E)$, parts (a) and (c) of (B_2) being (5.11) and Minkowski's inequality, respectively. [Regarding part (a), we do not distinguish between two L^p functions which are equal a.e.; thus, the zero element of $L^p(E)$ means any function equal to zero a.e. in E.]

To verify (B_3), suppose that $\{f_k\}$ is a Cauchy sequence in $L^p(E)$. If $p = \infty$, then $|f_k - f_m| \leq \|f_k - f_m\|_\infty$ except for a set $Z_{k,m}$ of measure zero. If $Z = \bigcup_{k,m} Z_{k,m}$, then Z has measure zero, and $|f_k - f_m| \leq \|f_k - f_m\|_\infty$ outside Z for all k and m. Hence, $\{f_k\}$ converges uniformly outside Z to a bounded limit f, and it follows that $\|f_k - f\|_\infty \to 0$. (Note that convergence in L^∞ is equivalent to uniform convergence outside a set of measure zero.)

In case $1 \leq p < \infty$, Tschebyshev's inequality (5.49) implies that

$$|\{\mathbf{x} \in E : |f_k(\mathbf{x}) - f_m(\mathbf{x})| > \varepsilon\}| \leq \varepsilon^{-p} \int_E |f_k - f_m|^p.$$

Hence, $\{f_k\}$ is a Cauchy sequence in measure. By (4.22) and (4.23), there is a subsequence $\{f_{k_j}\}$ and a function f such that $f_{k_j} \to f$ a.e. in E. Given $\varepsilon > 0$, there is a K such that

$$\left(\int_E |f_{k_j} - f_k|^p\right)^{1/p} = \|f_{k_j} - f_k\|_p < \varepsilon \text{ if } k_j,k > K.$$

Letting $k_j \to \infty$, we obtain by Fatou's lemma that $\|f - f_k\|_p \leq \varepsilon$ if $k > K$. Hence, $\|f - f_k\|_p \to 0$ as $k \to \infty$. Finally, since $\|f\|_p \leq \|f - f_k\|_p + \|f_k\|_p < +\infty$, it follows that $f \in L^p(E)$, which completes the proof.

A metric space X is said to be *separable* if it has a countable dense subset; that is, X is separable if there exists a countable set $\{x_k\}$ in X with the property that for every $x \in X$ and every $\varepsilon > 0$, there is an x_k with $d(x,x_k) < \varepsilon$. In the next theorem, we will show that L^p is separable if $1 \leq p < \infty$. Note that L^∞ is not separable: take $L^\infty(0,1)$, for example, and consider the functions $f_t(x) = \chi_{(0,t)}(x)$, $0 < t < 1$. There are an uncountable number of these, and $\|f_t - f_{t'}\|_\infty = 1$ if $t \neq t'$. See also Exercise 10.

(8.15) Theorem *If $1 \le p < \infty$, $L^p(E)$ is separable.*

Proof. Suppose first that $E = \mathbf{R}^n$, and consider a class of dyadic cubes in $\mathbf{R}^n$. Let D be the set of all (finite) linear combinations of characteristic functions of these cubes, the coefficients being complex numbers with rational real and imaginary parts. Then D is a countable subset of $L^p(\mathbf{R}^n)$. To see that D is dense, use the method of successively approximating more and more general functions: First, consider characteristic functions of open sets [every open set is the countable union of nonoverlapping dyadic cubes by (1.11)], of G_δ sets, and of measurable sets with finite measure; then consider simple functions whose supports have finite measure, nonnegative functions in $L^p(\mathbf{R}^n)$, and, finally, arbitrary functions in $L^p(\mathbf{R}^n)$. The details are left to the reader [cf. lemma (7.3)]. This proves the case $E = \mathbf{R}^n$.

For an arbitrary measurable E, let D' denote the restrictions to E of the functions in D. Then D' is dense in $L^p(E)$, $1 \le p < \infty$. In fact, given p and $f \in L^p(E)$, let $f_1 = f$ on E and $f_1 = 0$ off E. Then $f_1 \in L^p(\mathbf{R}^n)$, so that given $\varepsilon > 0$, there exists $g \in D$ with $(\int_{\mathbf{R}^n} |f_1 - g|^p \, d\mathbf{x})^{1/p} < \varepsilon$. Therefore, $(\int_E |f - g|^p d\mathbf{x})^{1/p} < \varepsilon$. This shows that D' is dense in $L^p(E)$ and completes the proof.

As we have already noted, Minkowski's inequality fails when $0 < p < 1$. Therefore, $\|\cdot\|_{p,E}$ is not a norm for such p. However, we still have the following facts.

(8.16) Theorem *If $0 < p < 1$, $L^p(E)$ is a complete, separable metric space, with distance defined by*

$$d(f,g) = \|f - g\|_{p,E}^p.$$

Proof. With $d(f,g)$ so defined, properties (M_1) and (M_2) of a metric space are clear. To verify (M_3), which is the triangle inequality, we first claim that

$$(a + b)^p \le a^p + b^p \text{ if } a,b \ge 0,\ 0 < p \le 1.$$

If both a and b are zero, this is obvious. If, say, $a \neq 0$, then dividing by a^p, we reduce the inequality to $(1 + t)^p \le 1 + t^p$, $t > 0$ $(t = b/a)$. This is clear since both sides are equal when $t = 0$ and the derivative of the right side majorizes that of the left for $t > 0$.

It follows that $|f(\mathbf{x}) - g(\mathbf{x})|^p \le |f(\mathbf{x}) - h(\mathbf{x})|^p + |h(\mathbf{x}) - g(\mathbf{x})|^p$ if $0 < p \le 1$. Integrating, we obtain $\|f - g\|_p^p \le \|f - h\|_p^p + \|h - g\|_p^p$, which is just the triangle inequality. The proofs that L^p is complete and separable with respect to $\|\cdot\|_p$ are the same as in (8.14) and (8.15).

It is worth noting that the triangle inequality is equivalent to the basic estimate

$$\|f + g\|_p^p \le \|f\|_p^p + \|g\|_p^p \qquad (0 < p \le 1). \tag{8.17}$$

The analogous results for series are listed in the following theorem.

(8.18) Theorem

(i) *If* $1 \le p \le \infty$, l^p *is a Banach space with* $\|a\| = \|a\|_p$. *For* $1 \le p < \infty$, l^p *is separable;* l^∞ *is not separable.*

(ii) *If* $0 < p < 1$, l^p *is a complete, separable metric space, with distance* $d(a,b) = \|a - b\|_p^p$.

Proof. We will show that l^p is complete and separable when $1 \le p < \infty$ and that l^∞ is not separable. The rest of the proof of (i) and the proof of (ii) are left to the reader.

Suppose that $1 \le p < \infty$, $a^{(i)} = \{a_k^{(i)}\} \in l^p$ for $i = 1, 2, \ldots,$ and $\|a^{(i)} - a^{(j)}\|_p \to 0$ as $i,j \to \infty$. Since $\|a^{(i)} - a^{(j)}\|_p \ge |a_k^{(i)} - a_k^{(j)}|$ for every k, it follows that $|a_k^{(i)} - a_k^{(j)}| \to 0$ for every k as $i,j \to \infty$. Let $a_k = \lim_{i\to\infty} a_k^{(i)}$ and $a = \{a_k\}$. We will show that $a \in l^p$ and $\|a^{(i)} - a\|_p \to 0$. Given $\varepsilon > 0$, there exists N such that

$$(\sum_k |a_k^{(i)} - a_k^{(j)}|^p)^{1/p} = \|a^{(i)} - a^{(j)}\|_p < \varepsilon \quad \text{if} \quad i,j > N.$$

Restricting the summation to $k \le M$ and letting $j \to \infty$, we obtain

$$\left(\sum_{k=1}^{M} |a_k^{(i)} - a_k|^p\right)^{1/p} \le \varepsilon \text{ for any } M, \text{ if } i > N.$$

Letting $M \to \infty$, we get $\|a^{(i)} - a\|_p \le \varepsilon$ if $i > N$; that is, $\|a^{(i)} - a\|_p \to 0$. The fact that $\|a\|_p \le \|a - a^{(i)}\|_p + \|a^{(i)}\|_p$ shows that $a \in l^p$.

To prove that l^p is separable when $p < \infty$, let D be the set of all sequences $\{d_k\}$ such that (a) the real and imaginary parts of d_k are rational, and (b) $d_k = 0$ for $k \ge N$ (N may vary from sequence to sequence). Then D is a countable subset of l^p. If $a = \{a_k\} \in l^p$ and $\varepsilon > 0$, choose N so that $\sum_{k=N+1}^{\infty} |a_k|^p < \varepsilon/2$. Choose $d_1, \ldots, d_N$ with rational real and imaginary parts such that $\sum_{k=1}^{N} |a_k - d_k|^p < \varepsilon/2$. Then $d = \{d_1, \ldots, d_N, 0, \ldots\}$ belongs to D and $\|a - d\|_p^p < \varepsilon$. It follows that D is dense in l^p, and therefore that l^p is separable.

To see that l^∞ is not separable, consider the subclass of sequences $a = \{a_k\}$ for which each a_k is 0 or 1. The number of such sequences is uncountable, and $\|a - a'\|_\infty = 1$ for any two different such sequences. Hence, l^∞ cannot be separable.

We know from Lusin's theorem that measurable functions have continuity properties. The next theorem gives a useful continuity property of functions in L^p.

(8.19) Theorem (*Continuity in L^p*) *If* $f \in L^p(\mathbf{R}^n)$, $1 \le p < \infty$, *then*

$$\lim_{|\mathbf{h}|\to 0} \|f(\mathbf{x} + \mathbf{h}) - f(\mathbf{x})\|_p = 0.$$

Proof. Let C_p denote the class of $f \in L^p$ such that $\|f(\mathbf{x} + \mathbf{h}) - f(\mathbf{x})\|_p \to 0$ as $|\mathbf{h}| \to 0$. We claim that (a) a finite linear combination of functions in C_p is in C_p, and (b) if $f_k \in C_p$ and $\|f_k - f\|_p \to 0$, then $f \in C_p$. Both of these facts follow easily from Minkowski's inequality; for (b), for example, note that

$$\begin{aligned}
&\|f(\mathbf{x} + \mathbf{h}) - f(\mathbf{x})\|_p \\
&\quad \le \|f(\mathbf{x} + \mathbf{h}) - f_k(\mathbf{x} + \mathbf{h})\|_p + \|f_k(\mathbf{x} + \mathbf{h}) - f_k(\mathbf{x})\|_p + \|f_k(\mathbf{x}) - f(\mathbf{x})\|_p \\
&\quad = \|f_k(\mathbf{x} + \mathbf{h}) - f_k(\mathbf{x})\|_p + 2\|f_k(\mathbf{x}) - f(\mathbf{x})\|_p.
\end{aligned}$$

Since $f_k \in C_p$, we have $\limsup_{|\mathbf{h}|\to 0} \|f(\mathbf{x} + \mathbf{h}) - f(\mathbf{x})\|_p \le 2\|f_k(\mathbf{x}) - f(\mathbf{x})\|_p$, and (b) follows by letting $k \to \infty$.

Clearly, the characteristic function of a cube belongs to C_p. Hence, in view of the fact that linear combinations of characteristic functions of cubes are dense in $L^p(\mathbf{R}^n)$ [see (8.15)], it follows from (a) and (b) that L^p is contained in C_p.

We remark without proof that this result is also true for $0 < p < 1$. (Use the same ideas for $\|\cdot\|_p^p$.) It fails, however, for $p = \infty$, as shown by the function $\chi = \chi_{(0,\infty)}(x)$ on $(-\infty,+\infty)$. In fact, $\chi \in L^\infty(-\infty,+\infty)$ but $\|\chi(x + h) - \chi(x)\|_\infty = 1$ for all $h \ne 0$.

5. The Space L^2; Orthogonality

For complex-valued measurable f, $f = f_1 + if_2$ with f_1 and f_2 real-valued and measurable, we have $\int_E f = \int_E f_1 + i\int_E f_2$ (see p. 76). We will use the fact that $|\int_E f| \le \int_E |f|$. (See Exercise 1.)

Among the L^p spaces, L^2 has the special property that the product of any two of its elements is integrable (Schwarz's inequality). This simple fact leads to some important extra structure in L^2 which we will now discuss.

Consider $L^2 = L^2(E)$, where E is a fixed subset of $\mathbf{R}^n$ of positive measure, and write $\|f\| = \|f\|_{2,E}$, $\int_E f = \int f$, etc. For $f,g \in L^2$, define the *inner product* of f and g by

(8.20)
$$\langle f,g\rangle = \int f\bar{g},$$

where $\bar{g}$ denotes the complex conjugate of g. Note that by Schwarz's inequality,

$$|\langle f,g\rangle| \le \|f\|\|g\|.$$

Moreover, the inner product has the following properties:

(a) $\langle g,f\rangle = \overline{\langle f,g\rangle}$.
(b) $\langle f_1 + f_2, g\rangle = \langle f_1,g\rangle + \langle f_2,g\rangle$, $\langle f, g_1 + g_2\rangle = \langle f,g_1\rangle + \langle f,g_2\rangle$.
(c) $\langle \alpha f,g\rangle = \alpha\langle f,g\rangle$, $\langle f,\alpha g\rangle = \bar{\alpha}\langle f,g\rangle$, $\alpha \in \mathbf{C}$.
(d) $\langle f,f\rangle^{1/2} = \|f\|$.

If $\langle f,g\rangle = 0$, then f and g are said to be *orthogonal*. A set $\{\phi_\alpha\}_{\alpha\in A}$ is *orthogonal* if any two of its elements are orthogonal; $\{\phi_\alpha\}$ is *orthonormal* if it is orthogonal and $\|\phi_\alpha\| = 1$ for all α. Note that if $\{\phi_\alpha\}$ is orthogonal and

$\|\phi_\alpha\| \neq 0$ for every α, then $\{\phi_\alpha/\|\phi_\alpha\|\}$ is orthonormal. Henceforth, we will assume that $\|\phi_\alpha\| \neq 0$ for all α for an orthogonal system $\{\phi_\alpha\}$. This implies that no element is zero and that no two elements are equal.

(8.21) Theorem *Any orthogonal system $\{\phi_\alpha\}$ in L^2 is countable.*

Proof. We may assume that $\{\phi_\alpha\}$ is orthonormal. Then for $\alpha \neq \beta$, we have

$$\|\phi_\alpha - \phi_\beta\|^2 = \int(\phi_\alpha - \phi_\beta)(\bar{\phi}_\alpha - \bar{\phi}_\beta) = \|\phi_\alpha\|^2 + \|\phi_\beta\|^2 = 2,$$

so that $\|\phi_\alpha - \phi_\beta\| = \sqrt{2}$. Since L^2 is separable, it follows that $\{\phi_\alpha\}$ must be countable.

A collection $\psi_1, \ldots, \psi_N$ is said to be *linearly independent* if $\sum_{k=1}^N a_k\psi_k(\mathbf{x}) = 0$ (a.e.) implies that every a_k is zero. An infinite collection of functions is called *linearly independent* if each finite subcollection is. No function in a linearly independent set can be zero a.e.

(8.22) Theorem *If $\{\psi_k\}$ is orthogonal, it is linearly independent.*

Proof. Suppose that $a_1\psi_{k_1} + \cdots + a_N\psi_{k_N} = 0$. Multiplying both sides by $\bar{\psi}_{k_1}$ and integrating, we obtain by orthogonality that $a_1 = 0$. Similarly, $a_2 = \cdots = a_N = 0$.

The converse of (8.22) is not true. However, the next result shows that if $\{\psi_k\}$ is linearly independent, then the system formed from suitable linear combinations of its elements is orthogonal.

(8.23) Theorem (*Gram-Schmidt Process*) *If $\{\psi_k\}$ is linearly independent, then the system $\{\phi_k\}$ defined by*

$$\begin{aligned}
\phi_1 &= \psi_1 \\
\phi_2 &= a_{21}\psi_1 + \psi_2 \\
&\vdots \\
\phi_k &= a_{k1}\psi_1 + \cdots + a_{k,k-1}\psi_{k-1} + \psi_k \\
&\vdots
\end{aligned}$$

is orthogonal for proper selection of the a_{ij}.

Proof. Having $\phi_1 = \psi_1$, we proceed by induction, assuming that $\phi_1, \ldots, \phi_{k-1}$ have been chosen as required. We will determine constants $b_{k1}, \ldots, b_{k,k-1}$ so that the function ϕ_k defined by

$$\phi_k = b_{k1}\phi_1 + \cdots + b_{k,k-1}\phi_{k-1} + \psi_k$$

is orthogonal to $\phi_1, \ldots, \phi_{k-1}$. If $j < k$,

$$\langle\phi_k,\phi_j\rangle = b_{kj}\langle\phi_j,\phi_j\rangle + \langle\psi_k,\phi_j\rangle$$

by orthogonality. Since $\langle\phi_j,\phi_j\rangle \neq 0$, b_{kj} can be chosen so that $\langle\phi_k,\phi_j\rangle = 0$,

$j < k$. Since each ϕ_j with $j < k$ is a linear combination of $\psi_1, \ldots, \psi_j$, the theorem follows.

When the ϕ_k are selected by the Gram-Schmidt process, we shall say that they are *generated* from the ψ_k. Note that the triangular character of the matrix in (8.23) means that each ψ_k can also be written as a linear combination of the $\phi_j, j \leq k$.

We call an orthogonal system $\{\phi_k\}$ *complete* if the only function which is orthogonal to every ϕ_k is zero; that is, $\{\phi_k\}$ is complete if $\langle f,\phi_k\rangle = 0$ for all k implies that $f = 0$ a.e. Thus, a complete orthogonal system is one which is maximal in the sense that it is not properly contained in any larger orthogonal system.

The *span* of a set of functions $\{\psi_k\}$ is the collection of all finite linear combinations of the ψ_k. In speaking of the span of $\{\psi_k\}$, we may always assume that $\{\psi_k\}$ is orthogonal by discarding any dependent functions and applying the Gram-Schmidt process to the resulting linearly independent set.

A set $\{\psi_k\}$ is called a *basis* for L^2 if its span is dense in L^2; that is, $\{\psi_k\}$ is a basis if given $f \in L^2$ and $\varepsilon > 0$, there exist N and $\{a_k\}$ such that $\|f - \sum_{k=1}^N a_k\psi_k\| < \varepsilon$. The a_k can always be chosen with rational real and imaginary parts. Any countable dense set in L^2 is of course a basis. It follows that L^2 has an orthogonal basis.

(8.24) Theorem *Any orthogonal basis in L^2 is complete. In particular, there exists a complete orthonormal basis for L^2.*

Proof. Let $\{\psi_k\}$ be an orthogonal basis for L^2. We may assume that $\{\psi_k\}$ is orthonormal. To show that it is complete, let $\langle f,\psi_k\rangle = 0$ for all k. Then $\langle f,f\rangle = \langle f, f - \sum_{k=1}^N a_k\psi_k\rangle$ for any N and a_k. By Schwarz's inequality, $|\langle f,f\rangle| \leq \|f\|\|f - \sum_{k=1}^N a_k\psi_k\|$, and so, since the term on the right can be chosen arbitrarily small, $\langle f,f\rangle = 0$. Therefore, $f = 0$ a.e., which completes the proof.

6. Fourier Series; Parseval's Formula

Let $\{\phi_k\}$ be any orthonormal system for L^2. If $f \in L^2$, the numbers defined by

$$c_k = c_k(f) = \langle f,\phi_k\rangle = \int_E f\bar{\phi}_k$$

are called the *Fourier coefficients* of f with respect to $\{\phi_k\}$. The series $\sum_k c_k\phi_k$ is called the *Fourier series* of f with respect to $\{\phi_k\}$, and denoted $S[f] = \sum_k c_k\phi_k$. We also write

$$f \sim \sum_k c_k\phi_k.$$

The first question we ask is how well $S[f]$, or more precisely, the sequence of its partial sums, approximates f. Fix N and let $L = \sum_{k=1}^{N} \gamma_k \phi_k$ be a linear combination of $\phi_1, \ldots, \phi_N$. We wish to know what choice of $\gamma_1, \ldots, \gamma_N$ makes $\|f - L\|$ a minimum. Note that since $\{\phi_k\}$ is orthonormal, $\|L\|^2 = \langle L,L \rangle = \sum_{k=1}^{N} |\gamma_k|^2$. Hence,

$$\|f - L\|^2 = \int \left(f - \sum_{k=1}^{N} \gamma_k \phi_k\right)\left(\bar{f} - \sum_{k=1}^{N} \bar{\gamma}_k \bar{\phi}_k\right)$$
$$= \|f\|^2 - \sum_{k=1}^{N} (\bar{\gamma}_k c_k + \gamma_k \bar{c}_k) + \sum_{k=1}^{N} |\gamma_k|^2,$$

where the c_k are the Fourier coefficients of f. Since

$$|c_k - \gamma_k|^2 = (c_k - \gamma_k)(\bar{c}_k - \bar{\gamma}_k) = |c_k|^2 - (\bar{\gamma}_k c_k + \gamma_k \bar{c}_k) + |\gamma_k|^2,$$

we obtain

$$\|f - L\|^2 = \|f\|^2 + \sum_{k=1}^{N} |c_k - \gamma_k|^2 - \sum_{k=1}^{N} |c_k|^2.$$

Therefore,

(8.25) $$\min_{\gamma_1, \ldots, \gamma_N} \|f - L\|^2 = \|f\|^2 - \sum_{k=1}^{N} |c_k|^2;$$

that is, the minimum is achieved when $\gamma_k = c_k$ for $k = 1, \ldots, N$, or equivalently, when L is the Nth partial sum of $S[f]$. Writing $s_N = s_N(f) = \sum_{k=1}^{N} c_k \phi_k$, we have from (8.25) that

(8.26) $$\|f - s_N\|^2 = \|f\|^2 - \sum_{k=1}^{N} |c_k|^2.$$

(8.27) Theorem *Let $\{\phi_k\}$ be an orthonormal system in L^2, and let $f \in L^2$.*

(i) *Of all linear combinations $\sum_1^N \gamma_k \phi_k$ with N fixed, the best L^2-approximation to f is given by the partial sum $s_N = \sum_1^N c_k \phi_k$ of the Fourier series of f.*

(ii) *(Bessel's inequality) The sequence $\{c_k\}$ of Fourier coefficients of f belongs to l^2 and*

$$\left(\sum_{k=1}^{\infty} |c_k|^2\right)^{1/2} \leq \|f\|.$$

Proof. Part (i) has been proved. Note that since $\|f - s_N\|^2 \geq 0$, Bessel's inequality follows from (8.26) by letting $N \to \infty$.

If f is a function for which equality holds in Bessel's inequality, that is, if

(8.28) $$\left(\sum_{k=1}^{\infty} |c_k|^2\right)^{1/2} = \|f\|,$$

then f is said to satisfy *Parseval's formula*. From (8.26), we immediately obtain the next result.

(8.29) Theorem *Parseval's formula holds for f if and only if $\|s_N - f\| \to 0$, i.e., if and only if $S[f]$ converges to f in L^2 norm.*

The following theorem is of great importance.

(8.30) Theorem (*Riesz-Fischer Theorem*) *Let $\{\phi_k\}$ be any orthonormal system, and let $\{c_k\}$ be any sequence in l^2. Then there is an $f \in L^2$ such that $S[f] = \sum c_k\phi_k$, i.e., such that $\{c_k\}$ is the sequence of Fourier coefficients of f with respect to $\{\phi_k\}$. Moreover, f can be chosen to satisfy Parseval's formula.*

Proof. Let $t_N = \sum_{k=1}^N c_k\phi_k$. Then if $M < N$,

$$\|t_N - t_M\|^2 = \left\|\sum_{M+1}^{N} c_k\phi_k\right\|^2 = \sum_{M+1}^{N} |c_k|^2.$$

The fact that $\{c_k\} \in l^2$ implies that $\{t_N\}$ is a Cauchy sequence in L^2. Since L^2 is complete, there is an $f \in L^2$ such that $\|f - t_N\| \to 0$. If $N \geq k$,

$$\int f\bar{\phi}_k = \int (f - t_N)\bar{\phi}_k + \int t_N\bar{\phi}_k = \int (f - t_N)\bar{\phi}_k + c_k.$$

Since the integral on the right is bounded in absolute value by $\|f - t_N\|\|\phi_k\| = \|f - t_N\|$, we obtain by letting $N \to \infty$ that $\int f\bar{\phi}_k = c_k$. Thus, $S[f] = \sum_k c_k\phi_k$, so that $t_N = s_N(f)$, and it follows from (8.29) that Parseval's formula holds for f. This completes the proof.

There is no guarantee that the Fourier coefficients of a function uniquely determine the function. However, if $\{\phi_k\}$ is complete, we can show that the correspondence between a function and its Fourier coefficients is unique; that is, if f and g have the same Fourier coefficients with respect to a complete system, then $f = g$ a.e. This is simple, since the vanishing of all the Fourier coefficients of $f - g$ implies that $f - g = 0$ a.e. An important related fact is the following.

(8.31) Theorem *Let $\{\phi_k\}$ be an orthonormal system. Then $\{\phi_k\}$ is complete if and only if Parseval's formula holds for every $f \in L^2$.*

Proof. Suppose that $\{\phi_k\}$ is complete. If $f \in L^2$, Bessel's inequality implies that its Fourier coefficients $\{c_k\}$ belong to l^2. Hence, by the Riesz-Fischer theorem, there exists a g in L^2 with $S[g] = \sum c_k\phi_k$ and $\|g\| = (\sum |c_k|^2)^{1/2}$. Since f and g have the same Fourier coefficients and $\{\phi_k\}$ is complete, we see that $f = g$ a.e. Hence, $\|f\| = \|g\| = (\sum |c_k|^2)^{1/2}$, which is Parseval's formula.

Conversely, suppose that Parseval's formula holds with respect to $\{\phi_k\}$ for every $f \in L^2$. If $\langle f,\phi_k\rangle = 0$ for all k, then $\|f\| = (\sum |\langle f,\phi_k\rangle|^2)^{1/2} = 0$. Therefore, $f = 0$ a.e., which proves that $\{\phi_k\}$ is complete.

Suppose that $\{\phi_k\}$ is orthonormal and complete and that $f,g \in L^2$. Let

$c_k = \langle f,\phi_k \rangle$, $d_k = \langle g,\phi_k \rangle$, $c = \{c_k\}$, $d = \{d_k\}$, $\|c\| = (\sum |c_k|^2)^{1/2}$ and $(c,d) = \sum c_k \bar{d}_k$. We claim that

(8.32) $$\langle f,g \rangle = (c,d).$$

To prove this, observe that by Parseval's formula, $\langle f + g, f + g \rangle = (c + d, c + d)$, or

$$\langle f,f \rangle + \langle g,g \rangle + 2 \operatorname{Re} \langle f,g \rangle = (c,c) + (d,d) + 2 \operatorname{Re} (c,d),$$

where $\operatorname{Re} z$ denotes the real part of z. Cancelling equal terms, we see that $\operatorname{Re} \langle f,g \rangle = \operatorname{Re} (c,d)$. Applying this to the function if gives $\operatorname{Re} \langle if,g \rangle = \operatorname{Re} (ic,d)$. But $\operatorname{Re} \langle if,g \rangle = \operatorname{Re} i\langle f,g \rangle = -\operatorname{Im} \langle f,g \rangle$. Similarly, $\operatorname{Re} (ic,d) = -\operatorname{Im} (c,d)$. Therefore, $\operatorname{Im} \langle f,g \rangle = \operatorname{Im} (c,d)$, and (8.32) is proved.

Another corollary of (8.31) is given in the next result. First, we make several definitions. Let X_1 and X_2 be metric spaces with metrics d_1 and d_2, respectively. Then X_1 and X_2 are said to be *isometric* if there is a mapping T of X_1 onto X_2 such that

$$d_1(f,g) = d_2(Tf,Tg)$$

for all $f,g \in X_1$. Such a T is called an *isometry*. Thus, an isometry is a mapping which preserves distances. An isometry is automatically one-to-one, and two isometric metric spaces may be regarded as the same space with a relabeling of the points. For example, two L^2 spaces, $L^2(E)$ and $L^2(E')$, are isometric if there is a mapping T of $L^2(E)$ onto $L^2(E')$ such that $\|f - g\|_{2,E} = \|Tf - Tg\|_{2,E'}$ for all $f,g \in L^2(E)$. The isometries we shall encounter will be *linear*, i.e., will satisfy

$$T(\alpha f + \beta g) = \alpha Tf + \beta Tg \text{ for all scalars } \alpha, \beta.$$

If T is a linear map of $L^2(E)$ onto $L^2(E')$, then since $Tf - Tg = T(f - g)$, it follows that T is an isometry if and only if

$$\|f\|_{2,E} = \|Tf\|_{2,E'}$$

for all $f \in L^2(E)$. Similarly, a linear map of $L^2(E)$ onto l^2 is an isometry if and only if $\|f\|_{2,E} = \|Tf\|_{l^2}$ for all $f \in L^2(E)$.

(8.33) Theorem *All spaces $L^2(E)$ are linearly isometric with l^2, and so with one another.*

Proof. For a given E, define a linear correspondence between $L^2(E)$ and l^2 by choosing a complete orthonormal system $\{\phi_k\}$ in $L^2(E)$ and mapping an $f \in L^2(E)$ onto the sequence $\{\langle f,\phi_k \rangle\}$ of its Fourier coefficients. This mapping is onto all of l^2 by the Riesz-Fischer theorem, and is an isometry by (8.31).

7. Hilbert Spaces

A set H is called a *Hilbert space over the complex numbers* $\mathbf{C}$ if it satisfies the following three conditions:

(H_1) H is a vector space over $\mathbf{C}$; that is, if $f,g \in H$ and $\alpha \in \mathbf{C}$, then $f + g \in H$ and $\alpha f \in H$.

(H_2) For every pair $f,g \in H$, there is a complex number (f,g), called the *inner product* of f and g, which satisfies

(a) $(g,f) = \overline{(f,g)}$.
(b) $(f_1 + f_2, g) = (f_1,g) + (f_2,g)$.
(c) $(\alpha f,g) = \alpha(f,g)$ for $\alpha \in \mathbf{C}$.
(d) $(f,f) \geq 0$, and $(f,f) = 0$ if and only if $f = 0$.

Notice that (a), (b) and (c) imply that $(f, g_1 + g_2) = (f,g_1) + (f,g_2)$ and $(f,\alpha g) = \bar{\alpha}(f,g)$. Define

$$\|f\| = (f,f)^{1/2}.$$

Before stating the third condition, we claim that

(8.34) $$|(f,g)| \leq \|f\|\|g\| \qquad (\textit{Schwarz's inequality}).$$

If $\|g\| = 0$, this is obvious. Otherwise, letting $\lambda = -(f,g)/\|g\|^2$, we obtain

$$0 \leq (f + \lambda g, f + \lambda g) = \|f\|^2 - 2\frac{|(f,g)|^2}{\|g\|^2} + \frac{|(f,g)|^2}{\|g\|^2} = \|f\|^2 - \frac{|(f,g)|^2}{\|g\|^2},$$

and Schwarz's inequality follows at once.

We will show that $\|\cdot\|$ is a norm on H by proving the triangle inequality. In fact,

$$\|f + g\|^2 = (f + g, f + g) = \|f\|^2 + 2\,\mathrm{Re}\,(f,g) + \|g\|^2.$$

Since $|\mathrm{Re}\,(f,g)| \leq |(f,g)| \leq \|f\|\|g\|$, it follows that the right side is at most $(\|f\| + \|g\|)^2$. Taking square roots, we obtain $\|f + g\| \leq \|f\| + \|g\|$, as desired. Hence, H is a normed linear space.

We also require

(H_3) H is complete with respect to $\|\cdot\|$.

In particular, a Hilbert space is a Banach space.

As for L^2 spaces, a linear map T of a Hilbert space H onto a Hilbert space H' is an isometry if and only if $\|f\|_H = \|Tf\|_{H'}$ for all $f \in H$.

A Hilbert space is called *infinite dimensional* if it cannot be spanned by a finite number of elements; hence, an infinite dimensional Hilbert space has an infinite linearly independent subset. The space L^2 with inner product

$(f,g) = \int f\bar{g}$ and the space l^2 with $(c,d) = \sum_k c_k\bar{d}_k$ are examples of separable infinite dimensional Hilbert spaces. In fact, there are essentially no other examples, as the following theorem shows.

(8.35) Theorem *All separable infinite dimensional Hilbert spaces are linearly isometric with l^2, and so with one another.*

Proof. The proof is a repetition of the ideas leading to (8.33), so we shall be brief. Let H be a separable infinite dimensional Hilbert space, and let $\{e'_k\}$ be a countable dense subset. Discarding those e'_k which are spanned by other e'_i, we obtain a linearly independent set $\{e_k\}$ with the same dense span as $\{e'_k\}$. Since H is infinite dimensional, $\{e_k\}$ is infinite. Using the Gram-Schmidt process, we may assume that $\{e_k\}$ is orthonormal: $(e_i,e_k) = 0$ for $i \neq k$ and $\|e_k\| = 1$ for all k. It follows that $\{e_k\}$ is complete; in fact, if $(f,e_k) = 0$ for all k, then

$$\left\| f - \sum_{k=1}^{N} a_k e_k \right\|^2 = \|f\|^2 + \sum_{k=1}^{N} |a_k|^2 \geq \|f\|^2.$$

If f were not zero, the span of the e_k could not be dense. Hence, $f = 0$, which shows that H has a complete orthonormal system $\{e_k\}$.

Next, we will show that Bessel's inequality and the Riesz-Fischer theorem hold for $\{e_k\}$. If $f \in H$, let $c_k = (f,e_k)$. Then

$$0 \leq \left\| f - \sum_{k=1}^{N} c_k e_k \right\|^2 = \|f\|^2 - \sum_{k=1}^{N} |c_k|^2.$$

Letting $N \to \infty$, we obtain Bessel's inequality $(\sum |c_k|^2)^{1/2} \leq \|f\|$. In particular, $\{c_k\}$ belongs to l^2.

To derive the Riesz-Fischer theorem, let $\{\gamma_k\}$ be a sequence in l^2, and set $t_N = \sum_{k=1}^{N} \gamma_k e_k$. Then

$$\|t_M - t_N\|^2 = \sum_{k=M+1}^{N} |\gamma_k|^2 \to 0 \text{ as } M,N \to \infty,\ M < N.$$

Since H is complete, there is a $g \in H$ such that $\|g - t_N\| \to 0$. We have

$$(g,e_k) = (g - t_N, e_k) + (t_N,e_k) = (g - t_N, e_k) + \gamma_k \qquad (k \leq N).$$

Letting $N \to \infty$, it follows from Schwarz's inequality that $(g,e_k) = \gamma_k$. Hence, $t_N = s_N(g)$ and $\|g - t_N\|^2 = \|g\|^2 - \sum_{k=1}^{N} |\gamma_k|^2$. Letting $N \to \infty$ in the last equation, we see that g satisfies Parseval's formula $\|g\| = (\sum |\gamma_k|^2)^{1/2}$. This gives the analogue of the Riesz-Fischer theorem.

Now let $f \in H$, and let $c_k = (f,e_k)$. Taking $\gamma_k = c_k$ above, we see by the completeness of $\{e_k\}$ that $g = f$, so that Parseval's formula holds: $\|f\| = (\sum |c_k|^2)^{1/2}$. The fact that H is linearly isometric with l^2 now follows as in the proof of (8.33).

Exercises

1. For complex-valued, measurable f, $f = f_1 + if_2$ with f_1 and f_2 real-valued and measurable, we have $\int_E f = \int_E f_1 + i\int_E f_2$. Prove that $\int_E f$ is finite if and only if $\int_E |f|$ is finite, and $|\int_E f| \le \int_E |f|$. (Note that $|\int_E f| = [(\int_E f_1)^2 + (\int_E f_2)^2]^{1/2}$, and use the fact that $(a^2 + b^2)^{1/2} = a\cos\alpha + b\sin\alpha$ for an appropriate α, while $(a^2 + b^2)^{1/2} \ge |a\cos\alpha + b\sin\alpha|$ for all α.)
2. Prove the converse of Hölder's inequality for $p = 1$ and ∞. Show also that for real-valued $f \notin L^p(E)$, there exists a function $g \in L^{p'}(E)$, $1/p + 1/p' = 1$, such that $fg \notin L^1(E)$. (Construct g of the form $\sum a_k g_k$ for appropriate a_k and g_k, with g_k satisfying $\int_E fg_k \to +\infty$.)
3. Prove Theorems (8.12) and (8.13). Show that Minkowski's inequality for series fails when $p < 1$.
4. Let f and g be real-valued, and let $1 < p < \infty$. Prove that equality holds in the inequality $|\int fg| \le \|f\|_p\|g\|_{p'}$ if and only if fg has constant sign a.e. and $|f|^p$ is a multiple of $|g|^{p'}$ a.e.

   If $\|f + g\|_p = \|f\|_p + \|g\|_p$ and $g \ne 0$ in Minkowski's inequality, show that f is a multiple of g.
5. For $1 \le p < \infty$ and $0 < |E| < +\infty$, define
$$N_p[f] = \left(\frac{1}{|E|}\int_E |f|^p\right)^{1/p}.$$
   Prove that if $p_1 < p_2$, then $N_{p_1}[f] \le N_{p_2}[f]$.

   Prove also that $N_p[f + g] \le N_p[f] + N_p[g]$, $(1/|E|)\int_E |fg| \le N_p[f]N_{p'}[g]$, $1/p + 1/p' = 1$, and that $\lim_{p\to\infty} N_p[f] = \|f\|_\infty$. Thus, N_p behaves like $\|\cdot\|_p$, but has the advantage of being monotone in p.
6. Prove the following generalization of Hölder's inequality. If $\sum_{i=1}^k 1/p_i = 1/r$, $p_i, r \ge 1$, then
$$\|f_1 \cdots f_k\|_r \le \|f_1\|_{p_1} \cdots \|f_k\|_{p_k}.$$
   (See also Exercise 12 of Chapter 7.)
7. Show that when $0 < p < 1$, the neighborhoods $\{f : \|f\|_p < \varepsilon\}$ of zero in $L^p(0,1)$ are not convex. (Let $f = \chi_{(0,\varepsilon^p)}$ and $g = \chi_{(\varepsilon^p, 2\varepsilon^p)}$. Show that $\|f\|_p = \|g\|_p = \varepsilon$, but that $\|\frac{1}{2}f + \frac{1}{2}g\|_p > \varepsilon$.)
8. Prove the following *integral version of Minkowski's inequality* for $1 \le p < \infty$:
$$\left[\int\left|\int f(x,y)\,dx\right|^p dy\right]^{1/p} \le \int\left[\int |f(x,y)|^p\,dy\right]^{1/p} dx.$$
   [For $1 < p < \infty$, note that the pth power of the left-hand side is majorized by $\iint [\int |f(z,y)|\,dz]^{p-1}|f(x,y)|\,dx\,dy$. Integrate first with respect to y and apply Hölder's inequality.]
9. If f is real-valued and measurable on E, define its *essential infimum* on E by
$$\operatorname*{ess\,inf}_E f = \sup\{\alpha : |\{\mathbf{x} \in E : f(\mathbf{x}) < \alpha\}| = 0\}.$$
   If $f \ge 0$, show that $\text{ess}_E \inf f = (\text{ess}_E \sup 1/f)^{-1}$.

10. Prove that $L^\infty(E)$ is not separable for any E with $|E| > 0$. (Construct a sequence of decreasing subsets of E whose measures strictly decrease. Consider the characteristic functions of the class of sets obtained by taking all possible unions of the differences of these subsets.)
11. If $f_k \to f$ in L^p, $1 \le p < \infty$, $g_k \to g$ pointwise, and $\|g_k\|_\infty \le M$ for all k, prove that $f_k g_k \to fg$ in L^p.
12. Let $f, \{f_k\} \in L^p$. Show that if $\|f - f_k\|_p \to 0$, then $\|f_k\|_p \to \|f\|_p$. Conversely, if $f_k \to f$ a.e. and $\|f_k\|_p \to \|f\|_p$, $1 \le p < \infty$, show that $\|f - f_k\|_p \to 0$.
13. Suppose that $f_k \to f$ a.e. and that $f_k, f \in L^p$, $1 < p < \infty$. If $\|f_k\|_p \le M < +\infty$, show that $\int f_k g \to \int fg$ for all $g \in L^{p'}$, $1/p + 1/p' = 1$.
14. Verify that the following systems are orthogonal:
    (a) $\{\frac{1}{2}, \cos x, \sin x, \ldots, \cos kx, \sin kx, \ldots\}$ on any interval of length 2π;
    (b) $\{e^{2\pi ikx/(b-a)}; k = 0, \pm 1, \pm 2, \ldots\}$ on (a,b).
15. If $f \in L^2(0,2\pi)$, show that

$$\lim_{k\to\infty} \int_0^{2\pi} f(x) \cos kx \, dx = \lim_{k\to\infty} \int_0^{2\pi} f(x) \sin kx \, dx = 0.$$

Prove that the same is true if $f \in L^1(0,2\pi)$ (This last statement is the *Riemann-Lebesgue lemma*. To prove it, approximate f in L^1 norm by L^2 functions. See (12.21).)
16. A sequence $\{f_k\}$ in L^p is said to *converge weakly* to a function f in L^p if $\int f_k g \to \int fg$ for all $g \in L^{p'}$. Prove that if $f_k \to f$ in L^p norm, $1 \le p \le \infty$, then $\{f_k\}$ converges weakly to f in L^p. Note by Exercise 15 that the converse is not true.
17. Suppose that $f_k, f \in L^2$ and that $\int f_k g \to \int fg$ for all $g \in L^2$ (that is, $\{f_k\}$ converges weakly to f in L^2). If $\|f_k\|_2 \to \|f\|_2$, show that $f_k \to f$ in L^2 norm.
18. Prove the *parallelogram law* for L^2:

$$\|f + g\|^2 + \|f - g\|^2 = 2\|f\|^2 + 2\|g\|^2.$$

Is this true for L^p when $p \ne 2$? The geometric interpretation is that the sum of the squares of the diagonals of a parallelogram equals the sum of the squares of the sides.
19. Prove that a finite dimensional Hilbert space is isometric with $\mathbf{R}^n$ for some n.
20. Construct a function in $L^1(-\infty, +\infty)$ which is not in $L^2(a,b)$ for any $a < b$. (Let $g(x) = x^{-1/2}$ on $(0,1)$ and $g(x) = 0$ elsewhere, so that $\int_{-\infty}^{+\infty} g = 2$. Consider the function $f(x) = \sum a_k g(x - r_k)$, where $\{r_k\}$ is the rational numbers and $\{a_k\}$ satisfies $a_k > 0$, $\sum a_k < +\infty$.)
21. If $f \in L^p(\mathbf{R}^n)$, $0 < p < \infty$, show that

$$\lim_{Q \searrow x} \frac{1}{|Q|} \int_Q |f(\mathbf{y}) - f(\mathbf{x})|^p \, d\mathbf{y} = 0 \text{ a.e.}$$

Note by Exercise 5 that this condition for a given p implies it for all smaller p.
22. Let $\{\phi_k\}$ be a complete orthonormal system in L^2, and let $m = \{m_k\}$ be a given sequence of numbers. If $f \in L^2$, $f \sim \sum c_k \phi_k$, define Tf by $Tf \sim \sum m_k c_k \phi_k$. Show that T is bounded on L^2, i.e., that there is a constant c independent of f such that $\|Tf\|_2 \le c\|f\|_2$ for all $f \in L^2$, if and only if $m \in l^\infty$. Show that the smallest possible choice for c is $\|m\|_{l^\infty}$.

Chapter 9

Approximations of the Identity; Maximal Functions

1. Convolutions

The convolution of two functions f and g which are measurable in $\mathbf{R}^n$ is defined by

$$(f * g)(\mathbf{x}) = \int_{\mathbf{R}^n} f(\mathbf{t})g(\mathbf{x} - \mathbf{t})\, dt \qquad (\mathbf{x} \in \mathbf{R}^n).$$

In Chapter 6, (6.14), we saw that $\|f * g\|_1 \le \|f\|_1\|g\|_1$. Moreover, according to (6.16), $\|f * g\|_1 = \|f\|_1\|g\|_1$ if f and g are nonnegative. In this section, we will study some additional properties of convolutions, beginning with the following theorem.

(9.1) Theorem *Let* $1 \le p \le \infty$, $f \in L^p(\mathbf{R}^n)$ *and* $g \in L^1(\mathbf{R}^n)$. *Then* $f * g \in L^p(\mathbf{R}^n)$ *and*

$$\|f * g\|_p \le \|f\|_p\|g\|_1.$$

Proof. Since $|f * g| \le |f| * |g|$, we may assume that f and g are nonnegative. We may also suppose that $1 < p \le \infty$, since when $p = 1$ the result is just (6.14). If $p = \infty$, then

$$(f * g)(\mathbf{x}) \le \int_{\mathbf{R}^n} \|f\|_\infty g(\mathbf{x} - \mathbf{t})\, dt = \|f\|_\infty \int_{\mathbf{R}^n} g(\mathbf{x} - \mathbf{t})\, dt = \|f\|_\infty\|g\|_1.$$

Therefore, $\|f * g\|_\infty \le \|f\|_\infty\|g\|_1$, as claimed.

If $1 < p < \infty$, we write

$$(f * g)(\mathbf{x}) = \int_{\mathbf{R}^n} [f(\mathbf{t})g(\mathbf{x} - \mathbf{t})^{1/p}]g(\mathbf{x} - \mathbf{t})^{1/p'}\, dt, \qquad \frac{1}{p} + \frac{1}{p'} = 1.$$

By Hölder's inequality with exponents p and p',

$$(f * g)(\mathbf{x}) \le \left(\int_{\mathbf{R}^n} f(\mathbf{t})^p g(\mathbf{x} - \mathbf{t})\, d\mathbf{t}\right)^{1/p} \left(\int_{\mathbf{R}^n} g(\mathbf{x} - \mathbf{t})\, d\mathbf{t}\right)^{1/p'}$$
$$= (f^p * g)(\mathbf{x})^{1/p} \|g\|_1^{1/p'}.$$

Now raise the first and last terms in this inequality to the pth power and integrate the result. Since $\int_{\mathbf{R}^n} (f^p * g)\, d\mathbf{x} = \|f\|_p^p \|g\|_1$ by (6.16), we obtain

$$\|f * g\|_p^p \le \|f\|_p^p \|g\|_1^{1+(p/p')} = \|f\|_p^p \|g\|_1^p.$$

The theorem follows by taking pth roots.

Theorem (9.1) is an important special case of the next result, whose proof is left to the reader. (See Exercise 2.)

(9.2) Theorem (***Young's Convolution Theorem***) *Let p and q satisfy $1 \le p,q \le \infty$ and $1/p + 1/q \ge 1$, and let r be defined by $1/r = 1/p + 1/q - 1$. If $f \in L^p(\mathbf{R}^n)$ and $g \in L^q(\mathbf{R}^n)$, then $f * g \in L^r(\mathbf{R}^n)$ and*

$$\|f * g\|_r \le \|f\|_p \|g\|_q.$$

Note that when $q = 1$, Young's theorem reduces to (9.1). See also Exercise 3.

A convolution $f * K$ with K fixed defines a transformation $T : f \to f * K$ which is called the *convolution operator with kernel K*. Theorem (9.1) states that a convolution operator with an integrable kernel maps functions in L^p into the same L^p. The next result shows an effect that convolution operators with smooth kernels have on L^p.

For a positive integer m, we denote by C^m the class of functions $f(\mathbf{x})$, $\mathbf{x} \in \mathbf{R}^n$, whose partial derivatives up to and including those of order m exist and are continuous. The subset of C^m of functions with compact support is denoted C_0^m. Similarly, C^∞ is the class of infinitely differentiable functions, and C_0^∞ is the corresponding subset of functions with compact support. (For the existence of such functions, see Exercise 4.) Finally, if $\alpha = (\alpha_1, \ldots, \alpha_n)$, where the α_k are nonnegative integers, then the αth partial derivative of f is denoted by

$$(D^\alpha f)(\mathbf{x}) = \left(\frac{\partial^\alpha f}{\partial \mathbf{x}^\alpha}\right)(\mathbf{x}) = \left(\frac{\partial^{\alpha_1}}{\partial x_1^{\alpha_1}} \cdots \frac{\partial^{\alpha_n}}{\partial x_n^{\alpha_n}} f\right)(\mathbf{x}).$$

(9.3) Theorem *If $1 \le p \le \infty$, $f \in L^p(\mathbf{R}^n)$ and $K \in C_0^m$, then $f * K \in C^m$ and*

$$D^\alpha(f * K)(\mathbf{x}) = (f * D^\alpha K)(\mathbf{x}),$$

$$\alpha = (\alpha_1, \ldots, \alpha_n),\ \alpha_1 + \cdots + \alpha_n \le m.$$

Proof. We first claim that if K is any continuous kernel with compact support, then $f * K$ is continuous. In fact,

$$\begin{aligned}
&|(f * K)(\mathbf{x}+\mathbf{h}) - (f * K)(\mathbf{x})| \\
&\quad = \left| \int_{\mathbf{R}^n} f(\mathbf{t})K(\mathbf{x}+\mathbf{h}-\mathbf{t})\,dt - \int_{\mathbf{R}^n} f(\mathbf{t})K(\mathbf{x}-\mathbf{t})\,dt \right| \\
&\quad = \left| \int_{\mathbf{R}^n} f(\mathbf{x}-\mathbf{t})[K(\mathbf{t}+\mathbf{h}) - K(\mathbf{t})]\,dt \right| \\
&\quad \le \left(\int_{\mathbf{R}^n} |f(\mathbf{x}-\mathbf{t})|^p\,dt \right)^{1/p} \left(\int_{\mathbf{R}^n} |K(\mathbf{t}+\mathbf{h}) - K(\mathbf{t})|^{p'}\,dt \right)^{1/p'} \\
&\quad = \|f\|_p \|K(\mathbf{t}+\mathbf{h}) - K(\mathbf{t})\|_{p'}.
\end{aligned}$$

The last expression tends to zero as $|\mathbf{h}| \to 0$ since K is uniformly continuous and has compact support. [Note that in case $p' < \infty$, this also follows from (8.19) since $K \in L^{p'}$. See also Exercise 3.]

Next, let $K \in C_0^m$, $m \ge 1$. Fix i, $i = 1, \ldots, n$, and let $\mathbf{h} = (0, \ldots, 0, h, 0, \ldots, 0)$, where h is in the ith coordinate position. Note that

$$\begin{aligned}
\frac{(f * K)(\mathbf{x}+\mathbf{h}) - (f * K)(\mathbf{x})}{h} &= \int_{\mathbf{R}^n} f(\mathbf{t}) \left\{ \frac{K(\mathbf{x}-\mathbf{t}+\mathbf{h}) - K(\mathbf{x}-\mathbf{t})}{h} \right\} dt \\
&= \int_{\mathbf{R}^n} f(\mathbf{t}) \frac{\partial K}{\partial x_i}(\mathbf{x}-\mathbf{t}+\mathbf{h}')\,dt,
\end{aligned}$$

by the mean-value theorem, where $\mathbf{h}' = (0, \ldots, 0, h', 0, \ldots, 0)$ for some h' depending on $\mathbf{x}$ and $\mathbf{t}$ which is between 0 and h. Hence, as $h \to 0$, $(\partial K/\partial x_i)(\mathbf{x}-\mathbf{t}+\mathbf{h}')$ converges to $(\partial K/\partial x_i)(\mathbf{x}-\mathbf{t})$ uniformly in $\mathbf{t}$. Since $\partial K/\partial x_i$ has compact support, it follows from the uniform convergence theorem that the last integral converges to $[f * (\partial K/\partial x_i)](\mathbf{x})$. Therefore, $\partial(f * K)(\mathbf{x})/\partial x_i$ exists and equals $[f * (\partial K/\partial x_i)](\mathbf{x})$, which is continuous by our earlier remarks. The proof of the theorem for $m = 1$ is now complete. The proof for $m = 2, 3, \ldots$ follows by repeated application of the case $m = 1$.

It follows from this result that $f * K \in C^\infty$ if $f \in L^p$, $1 \le p \le \infty$, and $K \in C_0^\infty$. If, in addition, f has compact support, then so has $f * K$. In fact, if S_1 is the support of K and S_2 is the support of f, then the formula $(f * K)(\mathbf{x}) = \int_{S_2} f(\mathbf{t})K(\mathbf{x}-\mathbf{t})\,dt$ implies that $(f * K)(\mathbf{x}) = 0$ unless there are points $\mathbf{t} \in S_2$ for which $\mathbf{x} - \mathbf{t} \in S_1$. Hence, the support of $f * K$ is contained in $\{\mathbf{x} : \mathbf{x} = \mathbf{s}_1 + \mathbf{s}_2, \mathbf{s}_1 \in S_1, \mathbf{s}_2 \in S_2\}$, and so is bounded. An application of this fact is given in Exercise 5.

(9.4) Theorem *If $f \in L(\mathbf{R}^n)$ and K is bounded and uniformly continuous on $\mathbf{R}^n$, then $f * K$ is bounded and uniformly continuous on $\mathbf{R}^n$.*

The proof is similar to the first part of the proof of (9.3) and is left as an exercise. See also Exercise 3.

2. Approximations of the Identity

Given K and $\varepsilon > 0$, let

$$K_\varepsilon(\mathbf{x}) = \varepsilon^{-n}K\left(\frac{\mathbf{x}}{\varepsilon}\right) = \varepsilon^{-n}K\left(\frac{x_1}{\varepsilon}, \ldots, \frac{x_n}{\varepsilon}\right).$$

For example, if $K(\mathbf{x}) = \chi_{\{|\mathbf{x}|<1\}}(\mathbf{x})$, then $K_\varepsilon(\mathbf{x}) = \varepsilon^{-n}\chi_{\{|\mathbf{x}|<\varepsilon\}}(\mathbf{x})$. In this case, taking successively smaller values of ε produces kernels with successively higher peaks and smaller supports. The effect on any positive K with compact support is roughly the same.

In general, K_ε has the following basic properties.

(9.5) Lemma *If $K \in L^1(\mathbf{R}^n)$ and $\varepsilon > 0$, then*

(i) $\int_{\mathbf{R}^n} K_\varepsilon = \int_{\mathbf{R}^n} K$.
(ii) $\int_{|\mathbf{x}|>\delta} |K_\varepsilon| \to 0$ as $\varepsilon \to 0$, for any fixed $\delta > 0$.

Proof. Part (i) follows immediately from the change of variables $\mathbf{y} = \mathbf{x}/\varepsilon$. (See Exercise 20, Chapter 5.) For part (ii), fix $\delta > 0$, and let $\mathbf{y} = \mathbf{x}/\varepsilon$. Then

$$\int_{|\mathbf{x}|>\delta} |K_\varepsilon(\mathbf{x})|\, d\mathbf{x} = \varepsilon^{-n}\int_{|\mathbf{x}|>\delta} \left|K\left(\frac{\mathbf{x}}{\varepsilon}\right)\right| d\mathbf{x} = \int_{|\mathbf{y}|>\delta/\varepsilon} |K(\mathbf{y})|\, d\mathbf{y}.$$

Since $K \in L$ and $\delta/\varepsilon \to +\infty$ as $\varepsilon \to 0$, it follows that the last integral tends to zero as $\varepsilon \to 0$. This completes the proof.

Note that for $K \geq 0$, property (i) above means that the areas under the graphs of K and K_ε are the same, while (ii) means that for small ε the bulk of the area under the graph of K_ε is concentrated in the region above a small neighborhood of the origin.

For any $K \in L$, we can expect from (ii) that the effect of letting $\varepsilon \to 0$ in the formula $(f * K_\varepsilon)(\mathbf{x}) = \int f(\mathbf{x} - \mathbf{t})K_\varepsilon(\mathbf{t})\, d\mathbf{t}$ will be to emphasize the values of $f(\mathbf{x} - \mathbf{t})$ corresponding to small $\mathbf{t}$. In fact, the next four theorems show that $(f * K_\varepsilon)(\mathbf{x}) \to f(\mathbf{x})$ in various senses (e.g., in norm or pointwise) as $\varepsilon \to 0$, if K is suitably restricted. A family $\{K_\varepsilon : \varepsilon > 0\}$ of kernels for which $f * K_\varepsilon \to f$ in some sense is called an *approximation of the identity*.

In what follows, we shall use the notation $f_\varepsilon(\mathbf{x})$ for the convolution $(f * K_\varepsilon)(\mathbf{x})$.

(9.6) Theorem *Let $f_\varepsilon = f * K_\varepsilon$, where $K \in L^1(\mathbf{R}^n)$ and $\int_{\mathbf{R}^n} K = 1$. If $f \in L^p(\mathbf{R}^n)$, $1 \leq p < \infty$, then*

$$\|f_\varepsilon - f\|_p \to 0 \text{ as } \varepsilon \to 0.$$

Proof. By (9.5)(i),

$$f(\mathbf{x}) = f(\mathbf{x})\int_{\mathbf{R}^n} K_\varepsilon(\mathbf{t})\, d\mathbf{t} = \int_{\mathbf{R}^n} f(\mathbf{x})K_\varepsilon(\mathbf{t})\, d\mathbf{t}.$$

Therefore,

$$|f_\varepsilon(\mathbf{x}) - f(\mathbf{x})| = \left|\int_{\mathbf{R}^n} [f(\mathbf{x} - \mathbf{t}) - f(\mathbf{x})]K_\varepsilon(\mathbf{t})\, d\mathbf{t}\right|$$
$$\leq \int_{\mathbf{R}^n} |f(\mathbf{x} - \mathbf{t}) - f(\mathbf{x})|\, |K_\varepsilon(\mathbf{t})|^{1/p} |K_\varepsilon(\mathbf{t})|^{1/p'}\, d\mathbf{t},$$

where $1/p + 1/p' = 1$ ($1/p' = 0$ if $p = 1$). Applying Hölder's inequality with exponents p and p', and then raising both sides to the pth power and integrating with respect to $\mathbf{x}$, we obtain

$$\int_{\mathbf{R}^n} |f_\varepsilon(\mathbf{x}) - f(\mathbf{x})|^p\, d\mathbf{x}$$
$$\leq \int_{\mathbf{R}^n} \left[\int_{\mathbf{R}^n} |f(\mathbf{x} - \mathbf{t}) - f(\mathbf{x})|^p |K_\varepsilon(\mathbf{t})|\, d\mathbf{t}\right]\left[\int_{\mathbf{R}^n} |K_\varepsilon(\mathbf{t})|\, d\mathbf{t}\right]^{p/p'} d\mathbf{x}$$
$$= \|K\|_1^{p/p'} \int_{\mathbf{R}^n} \left[\int_{\mathbf{R}^n} |f(\mathbf{x} - \mathbf{t}) - f(\mathbf{x})|^p |K_\varepsilon(\mathbf{t})|\, d\mathbf{t}\right] d\mathbf{x}.$$

Changing the order of integration in the last expression (which is justified since the integrand is nonnegative), we obtain

$$\|f_\varepsilon - f\|_p^p \leq \|K\|_1^{p/p'} \int_{\mathbf{R}^n} |K_\varepsilon(\mathbf{t})|\phi(\mathbf{t})\, d\mathbf{t},$$

where $\phi(\mathbf{t}) = \int_{\mathbf{R}^n} |f(\mathbf{x} - \mathbf{t}) - f(\mathbf{x})|^p\, d\mathbf{x} = \|f(\mathbf{x} - \mathbf{t}) - f(\mathbf{x})\|_p^p$. For $\delta > 0$, write

$$I_\varepsilon = \int_{\mathbf{R}^n} |K_\varepsilon(\mathbf{t})|\phi(\mathbf{t})\, d\mathbf{t} = \int_{|\mathbf{t}|<\delta} + \int_{|\mathbf{t}|\geq\delta} = A_{\varepsilon,\delta} + B_{\varepsilon,\delta}.$$

Given $\eta > 0$, we can choose δ so small that $\phi(\mathbf{t}) < \eta$ if $|\mathbf{t}| < \delta$ [note that $\phi(\mathbf{t}) \to 0$ as $|\mathbf{t}| \to 0$ by (8.19)]. Then

$$A_{\varepsilon,\delta} \leq \eta \int_{|\mathbf{t}|<\delta} |K_\varepsilon(\mathbf{t})|\, d\mathbf{t} \leq \eta\|K\|_1$$

for all ε. Moreover, ϕ is a bounded function by Minkowski's inequality [note that $\|\phi\|_\infty \leq (2\|f\|_p)^p$], so that $B_{\varepsilon,\delta}$ is less than a constant multiple of $\int_{|\mathbf{t}|\geq\delta} |K_\varepsilon(\mathbf{t})|\, d\mathbf{t}$, which tends to zero with ε. This proves that $I_\varepsilon \to 0$ as $\varepsilon \to 0$, and the theorem follows.

(9.7) Corollary *For $1 \leq p < \infty$, C_0^∞ is dense in $L^p(\mathbf{R}^n)$.*

Proof. Let $f \in L^p$, $1 \leq p < \infty$. Given $\eta > 0$, write $f = g + h$ where g has compact support and $\|h\|_p < \eta$. Choose a kernel $K \in C_0^\infty$ with $\int_{\mathbf{R}^n} K = 1$, and let $g_\varepsilon = g * K_\varepsilon$. Then $g_\varepsilon \in C_0^\infty$ and, by (9.6), $\|g - g_\varepsilon\|_p \to 0$. By Minkowski's inequality, $\|f - g_\varepsilon\|_p \leq \|g - g_\varepsilon\|_p + \|h\|_p < \|g - g_\varepsilon\|_p + \eta$. Choosing ε so that $\|g - g_\varepsilon\|_p < \eta$, we obtain $\|f - g_\varepsilon\|_p < 2\eta$, and the corollary follows.

The next result is a substitute for (9.6) in case $f \in L^\infty$.

(9.8) Theorem *Let $f_\varepsilon = f * K_\varepsilon$, where $f \in L^\infty(\mathbf{R}^n)$, $K \in L^1(\mathbf{R}^n)$ and $\int_{\mathbf{R}^n} K = 1$. Then $f_\varepsilon \to f$ as $\varepsilon \to 0$ at every point of continuity of f.*

Proof. As before,

$$|f_\varepsilon(\mathbf{x}) - f(\mathbf{x})| \le \int_{\mathbf{R}^n} |f(\mathbf{x} - \mathbf{t}) - f(\mathbf{x})||K_\varepsilon(\mathbf{t})|\,dt.$$

If f is continuous at $\mathbf{x}$, then given $\eta > 0$, there exists $\delta > 0$ such that $|f(\mathbf{x} - \mathbf{t}) - f(\mathbf{x})| < \eta$ if $|\mathbf{t}| < \delta$. Hence,

$$\int_{\mathbf{R}^n} |f(\mathbf{x} - \mathbf{t}) - f(\mathbf{x})||K_\varepsilon(\mathbf{t})|\,dt \le \eta \int_{|\mathbf{t}|<\delta} |K_\varepsilon(\mathbf{t})|\,dt + 2\|f\|_\infty \int_{|\mathbf{t}|\ge\delta} |K_\varepsilon(\mathbf{t})|\,dt.$$

Since $\int_{|\mathbf{t}|<\delta} |K_\varepsilon(\mathbf{t})|\,dt \le \|K\|_1$ and $\int_{|\mathbf{t}|>\delta} |K_\varepsilon(\mathbf{t})|\,dt \to 0$ as $\varepsilon \to 0$ for any fixed δ, it follows that $|f_\varepsilon(\mathbf{x}) - f(\mathbf{x})| \to 0$.

Before stating the next result, we introduce some useful notation. If $\psi(\mathbf{x})$ and $\phi(\mathbf{x})$ are defined in a neighborhood of $\mathbf{x}_0$ and if $\phi > 0$ there, we say that

$$\psi(\mathbf{x}) = O(\phi(\mathbf{x})) \text{ as } \mathbf{x} \to \mathbf{x}_0$$

if there is a constant c such that $|\psi(\mathbf{x})/\phi(\mathbf{x})| \le c$ near $\mathbf{x}_0$. If in addition, $\lim_{\mathbf{x}\to\mathbf{x}_0} \psi(\mathbf{x})/\phi(\mathbf{x}) = 0$, we say that

$$\psi(\mathbf{x}) = o(\phi(\mathbf{x})) \text{ as } \mathbf{x} \to \mathbf{x}_0.$$

In particular, the expressions $\psi(\mathbf{x}) = O(1)$ and $\psi(\mathbf{x}) = o(1)$ as $\mathbf{x} \to \mathbf{x}_0$ mean respectively that ψ is bounded and that $\psi \to 0$ as $\mathbf{x} \to \mathbf{x}_0$. The notation most commonly occurs when $\mathbf{x}_0 = 0$ or $\pm\infty$. A similar notation is used when $\mathbf{x}$ is a discontinuous variable, say a sequence of integers tending to $+\infty$. For example, $a_k = O(1)$ and $a_k = o(1)$ as $k \to +\infty$ mean respectively that $\{a_k\}$ is a bounded sequence and that $a_k \to 0$ as $k \to +\infty$.

The following theorem concerns the pointwise convergence of f_ε when $f \in L^1$. See Exercise 12 for the case $f \in L^p$.

(9.9) Theorem *Let $f_\varepsilon = f * K_\varepsilon$, where $f \in L^1(\mathbf{R}^n)$, $K \in L^1(\mathbf{R}^n) \cap L^\infty(\mathbf{R}^n)$, $\int_{\mathbf{R}^n} K = 1$, and $K(\mathbf{x}) = o(|\mathbf{x}|^{-n})$ as $|\mathbf{x}| \to +\infty$. Then $f_\varepsilon \to f$ as $\varepsilon \to 0$ at each point of continuity of f.*

Proof. If f is continuous at $\mathbf{x}$, then given $\eta > 0$, choose $\delta > 0$ such that $|f(\mathbf{x} - \mathbf{t}) - f(\mathbf{x})| < \eta$ if $|\mathbf{t}| < \delta$. As usual,

$$\begin{aligned}|f_\varepsilon(\mathbf{x}) - f(\mathbf{x})| &\le \eta \int_{|\mathbf{t}|<\delta} |K_\varepsilon(\mathbf{t})|\,dt + \int_{|\mathbf{t}|\ge\delta} |f(\mathbf{x} - \mathbf{t}) - f(\mathbf{x})||K_\varepsilon(\mathbf{t})|\,dt \\ &\le \eta\|K\|_1 + \int_{|\mathbf{t}|\ge\delta} |f(\mathbf{x} - \mathbf{t})||K_\varepsilon(\mathbf{t})|\,dt + |f(\mathbf{x})| \int_{|\mathbf{t}|\ge\delta} |K_\varepsilon(\mathbf{t})|\,dt.\end{aligned}$$

The last term on the right tends to zero with ε by (9.5). It is enough, therefore, to show that the second term tends to zero with ε. Write $|K(\mathbf{x})| = \mu(\mathbf{x})|\mathbf{x}|^{-n}$, where $\mu(\mathbf{x}) \to 0$ as $|\mathbf{x}| \to +\infty$. Then

$$\int_{|\mathbf{t}|\geq\delta} |f(\mathbf{x}-\mathbf{t})||K_\varepsilon(\mathbf{t})|\,dt = \int_{|\mathbf{t}|\geq\delta} |f(\mathbf{x}-\mathbf{t})|\mu\left(\frac{\mathbf{t}}{\varepsilon}\right)|\mathbf{t}|^{-n}\,dt$$
$$\leq \delta^{-n}\left\{\sup_{|\mathbf{t}|\geq\delta}\mu\left(\frac{\mathbf{t}}{\varepsilon}\right)\right\}\int_{|\mathbf{t}|\geq\delta}|f(\mathbf{x}-\mathbf{t})|\,dt.$$

Note that $\sup_{|\mathbf{t}|\geq\delta}\mu(\mathbf{t}/\varepsilon) \to 0$ as $\varepsilon \to 0$. Hence, since $\int_{|\mathbf{t}|\geq\delta}|f(\mathbf{x}-\mathbf{t})|\,dt \leq \|f\|_1$, the last expression above tends to zero with ε, and the theorem follows.

There are many classical kernels which satisfy the restrictions we have imposed. Three important examples are listed below for the case $n = 1$.

(9.10) The Poisson kernel. Let

$$K(x) = P(x) = \frac{1}{\pi}\frac{1}{1+x^2} \qquad [x \in (-\infty,+\infty)].$$

Then $P \in L^1(-\infty,+\infty) \cap L^\infty(-\infty,+\infty)$, $\int_{-\infty}^{+\infty} P = 1$, P is positive and $P(x) = o(|x|^{-1})$ as $|x| \to +\infty$. In fact, $P(x) = O(|x|^{-2})$ *as* $|x| \to +\infty$. We have

$$P_\varepsilon(x) = \frac{1}{\varepsilon}P\left(\frac{x}{\varepsilon}\right) = \frac{1}{\pi}\frac{\varepsilon}{\varepsilon^2+|x|^2}, \qquad \varepsilon > 0.$$

P_ε is called the *Poisson kernel*, and the convolution

$$f_\varepsilon(x) = (f * P_\varepsilon)(x) = \frac{1}{\pi}\int_{-\infty}^{+\infty} f(t)\frac{\varepsilon}{\varepsilon^2+(x-t)^2}\,dt$$

is called the *Poisson integral of f*.

Setting $\varepsilon = y$ and letting $f(x,y) = f_y(x)$, we obtain a function $f(x,y)$ defined in the upper half-plane $\{(x,y) : -\infty < x < +\infty, y > 0\}$. Notice that $y/(y^2+x^2)$ is the imaginary part of $-1/z$, $z = x + iy$, and so is harmonic in the upper half-plane; that is, $P_y(x)$ satisfies Laplace's equation

$$\left(\frac{\partial^2}{\partial x^2}+\frac{\partial^2}{\partial y^2}\right)P_y(x) = 0 \qquad \text{if } y > 0.$$

We leave it as an exercise to show that if $f \in L^p$, $1 \leq p \leq \infty$, then

$$\left(\frac{\partial^2}{\partial x^2}+\frac{\partial^2}{\partial y^2}\right)f(x,y) = \int_{-\infty}^{+\infty} f(t)\left(\frac{\partial^2}{\partial x^2}+\frac{\partial^2}{\partial y^2}\right)P_y(x-t)\,dt,$$

so that $(\partial^2/\partial x^2 + \partial^2/\partial y^2)f(x,y) = 0$ for $y > 0$. Hence, $f(x,y)$ is also harmonic in the upper half-plane.

If f is integrable on $(-\infty,+\infty)$, it follows from (9.9) that $f(x,y) \to f(x)$ as

$y \to 0$ wherever f is continuous. Thus, $f(x,y)$ solves the *Dirichlet problem* for the upper half-plane; that is, if $f(x)$ is continuous and integrable on $(-\infty,+\infty)$, then $f(x,y)$ defines a function which is harmonic in the upper half-plane and which tends to $f(x)$ as $y \to 0$.

(9.11) The Fejér kernel Let

$$K(x) = \frac{1}{\pi}\left(\frac{\sin x}{x}\right)^2 \qquad [x \in (-\infty,+\infty)].$$

Then K satisfies the same conditions as in (9.10), and $K_\varepsilon(x) = (1/\pi)\,[\varepsilon \sin^2 (x/\varepsilon)/x^2]$. Setting $w = 1/\varepsilon$, we obtain the *Fejér kernel*

$$F(x,w) = \frac{1}{\pi}\frac{\sin^2 wx}{wx^2}.$$

If $f \in L(-\infty,+\infty)$ and if f is continuous at x, then by (9.9)

$$\lim_{w \to +\infty} \frac{1}{\pi}\int_{-\infty}^{+\infty} f(x-t)\frac{\sin^2 wt}{wt^2}\,dt = f(x).$$

(9.12) The Gauss-Weierstrass kernel The function

$$K(x) = \frac{1}{\sqrt{\pi}}e^{-x^2} \qquad [x \in (-\infty,+\infty)]$$

also satisfies all the required conditions (see Exercise 11, Chapter 6). Here, $K_\varepsilon(x) = (1/\sqrt{\pi}\varepsilon)e^{-x^2/\varepsilon^2}$, and letting $\varepsilon = \sqrt{y}$, $y > 0$, we obtain the *Gauss-Weierstrass kernel*

$$W(x,y) = \frac{1}{\sqrt{\pi y}}e^{-x^2/y}.$$

If f is integrable on $(-\infty,+\infty)$ and continuous at x, then

$$\lim_{y \to 0+} \frac{1}{\sqrt{\pi y}}\int_{-\infty}^{+\infty} f(x-t)e^{-t^2/y}\,dt = f(x).$$

Notice that $W(x,y)$ satisfies the heat equation

$$\frac{\partial^2}{\partial x^2}W = 4\frac{\partial}{\partial y}W.$$

If we strengthen the condition $K(x) = o(|x|^{-n})$, $|x| \to +\infty$, used in (9.9), we can obtain the convergence of f_ε to f almost everywhere. The following result is fairly typical of theorems of this kind. Its hypotheses are met by any of the three examples just listed.

(9.13) Theorem *Suppose that $f \in L(\mathbf{R}^n)$, K is bounded, $K(\mathbf{x}) = O(|\mathbf{x}|)^{-n-\lambda})$*

as $|\mathbf{x}| \to +\infty$ *for some* $\lambda > 0$, *and* $\int_{\mathbf{R}^n} K = 1$. *If* $f_\varepsilon = f * K_\varepsilon$, *then* $f_\varepsilon \to f$ *at each point of the Lebesgue set of* f.

Proof. Let $\mathbf{x}_0$ be a point of the Lebesgue set of f [see (7.14)], so that $\rho^{-n} \int_{|\mathbf{x}|<\rho} |f(\mathbf{x}_0 + \mathbf{x}) - f(\mathbf{x}_0)| \, d\mathbf{x} \to 0$ as $\rho \to 0$. By considering the function $f(\mathbf{x}_0 + \mathbf{x})$, we may assume that $\mathbf{x}_0 = 0$. Since the hypothesis on K implies that $K(\mathbf{x}) = o(|\mathbf{x}|^{-n})$, the conclusion follows from (9.9) if f is continuous at 0. Hence, subtracting from f a continuous function with compact support which equals $f(0)$ at 0, we may suppose that $f(0) = 0$.

The hypotheses $|K(\mathbf{x})| \le M$ and $K(\mathbf{x}) = O(|\mathbf{x}|^{-n-\lambda})$ can be combined into a single estimate:

$$|K(\mathbf{x})| \le \frac{M_1}{(1 + |\mathbf{x}|)^{n+\lambda}}.$$

Hence,

$$|K_\varepsilon(\mathbf{x})| \le M_1 \frac{\varepsilon^\lambda}{(\varepsilon + |\mathbf{x}|)^{n+\lambda}}.$$

Therefore,

$$|f_\varepsilon(0)| \le M_1 \int_{\mathbf{R}^n} |f(\mathbf{x})| \frac{\varepsilon^\lambda}{(\varepsilon + |\mathbf{x}|)^{n+\lambda}} \, d\mathbf{x},$$

and it remains to show that the integral tends to zero. We will use the following lemma, which is of some independent interest.

(9.14) Lemma *Suppose that* $f(\mathbf{x})$ *is integrable over a spherical shell* $a \le |\mathbf{x}| \le b$ *and that* $\phi(\rho)$ *is continuous for* $a \le \rho \le b$, $0 \le a < b < +\infty$. *Let* $F(\rho) = \int_{a \le |\mathbf{x}| \le \rho} f(\mathbf{x}) \, d\mathbf{x}$ *for* $a \le \rho \le b$. *Then*

$$\int_{a \le |\mathbf{x}| \le b} f(\mathbf{x}) \phi(|\mathbf{x}|) \, d\mathbf{x} = \int_a^b \phi(\rho) \, dF(\rho),$$

the integral on the right being a Riemann-Stieltjes integral.

Proof. Note that this reduces to the formula in (7.32)(i) in case $n = 1$. In any case, writing $f = f^+ - f^-$, we see that F is the difference of two bounded increasing functions. Hence, F is of bounded variation on $[a,b]$ and $\int_a^b \phi \, dF$ is well-defined. We may assume that $f \ge 0$. Let $I = \int_{a \le |\mathbf{x}| \le b} f(\mathbf{x}) \phi(|\mathbf{x}|) \, d\mathbf{x}$, and let $\{a = \rho_0 < \rho_1 < \cdots < \rho_k = b\}$ be a partition of $[a,b]$. Then

$$I = \sum_{i=1}^{k} \int_{\rho_{i-1} \le |\mathbf{x}| \le \rho_i} f(\mathbf{x}) \phi(|\mathbf{x}|) \, d\mathbf{x},$$

and since $f \ge 0$,

$$\sum_{i=1}^{k} m_i \int_{\rho_{i-1} \le |\mathbf{x}| \le \rho_i} f(\mathbf{x}) \, d\mathbf{x} \le I \le \sum_{i=1}^{k} M_i \int_{\rho_{i-1} \le |\mathbf{x}| \le \rho_i} f(\mathbf{x}) \, d\mathbf{x},$$

where m_i and M_i are respectively the minimum and maximum of ϕ in $[\rho_{i-1},\rho_i]$. This can be rewritten

$$\sum_{i=1}^{k} m_i[F(\rho_i) - F(\rho_{i-1})] \le I \le \sum_{i=1}^{k} M_i[F(\rho_i) - F(\rho_{i-1})].$$

By (2.24), the extreme terms in this inequality both converge to $\int_a^b \phi(\rho)\, dF(\rho)$ as the norm of the partition tends to zero, and the lemma follows.

Returning to the proof of (9.13), let $F(\rho) = \int_{|\mathbf{x}|\le\rho} |f(\mathbf{x})|\, d\mathbf{x}$. The hypotheses that $\mathbf{x}_0 = 0$ is a Lebesgue point of f and that $f(0) = 0$ imply that given $\eta > 0$, there is a $\delta > 0$ such that $F(\rho) < \eta\rho^n$ if $\rho \le \delta$. Write

$$\int_{\mathbf{R}^n} |f(\mathbf{x})| \frac{\varepsilon^\lambda}{(\varepsilon + |\mathbf{x}|)^{n+\lambda}}\, d\mathbf{x} = \int_{|\mathbf{x}|\le\delta} + \int_{|\mathbf{x}|>\delta} = A + B.$$

Taking $\phi(\rho) = \varepsilon^\lambda/(\varepsilon + \rho)^{n+\lambda}$ and $[a,b] = [0,\delta]$ in (9.14), we have

$$A = \int_0^\delta \frac{\varepsilon^\lambda}{(\varepsilon + \rho)^{n+\lambda}}\, dF(\rho).$$

Integrating by parts and observing that $F(0) = 0$, we obtain

$$A = \frac{\varepsilon^\lambda}{(\varepsilon + \delta)^{n+\lambda}} F(\delta) + (n + \lambda) \int_0^\delta F(\rho) \frac{\varepsilon^\lambda}{(\varepsilon + \rho)^{n+\lambda+1}}\, d\rho.$$

The first term on the right tends to zero as $\varepsilon \to 0$. The definition of δ and the change of variables $\rho = \varepsilon t$ show that the second term is at most

$$(n + \lambda)\eta \int_0^\delta \rho^n \frac{\varepsilon^\lambda}{(\varepsilon + \rho)^{n+\lambda+1}}\, d\rho = (n + \lambda)\eta \int_0^{\delta/\varepsilon} \frac{t^n}{(1 + t)^{n+\lambda+1}}\, dt.$$

Hence,

$$\limsup_{\varepsilon\to 0} A \le (n + \lambda)\eta \int_0^\infty \frac{t^n}{(1 + t)^{n+\lambda+1}} = c\eta,$$

where $c(= c_{\lambda,n})$ is finite since $\lambda > 0$.

Finally, to estimate B, note that if $|\mathbf{x}| > \delta$ then $\varepsilon + |\mathbf{x}| > \delta$, so that

$$B \le \frac{\varepsilon^\lambda}{\delta^{n+\lambda}} \int_{|\mathbf{x}|>\delta} |f(\mathbf{x})|\, d\mathbf{x} \le \frac{\varepsilon^\lambda}{\delta^{n+\lambda}} \|f\|_1.$$

Hence, $\lim_{\varepsilon\to 0} B = 0$. Combining these estimates, we obtain $\limsup_{\varepsilon\to 0} (A + B) \le c\eta$, and the theorem follows by letting $\eta \to 0$. See Exercise 12 for the case $f \in L^p$, $p > 1$.

The kernels $\{K_\varepsilon\}$ for K satisfying the kinds of conditions above are examples of approximations of the identity.

3. The Hardy-Littlewood Maximal Function

Let f^* denote the Hardy-Littlewood maximal function of f:

$$f^*(\mathbf{x}) = \sup \frac{1}{|Q|} \int_Q |f(\mathbf{y})| \, d\mathbf{y},$$

where the supremum is taken over all cubes Q with center $\mathbf{x}$ and edges parallel to the coordinate axes [see (7.5)].

We observed in Chapter 7, p. 105, that f^* is not integrable over $\mathbf{R}^n$ (unless $f = 0$ a.e.), but does satisfy the weak-type condition

(9.15) $$|\{\mathbf{x} \in \mathbf{R}^n : f^*(\mathbf{x}) > \alpha\}| \le \frac{c}{\alpha} \|f\|_1 \qquad (\alpha > 0),$$

where c depends only on n [lemma (7.9)]. The behavior of f^* on the other L^p spaces, $1 < p \le \infty$, turns out to be better. For example, it is clear from the definition of f^* that $f^*(\mathbf{x}) \le \|f\|_\infty$ for all x. Thus, f^* is bounded if f is, and $\|f^*\|_\infty \le \|f\|_\infty$. The following theorem describes the behavior of f^* when $f \in L^p$.

(9.16) Theorem *Let $1 < p \le \infty$ and $f \in L^p(\mathbf{R}^n)$. Then $f^* \in L^p(\mathbf{R}^n)$ and*

$$\|f^*\|_p \le c\|f\|_p,$$

where c depends only on n and p.

Proof. We may assume that $1 < p < \infty$. The idea is to obtain information for L^p by interpolating between the known results for L^1 and L^∞. Thus, for $\alpha > 0$, let

$$\omega(\alpha) = |\{\mathbf{x} \in \mathbf{R}^n : f^*(\mathbf{x}) > \alpha\}|$$

denote the distribution function of f^*. Fix $\alpha > 0$ and define a function g by $g(\mathbf{x}) = f(\mathbf{x})$ when $|f(\mathbf{x})| \ge \alpha/2$ and $g(\mathbf{x}) = 0$ otherwise. Then since $|f(\mathbf{x})| \le |g(\mathbf{x})| + \alpha/2$,

$$f^*(\mathbf{x}) \le \sup \frac{1}{|Q|} \int_Q |g(\mathbf{y})| \, d\mathbf{y} + \frac{\alpha}{2} = g^*(\mathbf{x}) + \frac{\alpha}{2}.$$

In particular,

$$\{\mathbf{x} \in \mathbf{R}^n : f^*(\mathbf{x}) > \alpha\} \subset \left\{\mathbf{x} \in \mathbf{R}^n : g^*(\mathbf{x}) > \frac{\alpha}{2}\right\},$$

so that, by (9.15),

$$\begin{aligned} \omega(\alpha) &\le \left|\left\{\mathbf{x} \in \mathbf{R}^n : g^*(\mathbf{x}) > \frac{\alpha}{2}\right\}\right| \\ &\le \frac{2c}{\alpha} \|g\|_1 = \frac{2c}{\alpha} \int_{\{\mathbf{x} \in \mathbf{R}^n : |f(\mathbf{x})| \ge \alpha/2\}} |f(\mathbf{x})| \, d\mathbf{x}. \end{aligned}$$

We have the formula $\int_{\mathbf{R}^n} f^{*p}\,d\mathbf{x} = p\int_0^\infty \alpha^{p-1}\omega(\alpha)\,d\alpha$, which was stated on two occasions: Exercise 16, Chapter 5, and Exercise 5, Chapter 6. Hence

$$\int_{\mathbf{R}^n} f^{*p}\,d\mathbf{x} \le p\int_0^\infty \alpha^{p-1}\left[\frac{2c}{\alpha}\int_{\{\mathbf{x}\in\mathbf{R}^n:|f(\mathbf{x})|\ge\alpha/2\}} |f(\mathbf{x})|\,d\mathbf{x}\right]d\alpha.$$

Interchanging the order of integration in the expression on the right (which is justified since the integrand is nonnegative), we obtain

$$\int_{\mathbf{R}^n} f^{*p}\,d\mathbf{x} \le 2cp\int_{\mathbf{R}^n} |f(\mathbf{x})|\left(\int_0^{2|f(\mathbf{x})|} \alpha^{p-2}\,d\alpha\right)d\mathbf{x}.$$

Since $p-2 > -1$ (i.e., $p > 1$), the inner integral equals $(2|f(\mathbf{x})|)^{p-1}/(p-1)$ a.e. (wherever $f(\mathbf{x})$ is finite), so that

$$\int_{\mathbf{R}^n} f^{*p}\,d\mathbf{x} \le \frac{2^p pc}{p-1}\int_{\mathbf{R}^n} |f(\mathbf{x})|^p\,d\mathbf{x} = \frac{2^p pc}{p-1}\|f\|_p^p.$$

Taking pth roots, we see that $\|f^*\|_p \le A_p\|f\|_p$, where $A_p^p = 2^p pc/(p-1)$. This completes the proof. Note that the constant A_p tends to $+\infty$ as $p \to 1$, and is bounded as $p \to \infty$.

The Hardy-Littlewood maximal function plays an important role in many parts of analysis concerned with operator theory and differentiation. It arose naturally in Chapter 7 in connection with Lebesgue's differentiation theorem. As another illustration of its usefulness, we have the following result.

(9.17) Theorem *Let $K(\mathbf{x})$ be nonnegative and integrable on $\mathbf{R}^n$, and suppose that $K(\mathbf{x})$ depends only on $|\mathbf{x}|$ and decreases as $|\mathbf{x}|$ increases [that is, $K(\mathbf{x}) = \phi(|\mathbf{x}|)$, where $\phi(t)$, $t > 0$, is monotone decreasing]. Then*

$$\sup_{\varepsilon>0} |(f * K_\varepsilon)(\mathbf{x})| \le cf^*(\mathbf{x}),$$

with c independent of f.

Proof. We first remark that there is a constant c depending only on n such that

$$\sup_{\delta>0} \delta^{-n}\int_{|\mathbf{y}|<\delta} |f(\mathbf{x}-\mathbf{y})|\,d\mathbf{y} \le cf^*(\mathbf{x}).$$

This follows by enclosing the ball $|\mathbf{y}| < \delta$ in the cube with center 0 and edge 2δ and observing that the ratio of the two volumes is bounded independent of δ.

To prove the result, we will use a method based on Fubini's theorem. Fix ε, and let $E = \{(\mathbf{y},t) : \mathbf{y}\in\mathbf{R}^n,\ t > 0,\ K_\varepsilon(\mathbf{y}) > t\}$. Then E is a measurable subset of $\mathbf{R}^{n+1}$ by (5.1), and

$$K_\varepsilon(\mathbf{y}) = \int_0^{K_\varepsilon(\mathbf{y})} dt = \int_0^\infty \chi_E(\mathbf{y},t)\,dt.$$

Hence,

$$|(f * K_\varepsilon)(\mathbf{x})| = \left|\int_{\mathbf{R}^n} f(\mathbf{x}-\mathbf{y})K_\varepsilon(\mathbf{y})\,d\mathbf{y}\right| \le \int_{\mathbf{R}^n} |f(\mathbf{x}-\mathbf{y})|\left[\int_0^\infty \chi_E(\mathbf{y},t)\,dt\right]d\mathbf{y}.$$

Changing the order of integration in the last expression, we obtain

$$\begin{aligned}|(f * K_\varepsilon)(\mathbf{x})| &\le \int_0^\infty \left[\int_{\mathbf{R}^n} |f(\mathbf{x}-\mathbf{y})|\chi_E(\mathbf{y},t)\,d\mathbf{y}\right]dt\\ &= \int_0^\infty \left[\int_{\{\mathbf{y}:K_\varepsilon(\mathbf{y})>t\}} |f(\mathbf{x}-\mathbf{y})|\,d\mathbf{y}\right]dt.\end{aligned}$$

Let $E_t = \{\mathbf{y} : K_\varepsilon(\mathbf{y}) > t\}$, $t > 0$. Since $K(\mathbf{y})$ depends only on $|\mathbf{y}|$ and decreases as $|\mathbf{y}|$ increases, E_t is a ball with center 0. Hence, by our earlier remark,

$$\begin{aligned}|(f * K_\varepsilon)(\mathbf{x})| &\le \int_0^\infty |E_t|\left[\frac{1}{|E_t|}\int_{E_t} |f(\mathbf{x}-\mathbf{y})|\,d\mathbf{y}\right]dt\\ &\le \int_0^\infty |E_t| cf^*(\mathbf{x})\,dt = cf^*(\mathbf{x})\int_0^\infty |E_t|\,dt.\end{aligned}$$

Finally, note that $|E_t|$ is the distribution function of K_ε, so that

$$\int_0^\infty |E_t|\,dt = \|K_\varepsilon\|_1 = \|K\|_1.$$

Therefore, $|(f * K_\varepsilon)(\mathbf{x})| \le c\|K\|_1 f^*(\mathbf{x})$, and the theorem follows by taking the sup over $\varepsilon > 0$.

We leave it to the reader to show that (9.17) can also be derived from (9.14); see, e.g., the proof of (12.61) in Chapter 12.

In particular, for the kernel $K(\mathbf{x}) = 1/(1 + |\mathbf{x}|^{n+\lambda})$, $\lambda > 0$, Theorem (9.17) gives

$$\textbf{(9.18)}\qquad \sup_{\varepsilon>0}\left|\int_{\mathbf{R}^n} f(\mathbf{x}-\mathbf{y})\frac{\varepsilon^\lambda}{\varepsilon^{n+\lambda} + |\mathbf{y}|^{n+\lambda}}\,d\mathbf{y}\right| \le cf^*(\mathbf{x}),$$

a fact which will be used in the next section.

Note that the conclusion of (9.17) is valid for any K which is majorized in absolute value by a kernel satisfying the hypothesis of (9.17). This includes any K satisfying the hypothesis of (9.13).

4. The Marcinkiewicz Integral

We recall from Chapter 6, (6.17), that if F is a closed subset of a bounded open interval (a,b) in $\mathbf{R}^1$, and if $\delta(x)$ denotes the distance from x to F, then the Marcinkiewicz integral

$$M_\lambda(x) = \int_a^b \frac{\delta^\lambda(y)}{|x-y|^{1+\lambda}}\,dy \qquad (\lambda > 0)$$

is integrable over F. More generally, in Exercise 7 of Chapter 6, we considered the expression

$$\int_{\mathbf{R}^1} \frac{\delta^\lambda(y)f(y)}{|x-y|^{1+\lambda}}\,dy \qquad (\lambda > 0),$$

where f is nonnegative and integrable over the complement of F. If $f = \chi_{(a,b)}$, this reduces to $M_\lambda(x)$.

Now consider the integral

$$J_\lambda(f)(\mathbf{x}) = \int_{\mathbf{R}^n} \frac{\delta^\lambda(\mathbf{y})f(\mathbf{y})}{|\mathbf{x}-\mathbf{y}|^{n+\lambda}}\,d\mathbf{y} \qquad (\mathbf{x} \in \mathbf{R}^n),$$

and the modified form

$$H_\lambda(f)(\mathbf{x}) = \int_{\mathbf{R}^n} \frac{\delta^\lambda(\mathbf{y})f(\mathbf{y})}{|\mathbf{x}-\mathbf{y}|^{n+\lambda} + \delta^{n+\lambda}(\mathbf{x})}\,d\mathbf{y}.$$

Here again, $\lambda > 0$, $\delta(\mathbf{x})$ denotes the distance from $\mathbf{x}$ to a closed set $F \subset \mathbf{R}^n$, and f is nonnegative and measurable on $\mathbf{R}^n$. Notice that $H_\lambda(f)$ and $J_\lambda(f)$ are equal in F since δ is zero there. For the same reason, $J_\lambda(f)$ and $H_\lambda(f)$ are independent of the values of f on F. Therefore, we may assume for simplicity that $f = 0$ on F.

We will prove below that if $f \in L^p(\mathbf{R}^n - F)$, $1 \le p < \infty$, then $H_\lambda(f) \in L^p(\mathbf{R}^n)$. This implies the basic fact that $J_\lambda(f) \in L^p(F)$. (In general, $J_\lambda(f)$ diverges outside F: see, for example, Exercise 9 of Chapter 6.) For the proof, it will be convenient to consider one more modification of $J_\lambda(f)$, namely,

$$H'_\lambda(f)(\mathbf{x}) = \int_{\mathbf{R}^n} \frac{\delta^\lambda(\mathbf{y})f(\mathbf{y})}{|\mathbf{x}-\mathbf{y}|^{n+\lambda} + \delta^{n+\lambda}(\mathbf{y})}\,d\mathbf{y}.$$

As we move from a point $\mathbf{x}$ to another point $\mathbf{y}$, the distance from F does not increase by more than $|\mathbf{x}-\mathbf{y}|$. Hence,

$$|\delta(\mathbf{x}) - \delta(\mathbf{y})| \le |\mathbf{x}-\mathbf{y}|.$$

It follows that $\delta(\mathbf{y}) \le |\mathbf{x}-\mathbf{y}| + \delta(\mathbf{x})$, so that we have

$$\delta^{n+\lambda}(\mathbf{y}) \le 2^{n+\lambda}[|\mathbf{x}-\mathbf{y}|^{n+\lambda} + \delta^{n+\lambda}(\mathbf{x})],$$

as well as a similar inequality with $\mathbf{x}$ and $\mathbf{y}$ interchanged. We immediately obtain that

$$2^{-n-\lambda-1}H'_\lambda(f)(\mathbf{x}) \le H_\lambda(f)(\mathbf{x}) \le 2^{n+\lambda+1}H'_\lambda(f)(\mathbf{x}).$$

Thus, inequalities for H'_λ lead to ones for H_λ, but H'_λ is easier to deal with.

(9.19) Theorem *If $f \in L^p(\mathbf{R}^n)$, $1 \le p < \infty$, and $\lambda > 0$, then $H_\lambda(f) \in L^p(\mathbf{R}^n)$ and*

$$\|H_\lambda(f)\|_p \le c\|f\|_p,$$

where c is independent of f. In particular, $\|J_\lambda(f)\|_{p,F} \le c\|f\|_p$.

Proof. Fix p, $1 \le p < \infty$, and let g be any nonnegative function with $\|g\|_{p'} \le 1$, where $1/p + 1/p' = 1$. By interchanging the order of integration, we obtain

$$\int_{\mathbf{R}^n} H'_\lambda(f)(\mathbf{x})g(\mathbf{x})\,d\mathbf{x} = \int_{\mathbf{R}^n} f(\mathbf{y})\delta^\lambda(\mathbf{y})\left[\int_{\mathbf{R}^n} \frac{g(\mathbf{x})}{|\mathbf{x}-\mathbf{y}|^{n+\lambda} + \delta^{n+\lambda}(\mathbf{y})}\,d\mathbf{x}\right] d\mathbf{y}.$$

The outer integration on the right can be restricted to $\mathbf{R}^n - F$ without changing the value of the integral. However, if $\mathbf{y} \in \mathbf{R}^n - F$, then $\delta(\mathbf{y}) > 0$, and it follows from (9.18) that the inner integral on the right is bounded by $c\delta^{-\lambda}(\mathbf{y})g^*(\mathbf{y})$. Combining this estimate with Hölder's inequality and (9.16), we obtain

$$\begin{aligned}\int_{\mathbf{R}^n} H'_\lambda(f)(\mathbf{x})g(\mathbf{x})\,d\mathbf{x} &\le c\int_{\mathbf{R}^n} f(\mathbf{y})g^*(\mathbf{y})\,d\mathbf{y}\\ &\le c\|f\|_p\|g^*\|_{p'}\\ &\le c_1\|f\|_p\|g\|_{p'} \le c_1\|f\|_p.\end{aligned}$$

By (8.9), the supremum of the left side for all such g is $\|H'_\lambda(f)\|_p$, so that $\|H'_\lambda(f)\|_p \le c_1\|f\|_p$, and the theorem follows.

Exercises

1. Use Minkowski's integral inequality (see Exercise 8, Chapter 8) to prove (9.1).
2. Prove Young's theorem (9.2). [For $f,g \ge 0$ and $p,q,r < \infty$, write

$$(f*g)(\mathbf{x}) = \int f(\mathbf{t})^{p/r}g(\mathbf{x}-\mathbf{t})^{q/r}\cdot f(\mathbf{t})^{p(1/p-1/r)}\cdot g(\mathbf{x}-\mathbf{t})^{q(1/q-1/r)}\,d\mathbf{t},$$

and apply Hölder's inequality for three functions (Exercise 6, Chapter 8) with exponents r, p_1, and p_2, where $1/p_1 = 1/p - 1/r$, $1/p_2 = 1/q - 1/r$.]
3. Show that if $f \in L^p(\mathbf{R}^n)$ and $K \in L^{p'}(\mathbf{R}^n)$, $1 \le p \le \infty$, $1/p + 1/p' = 1$, then $f*K$ is bounded and continuous in $\mathbf{R}^n$.
4. (a) Show that the function h defined by $h(x) = e^{-1/x^2}$ for $x > 0$ and $h(x) = 0$ for $x \le 0$ is in C^∞.
   (b) Show that the function $g(x) = h(x-a)h(b-x)$, $a < b$, is C^∞ with support $[a,b]$.
   (c) Construct a function in $C_0^\infty(\mathbf{R}^n)$ whose support is a ball or an interval.

5. Let G and G_1 be bounded open subsets of $\mathbf{R}^n$ such that $\bar{G}_1 \subset G$. Construct a function $h \in C_0^\infty$ such that $h = 1$ in G_1 and $h = 0$ outside G. (Choose an open G_2 such that $\bar{G}_1 \subset G_2$, $\bar{G}_2 \subset G$. Let $h = \chi_{G_2} * K$ for a $K \in C^\infty$ with suitably small support and $\int K = 1$.)
6. Prove theorem (9.4).
7. Let $f \in L^p(-\infty,+\infty)$, $1 \le p \le \infty$. Show that the Poisson integral of f, $f(x,y)$, is harmonic in the upper half-plane $y > 0$. [Show that $((\partial^2/\partial x^2) + (\partial^2/\partial y^2))f(x,y) = \int_{-\infty}^{+\infty} f(t)((\partial^2/\partial x^2) + (\partial^2/\partial y^2))P_y(x - t)\,dt$.]
8. (*Schur's lemma*) For $s,t \ge 0$, let $K(s,t)$ satisfy $K \ge 0$ and $K(\lambda s,\lambda t) = \lambda^{-1}K(s,t)$ for all $\lambda > 0$, and suppose that $\int_0^\infty t^{-1/p}K(1,t)\,dt = \gamma < +\infty$ for some $p, 1 \le p \le \infty$. For example, $K(s,t) = 1/(s + t)$ has these properties. Show that if

$$(Tf)(s) = \int_0^\infty f(t)K(s,t)\,dt \qquad (f \ge 0),$$

then $\|Tf\|_p \le \gamma\|f\|_p$. [Note that $K(s,t) = s^{-1}K(1,t/s)$, so that $(Tf)(s) = \int_0^\infty f(ts)K(1,t)\,dt$. Now apply Minkowski's integral inequality (see Exercise 8, Chapter 8).]
9. The maximal function is defined as $f^*(\mathbf{x}) = \sup |Q|^{-1}\int_Q |f|$, where the supremum is taken over cubes Q with center $\mathbf{x}$. Let $f^{**}(\mathbf{x})$ be defined similarly, but with the supremum taken over all Q *containing* $\mathbf{x}$. Thus, $f^*(\mathbf{x}) \le f^{**}(\mathbf{x})$. Show that there is a positive constant c depending only on the dimension such that $f^{**}(\mathbf{x}) \le cf^*(\mathbf{x})$.
10. Let $T : f \to Tf$ be a function transformation which is *sublinear*; that is, T has the property that if Tf_1 and Tf_2 are defined, then so is $T(f_1 + f_2)$, and

$$|T(f_1 + f_2)(\mathbf{x})| \le |(Tf_1)(\mathbf{x})| + |(Tf_2)(\mathbf{x})|.$$

Suppose also that there are constants c_1 and c_2 such that T satisfies $\|Tf\|_\infty \le c_1\|f\|_\infty$ and $|\{\mathbf{x}: |(Tf)(\mathbf{x})| > \alpha\}| \le c_2\alpha^{-1}\|f\|_1$, $\alpha > 0$. Show that for $1 < p < \infty$, there is a constant c_3 such that $\|Tf\|_p \le c_3\|f\|_p$. This is a special case of an interpolation result due to Marcinkiewicz. (An example of such a T is the maximal function operator $Tf = f^*$, and the proof in the general case is like that for f^*.)
11. Generalize theorem (9.6) as follows: Let $f_\varepsilon = f * K_\varepsilon$, $K \in L^1(\mathbf{R}^n)$ and $\int_{\mathbf{R}^n} K = \gamma$. If $f \in L^p(\mathbf{R}^n)$, $1 \le p < \infty$, show that $\|f_\varepsilon - \gamma f\|_p \to 0$. Derive analogous results for (9.8), (9.9), and (9.13). [The case $\gamma \ne 0$ follows from the case $\gamma = 1$ by considering $K(\mathbf{x})/\gamma$.]
12. Show that the conclusions of (9.9) and (9.13) remain true if the assumption that $f \in L^1$ is replaced by $f \in L^p$, $p > 1$.
13. Let $f \in L^p(0,1)$, $1 \le p < \infty$, and for each $k = 1, 2, \ldots$, define a function f_k on $(0,1)$ by letting $I_{k,j} = \{x: (j - 1)2^{-k} \le x < j2^{-k}\}$, $j = 1, \ldots, 2^k$, and setting $f_k(x)$ equal to $|I_{k,j}|^{-1}\int_{I_{k,j}} f$ for $x \in I_{k,j}$. Prove that $f_k \to f$ in $L^p(0,1)$ norm. [Exercise 18 of Chapter 7 may be helpful for the case $p = 1$.]

Chapter 10

Abstract Integration

In the preceding chapters, we developed a theory of integration based on a theory of measurable sets. The notion of the measure of a set was in turn based on the primitive and classical notion of the measure (or volume) of an interval in $\mathbf{R}^n$; this led almost automatically by the process of covering to the notion of measure for more general sets.

In this chapter, we follow an alternate approach. We will consider a family of sets in $\mathbf{R}^n$ and assume that they all have "measures", that is, assume that with each member of the family, we can associate a nonnegative number satisfying elementary and natural requirements which justify calling it a measure. Starting with this assumption, we will develop a theory of integration which follows the pattern of Lebesgue integration. The advantage of this method is that it can be applied not only to $\mathbf{R}^n$, but also to general abstract spaces with much less geometric structure than $\mathbf{R}^n$. Thus, it is important for applications. There are new questions which arise in the abstract setting, but many of the theorems and proofs are practically the same as those for Lebesgue measure in $\mathbf{R}^n$. In such cases, we will usually refer to earlier chapters for proofs.

It is natural to ask how we can construct such measures. One possible approach is to start with the more elementary notion of an "outer measure" in an abstract space, and, as in the case of $\mathbf{R}^n$ discussed in Chapter 3, select a subclass of sets on which the outer measure has additional properties, qualifying it as a measure. This idea will be developed in Chapter 11.

1. Additive Set Functions; Measures

Let $\mathscr{S}$ be a fixed set, and let Σ be a *σ-algebra* of subsets of $\mathscr{S}$; that is, let Σ satisfy

(a) $\mathscr{S} \in \Sigma$.

(b) If $E \in \Sigma$, then its complement CE $(= \mathscr{S} - E) \in \Sigma$ (i.e., Σ is closed under complements).

(c) If $E_k \in \Sigma$ for $k = 1, 2, \ldots$, then $\bigcup E_k \in \Sigma$ (i.e., Σ is closed under countable unions).

It is easy to see that the definition is unchanged if condition (a) is replaced by the assumption that Σ be nonempty; see also p. 39. Another widely used term for a σ-algebra is a *countably additive family of sets.*

Immediate consequences of the definition are that the following sets belong to Σ:

(1) The empty set $\varnothing$ $(= C\mathscr{S})$;

(2) $\bigcap E_k$ if $E_k \in \Sigma$, $k = 1, 2, \ldots$;

(3) $\limsup E_k$ $(= \bigcap_{m=1}^{\infty} \bigcup_{k=m}^{\infty} E_k)$ and $\liminf E_k$ $(= \bigcup_{m=1}^{\infty} \bigcap_{k=m}^{\infty} E_k)$ if each $E_k \in \Sigma$;

(4) $E_1 - E_2$ $(= E_1 \cap CE_2)$ if $E_1, E_2 \in \Sigma$.

We recall the basic fact that the collection of Lebesgue measurable subsets of $\mathbf{R}^n$ is a σ-algebra: see (3.20). In general, the elements E of a σ-algebra Σ are called Σ-*measurable* sets, or simply *measurable sets* if it is clear from context what Σ is.

If Σ is a σ-algebra, then a real-valued function $\phi(E)$, $E \in \Sigma$, is called an *additive set function* on Σ if

(i) $\phi(E)$ is finite for every $E \in \Sigma$, and

(ii) $\phi(\bigcup E_k) = \Sigma\phi(E_k)$ for every countable family $\{E_k\}$ of disjoint sets in Σ.

Since $\bigcup E_k$ is independent of the order of the E_k's, the series in (ii) converges absolutely.

We obtain a simple example of a set function by letting Σ be the σ-algebra of all subsets of $\mathscr{S}$ and defining $\phi(E) = \chi_E(x_0)$ for a fixed $x_0 \in \mathscr{S}$. As another example, let Σ be the collection of all Lebesgue measurable subsets of $\mathbf{R}^n$, and define $\phi(E) = \int_E f$, where $f \in L(\mathbf{R}^n)$.

A function $\mu(E)$ defined for E in Σ is called a *measure* on Σ if

(i) $0 \le \mu(E) \le +\infty$, and

(ii) $\mu(\bigcup E_k) = \Sigma\mu(E_k)$ for every countable family $\{E_k\}$ of disjoint sets in Σ.

The choices $\mu \equiv 0$ or $\mu \equiv +\infty$ are always possible, but of little interest.

If μ is a measure on Σ, then the triplet $(\mathscr{S},\Sigma,\mu)$ is called a *measure space.* For example, Lebesgue measure together with the class of Lebesgue measurable subsets of $\mathbf{R}^n$ is a measure space. As another example, let $\mathscr{S}$ be any countable set, $\mathscr{S} = \{x_k\}$, and let $\{a_k\}$ be a sequence of nonnegative numbers. Let Σ be the family of all subsets of $\mathscr{S}$, and define $\delta(E) = \sum a_{k_j}$ if $E = \{x_{k_j}\}$. Then $(\mathscr{S},\Sigma,\delta)$ is a measure space. Such a space is called a *discrete* measure space.

The distinction between a measure and an additive set function is that a measure is nonnegative, but may be infinite, while an additive set function may take both positive and negative values, but is finite. There are similarities between many of the properties of set functions and measures. If $E_1 \subset E_2$ and μ is a measure, then $\mu(E_2 - E_1) + \mu(E_1) = \mu(E_2)$, so that $\mu(E_2 - E_1) = \mu(E_2) - \mu(E_1)$ if $\mu(E_1)$ is finite. If ϕ is an additive set function, then the formula $\phi(E_2 - E_1) = \phi(E_2) - \phi(E_1)$, $E_1 \subset E_2$, always holds. Choosing $E_1 = E_2$, we see that $\phi(\varnothing) = 0$ for an additive set function, and also $\mu(\varnothing) = 0$ for a measure, unless $\mu(E) = +\infty$ for all E. Moreover, if $E_1 \subset E_2$, then $\mu(E_1) \leq \mu(E_2)$ even if $\mu(E_1) = +\infty$, and $\phi(E_1) \leq \phi(E_2)$ if $\phi \geq 0$.

The next few results concern limit properties and a basic decomposition for additive set functions. Both $\mathscr{S}$ and Σ are fixed.

(10.1) Theorem *If $\{E_k\}$ is a monotone sequence of sets in Σ (that is, if $E_k \nearrow E$ or $E_k \searrow E$) and ϕ is an additive set function, then $\phi(E) = \lim_{k\to\infty} \phi(E_k)$.*

Proof. If $E_k \nearrow E$, then $E = \bigcup E_k = E_1 \cup (E_2 - E_1) \cup (E_3 - E_2) \cup \cdots$. Hence,

$$\phi(E) = \phi(E_1) + \sum_{k=2}^{\infty} \phi(E_k - E_{k-1}) = \phi(E_1) + \lim_{N\to\infty} \sum_{k=2}^{N} [\phi(E_k) - \phi(E_{k-1})]$$
$$= \lim_{N\to\infty} \phi(E_N).$$

On the other hand, if $E_k \searrow E$, then $\mathscr{S} - E_k \nearrow \mathscr{S} - E$. Therefore, by the case already considered, we have $\phi(\mathscr{S} - E_k) \to \phi(\mathscr{S} - E)$. Since $\phi(\mathscr{S} - E_k) = \phi(\mathscr{S}) - \phi(E_k)$ and $\phi(\mathscr{S} - E) = \phi(\mathscr{S}) - \phi(E)$, the result follows.

The next theorem is similar to Fatou's lemma (5.17).

(10.2) Theorem *Let ϕ be a nonnegative additive set function, and let $\{E_k\}$ be any sequence of sets in Σ. Then*

$$\phi(\liminf E_k) \leq \liminf_{k\to\infty} \phi(E_k) \leq \limsup_{k\to\infty} \phi(E_k) \leq \phi(\limsup E_k).$$

Proof. The sets $H_m = \bigcap_{k=m}^{\infty} E_k$ increase to $\liminf E_k$. Therefore, by the preceding theorem, $\phi(\liminf E_k) = \lim \phi(H_m)$. Since $H_m \subset E_m$, we have $\phi(H_m) \leq \phi(E_m)$ and $\lim \phi(H_m) \leq \liminf \phi(E_m)$. Therefore, $\phi(\liminf E_k) \leq \liminf \phi(E_m)$, which proves the first inequality. The proof of the third one is similar, and the second is obvious.

If $E \in \Sigma$, the collection of sets $E \cap A$ as A ranges over Σ forms a σ-algebra Σ' of subsets of E. In fact, Σ' is just the collection of all Σ-measurable subsets of E. If ψ is an additive set function on Σ, then its restriction to Σ' is additive on Σ'. On the other hand, if ϕ is an additive set function on Σ', then the function defined by $\psi(A) = \phi(A \cap E)$ is additive on Σ.

Now, let ϕ be an additive set function on the measurable subsets of a set $E \in \Sigma$. Define

$$\overline{V}(E) = \overline{V}(E;\phi) = \sup_{\substack{A \subset E \\ A \in \Sigma}} \phi(A), \qquad \underline{V}(E) = \underline{V}(E;\phi) = -\inf_{\substack{A \subset E \\ A \in \Sigma}} \phi(A), \tag{10.3}$$

$$V(E) = V(E;\phi) = \overline{V}(E) + \underline{V}(E)$$

to be the *upper*, *lower* and *total variation* of ϕ on E, respectively. Note that all three are nonnegative since $\phi(\varnothing) = 0$. Moreover, as is easy to see from the definitions, each variation is monotone increasing with E; that is, if $E_1 \subset E_2$, then $\overline{V}(E_1) \leq \overline{V}(E_2)$, etc. We will show that if ϕ is an additive set function on Σ, then so are $\overline{V}$, $\underline{V}$, and V. The first step in proving this is the following lemma.

(10.4) Lemma *If ϕ is an additive set function on Σ, then each of its three variations is countably subadditive; that is, if $E_k \in \Sigma$, $k = 1, 2, \ldots$, then*

$$\overline{V}(\bigcup E_k) \leq \sum \overline{V}(E_k),$$

with similar formulas for $\underline{V}$ and V.

Proof. Let $H_1 = E_1$, $H_2 = E_2 - E_1$, $H_3 = E_3 - E_2 - E_1, \ldots$. Then the H_k are disjoint and $\bigcup E_k = \bigcup H_k$. If $A \in \Sigma$ and $A \subset \bigcup E_k$, then $A = \bigcup (A \cap H_k)$ and $\phi(A) = \Sigma \phi(A \cap H_k)$. Therefore, since $A \cap H_k \subset E_k$, we have $\phi(A) \leq \sum \overline{V}(E_k)$. Hence,

$$\overline{V}(\bigcup E_k) = \sup_{A \subset \bigcup E_k, A \in \Sigma} \phi(A) \leq \sum \overline{V}(E_k),$$

which proves the result for $\overline{V}$. The proof for $\underline{V}$ is similar, and the result for V follows by adding.

(10.5) Lemma *If ϕ is an additive set function on Σ, then its variations $\overline{V}(E)$, $\underline{V}(E)$, and $V(E)$ are finite for every $E \in \Sigma$.*

Proof. It is enough to show the result for V. Suppose that $V(E) = +\infty$ for some E. We claim that there would then exist sets $E_k \in \Sigma$, $k = 1, 2, \ldots$, such that $E_k \searrow$ and $|\phi(E_k)| \geq k - 1$. To see this, we argue by induction. Let $E_1 = E$, and suppose that $E_1 \supset E_2 \supset \cdots \supset E_N$ have been constructed with $|\phi(E_k)| \geq k - 1$ and $V(E_k) = +\infty$ for $k = 1, \ldots, N$. Since $V(E_N) = +\infty$, there exists $A \in \Sigma$ such that $A \subset E_N$ and $|\phi(A)| \geq |\phi(E_N)| + N$. If $V(A) = +\infty$, let $E_{N+1} = A$, noting that $|\phi(A)| \geq N$. If $V(A) < +\infty$, let $E_{N+1} = E_N - A$. Then $V(E_{N+1}) = +\infty$ since by (10.4) we have $V(E_N) \leq V(E_{N+1}) + V(A)$. Furthermore,

$$|\phi(E_{N+1})| = |\phi(E_N) - \phi(A)| \geq |\phi(A)| - |\phi(E_N)| \geq N.$$

This establishes the existence of sets E_k with the desired properties. Thus, by (10.1), we obtain $|\phi(\bigcap E_k)| = \lim |\phi(E_k)| = +\infty$, contradicting the finiteness of ϕ.

The final step in proving that the variations are additive set functions is given in the next lemma.

(10.6) Lemma *If ϕ is an additive set function on Σ and $\{E_k\}$ is a sequence of disjoint sets in Σ, then $\overline{V}(\bigcup E_k) = \sum \overline{V}(E_k)$. Similar formulas hold for $\underline{V}$ and V.*

Proof. By (10.4), we have $\overline{V}(\bigcup E_k) \leq \sum \overline{V}(E_k)$. To show the opposite inequality, given $\varepsilon > 0$, choose $A_k \subset E_k$ with $\overline{V}(E_k) \leq \phi(A_k) + \varepsilon 2^{-k}$. This is possible since $\overline{V}(E_k)$ is finite by the previous lemma. Since the E_k are disjoint so are the A_k, and we obtain

$$\sum \overline{V}(E_k) \leq \sum \phi(A_k) + \varepsilon = \phi(\bigcup A_k) + \varepsilon \leq \overline{V}(\bigcup E_k) + \varepsilon.$$

Since ε is an arbitrary positive number, the result for $\overline{V}$ follows. The analogous formula for $\underline{V}$ is proved similarly, and the one for V follows by adding.

Combining (10.5) and (10.6), we immediately obtain the next theorem.

(10.7) Theorem *If ϕ is an additive set function on Σ, then so are its variations $\overline{V}$, $\underline{V}$, and V.*

The result which follows is basic and gives a decomposition of an additive set function into the difference of two *nonnegative* additive set functions. It may be compared to (2.6).

(10.8) Theorem (*Jordan Decomposition*) *If ϕ is an additive set function on Σ, then*

$$\phi(E) = \overline{V}(E) - \underline{V}(E) \qquad (E \in \Sigma).$$

Proof. If $A \subset E$ and $A \in \Sigma$, then $\phi(E) = \phi(A) + \phi(E - A) = \phi(A) - (-\phi(E - A))$. Since $\phi(E)$ is fixed, we have $\phi(A_k) \to \overline{V}(E)$ for a sequence of sets $A_k \in \Sigma$ if and only if $-\phi(E - A_k) \to \underline{V}(E)$. Hence, $\phi(E) = \overline{V}(E) - \underline{V}(E)$, as claimed.

A sequence $\{E_k\}$ of sets is said to *converge* if $\limsup E_k = \liminf E_k$. Thus, $\{E_k\}$ converges if each point which belongs to infinitely many E_k belongs to all E_k from some k on. If $\{E_k\}$ converges, it is said to *converge to* the set $E = \limsup E_k = \liminf E_k$.

As a simple corollary of the Jordan decomposition, we obtain the following result.

(10.9) Corollary *If a sequence of sets $\{E_k\}$ of Σ converges to E, and if ϕ is an additive set function on Σ, then* $\lim_{k\to\infty} \phi(E_k) = \phi(E)$.

Proof. If $\phi \geq 0$, we may apply (10.2). Since the extreme terms there both equal $\phi(E)$, it follows that all four equal $\phi(E)$. Hence, $\lim \phi(E_k)$ exists and equals $\phi(E)$. For arbitrary ϕ, the result therefore holds for $\overline{V}$ and $\underline{V}$, and so, by the Jordan decomposition, for ϕ itself.

Let $(\mathscr{S},\Sigma,\mu)$ be a measure space. We have already observed that μ satisfies $\mu(E_1) \le \mu(E_2)$ if $E_1 \subset E_2$, $E_1,E_2 \in \Sigma$. Another basic property of μ is given in the next theorem.

(10.10) Theorem *Let $(\mathscr{S},\Sigma,\mu)$ be a measure space, and let $\{E_k\}$ be any sequence of measurable sets. Then*

$$\mu(\bigcup E_k) \le \sum \mu(E_k).$$

Proof. Write $\bigcup E_k$ as a disjoint union as follows:

$$\bigcup E_k = E_1 \cup (E_2 - E_1) \cup (E_3 - E_2 - E_1) \cup \cdots.$$

Then

$$\mu(\bigcup E_k) = \mu(E_1) + \mu(E_2 - E_1) + \mu(E_3 - E_2 - E_1) + \cdots$$
$$\le \mu(E_1) + \mu(E_2) + \mu(E_3) + \cdots = \sum \mu(E_k),$$

which completes the proof.

By definition, a measure is countably additive on disjoint measurable sets [cf. (3.23)]. The next result shows that it shares another basic property of Lebesgue measure [see (3.26)].

(10.11) Theorem *Let $(\mathscr{S},\Sigma,\mu)$ be a measure space, and let $\{E_k\}$ be a sequence of measurable sets.*

(i) *If $E_k \nearrow E$, then $\lim_{k\to\infty} \mu(E_k) = \mu(E)$.*
(ii) *If $E_k \searrow E$ and $\mu(E_{k_0}) < +\infty$ for some k_0, then $\lim_{k\to\infty} \mu(E_k) = \mu(E)$.*

Proof. Suppose that $E_k \nearrow E$. If $\mu(E_k) < +\infty$ for all k, we may use the same argument used to prove the first part of (10.1). If $\mu(E_k) = +\infty$ for some k, then $\lim \mu(E_k) = \mu(E) = +\infty$. To prove the second part, we may assume that $k_0 = 1$ and use the argument for (10.1).

(10.12) Corollary *Let $(\mathscr{S},\Sigma,\mu)$ be a measure space and let $\{E_k\}$ be a sequence of measurable sets. Then*

(i) *$\mu(\liminf E_k) \le \liminf_{k\to\infty} \mu(E_k)$.*
(ii) *If $\mu(\bigcup_{k_0}^{\infty} E_k) < +\infty$ for some k_0, then $\mu(\limsup E_k) \ge \limsup_{k\to\infty} \mu(E_k)$.*

Proof. Part (ii) is an immediate corollary of (10.2). For part (i), let $A_m = \bigcap_{k=m}^{\infty} E_k$, $m = 1, 2, \ldots.$ Then $A_m \nearrow \liminf E_k$, and by (10.11), $\mu(\liminf E_k) = \lim_{m\to\infty} \mu(A_m)$. Since $A_m \subset E_m$, we have $\mu(A_m) \le \mu(E_m)$ and $\lim_{m\to\infty} \mu(A_m) \le \liminf_{m\to\infty} \mu(E_m)$. The result follows by combining inequalities.

2. Measurable Functions; Integration

We will now develop the notions of measurable functions and integration in a measure space. These will be used later in the chapter to prove several important results for set functions.

Let Σ be a fixed σ-algebra of subsets of $\mathscr{S}$, and let $f(x)$ be a real-valued function defined for x in a measurable set E. (As usual, f may take the values $\pm\infty$.) Then f is said to be *Σ-measurable*, or simply *measurable*, if $\{x \in E : f(x) > a\}$ is measurable for $-\infty < a < +\infty$. We will state some familiar results whose proofs depend only on the fact that the class of measurable sets forms a σ-algebra. The proofs are therefore similar to those in Chapter 4 for Lebesgue measurable functions, and details are left to the reader.

(10.13) Theorem

(i) *If f and g are measurable on a set $E \in \Sigma$, then so are $f + g$, cf for real c, $\phi(f)$ if ϕ is continuous on $\mathbf{R}^1$, f^+, f^-, $|f|^p$ for $p > 0$, fg, and $1/f$ if $f \neq 0$ in E.*

(ii) *If $\{f_k\}$ are measurable on $E \in \Sigma$, then so are $\sup_k f_k$, $\inf_k f_k$, $\limsup_{k\to\infty} f_k$, $\liminf_{k\to\infty} f_k$, and, if it exists, $\lim_{k\to\infty} f_k$.*

(iii) *If f is a simple function taking values $v_1, \ldots, v_N$ on disjoint sets $E_1, \ldots, E_N$, respectively, then f is measurable if and only if each E_k is measurable. In particular, χ_E is measurable if and only if E is.*

(iv) *If f is nonnegative and measurable on $E \in \Sigma$, then there exist nonnegative, simple measurable $f_k \nearrow f$ on E.*

If $(\mathscr{S},\Sigma,\mu)$ is a measure space, a measurable set E is said to have *μ-measure zero*, or *measure zero*, if $\mu(E) = 0$. A property is said to hold *almost everywhere in E with respect to μ*, or *a.e.* (μ), if it holds in E except at most for a subset of measure zero.

We have the following analogue of Egorov's theorem.

(10.14) Theorem (*Egorov's Theorem*) *Let $(\mathscr{S},\Sigma,\mu)$ be a measure space, and let E be a measurable set with $\mu(E) < +\infty$. Let $\{f_k\}$ be a sequence of measurable functions on E such that each f_k is finite a.e. in E and $\{f_k\}$ converges a.e. in E to a finite limit. Then, given $\varepsilon > 0$, there is a measurable set $A \subset E$ with $\mu(E - A) < \varepsilon$ such that $\{f_k\}$ converges uniformly on A.*

In general, we cannot choose A to be closed; in fact, $\mathscr{S}$ has very little structure, and the notion of a closed set may not even be defined. The proof is similar to that for Lebesgue measure, and is left as an exercise.

Let f be nonnegative on a measurable set E. Define the *integral of f over E with respect to μ* by

$$\text{(10.15)} \qquad \int_E f\,d\mu = \sup \sum_j [\inf_{x \in E_j} f(x)]\mu(E_j) \quad (f \geq 0),$$

where the supremum is taken over all decompositions $E = \bigcup E_j$ of E into the union of a *finite* number of disjoint measurable sets E_j. We adopt the convention $0 \cdot \infty = \infty \cdot 0 = 0$ for the terms of the sum in (10.15). By (5.8), the definition reduces to the usual Lebesgue integral in case $\mathscr{S} = \mathbf{R}^n$, Σ is the class of Lebesgue measurable sets, μ is Lebesgue measure and f is Lebesgue measurable. Although definition (10.15) does not require the measurability of f, many of the familiar properties of the integral are valid only for measurable functions. All functions considered in the rest of this section are assumed to be measurable.

(10.16) Theorem *Let $(\mathscr{S},\Sigma,\mu)$ be a measure space, and let f be a nonnegative, simple measurable function defined on a measurable set E. If f takes values $v_1, \ldots, v_N$ on disjoint $E_1, \ldots, E_N$, then*

$$\int_E f\,d\mu = \sum v_j\mu(E_j).$$

Proof. Since f is measurable, each E_j is measurable by (10.13)(iii). Clearly, $\int_E f\,d\mu \geq \sum v_j\mu(E_j)$. On the other hand, consider any decomposition $E = \bigcup A_k$ of E into a finite number of disjoint measurable sets, and let $w_k = \inf_{A_k} f$. If $A_k \cap E_j$ is not empty, then $w_k \leq v_j$. Therefore, by the additivity of μ,

$$\begin{aligned}\sum w_k\mu(A_k) &= \sum_j \sum_k w_k\mu(A_k \cap E_j)\\ &\leq \sum_j v_j \sum_k \mu(A_k \cap E_j) = \sum v_j\mu(E_j).\end{aligned}$$

Taking the supremum over all such decompositions, we obtain $\int_E f\,d\mu \leq \sum v_j\mu(E_j)$, which completes the proof.

Note that the last theorem holds even if some of the v_j are $+\infty$.

(10.17) Theorem *Let $(\mathscr{S},\Sigma,\mu)$ be a measure space, and let f and g be measurable functions defined on a set $E \in \Sigma$.*

(i) *If $0 \leq f \leq g$ on E, then $\int_E f\,d\mu \leq \int_E g\,d\mu$.*
(ii) *If $f \geq 0$ on E and $\mu(E) = 0$, then $\int_E f\,d\mu = 0$.*

Proof. Both parts follow immediately from the definition (10.15). For part (ii), note that $\mu(E_j) = 0$ whenever $E_j \subset E$ and $E_j \in \Sigma$. Hence, each term of the sum in (10.15) is zero.

In order to investigate further the properties of the integral, we need the next two lemmas. In these and the results which follow, the measure space $(\mathscr{S},\Sigma,\mu)$ is fixed.

(10.18) Lemma

(i) *If f and g are nonnegative, simple measurable functions on E, and if c is a nonnegative constant, then* $\int_E (f+g)\,d\mu = \int_E f\,d\mu + \int_E g\,d\mu$ *and* $\int_E cf\,d\mu = c\int_E f\,d\mu$.

(ii) *If f is a nonnegative, simple measurable function on E, and $E = E_1 \cup E_2$ is the union of two disjoint measurable sets, then* $\int_E f\,d\mu = \int_{E_1} f\,d\mu + \int_{E_2} f\,d\mu$.

The proof of the first part of (i) is like the first part of the proof of (5.14). The second part of (i) follows immediately from (10.16). The details and the proof of (ii) are left as an exercise.

(10.19) Lemma *Let f_k, $k = 1, 2, \ldots$, and g be nonnegative, simple measurable functions defined on a set $E \in \Sigma$. If $f_k \nearrow$ and $\lim_{k\to\infty} f_k \geq g$ on E, then*

$$\lim_{k\to\infty} \int_E f_k\,d\mu \geq \int_E g\,d\mu.$$

Proof. Suppose that g takes values $v_1, \ldots, v_m$ on disjoint sets $E_1, \ldots, E_m$. By (10.18)(ii), it is enough to show that

$$\lim_{k\to\infty} \int_{E_j} f_k\,d\mu \geq \int_{E_j} g\,d\mu \text{ for each } j.$$

We thus reduce the proof to the case when g is constant on E, that is, $g = v \geq 0$ on E. If $v = 0$, the result is obvious. Suppose then that $0 < v < +\infty$, and let $0 < \varepsilon < v$ and $A_k = \{x \in E : f_k(x) \geq v - \varepsilon\}$, $k = 1, 2, \ldots$. Since $f_k \nearrow$, we have $A_k \nearrow E$, so that $\mu(A_k) \to \mu(E)$. Moreover,

$$\int_E f_k\,d\mu \geq \int_{A_k} f_k\,d\mu \geq (v - \varepsilon)\mu(A_k).$$

Therefore, $\lim_{k\to\infty} \int_E f_k\,d\mu \geq (v-\varepsilon)\mu(E)$. Letting $\varepsilon \to 0$ and observing that $v\mu(E) = \int_E g\,d\mu$, we obtain the desired result.

The next theorem is helpful in deriving properties of $\int f\,d\mu$ for arbitrary nonnegative f from those for simple f.

(10.20) Theorem *Let $\{f_k\}$ be a sequence of nonnegative, simple measurable functions defined on $E \in \Sigma$. If $f_k \nearrow f$ on E, then* $\int_E f_k\,d\mu \to \int_E f\,d\mu$.

Proof. Clearly, $\lim_{k\to\infty} \int_E f_k\,d\mu \leq \int_E f\,d\mu$. To show the opposite inequality, consider a partition $E = \bigcup E_j$ of E into a finite number of disjoint measurable sets E_j, and let $v_j = \inf_{E_j} f$ and $\sigma = \sum v_j\mu(E_j)$. The function g defined by $g = \sum v_j\chi_{E_j}$ is nonnegative and measurable, and $\int_E g\,d\mu = \sigma$. Since $\lim_{k\to\infty} f_k \geq g$, we have $\lim_{k\to\infty} \int_E f_k\,d\mu \geq \sigma$ by (10.19). Taking the supremum of such σ over all partitions of E, we obtain the desired result.

As a corollary, we have the following theorem.

(10.21) Theorem *Let f and g be nonnegative measurable functions defined on $E \in \Sigma$, and let c be a nonnegative constant. Then*

(i) $\int_E (f+g)\,d\mu = \int_E f\,d\mu + \int_E g\,d\mu$ *and* $\int_E cf\,d\mu = c\int_E f\,d\mu$.
(ii) *If $E = E_1 \cup E_2$, where E_1 and E_2 are disjoint and measurable, then* $\int_E f\,d\mu = \int_{E_1} f\,d\mu + \int_{E_2} f\,d\mu$.

Proof. By (10.13)(iv), choose simple measurable f_k and g_k such that $0 \le f_k \nearrow f$ and $0 \le g_k \nearrow g$. Then $f_k + g_k$ is simple and measurable, and $0 \le f_k + g_k \nearrow f + g$. Therefore, by (10.20) and (10.18),

$$\int_E (f+g)\,d\mu = \lim_{k\to\infty}\int_E (f_k+g_k)\,d\mu = \lim_{k\to\infty}\left(\int_E f_k\,d\mu + \int_E g_k\,d\mu\right)$$
$$= \int_E f\,d\mu + \int_E g\,d\mu.$$

This proves the first part of (i); the other parts are proved similarly.

If f is any real-valued measurable function defined on a measurable set E, we define its *integral with respect to μ* by

$$\text{(10.22)}\qquad \int_E f(x)\,d\mu(x) = \int_E f\,d\mu = \int_E f^+\,d\mu - \int_E f^-\,d\mu,$$

provided not both integrals on the right are $+\infty$.

We say that f is *integrable with respect to μ*, or *μ-integrable*, over E if $\int_E f\,d\mu$ exists and is finite. When this is the case, we write $f \in L(E;d\mu)$ or $f \in L(E;\mu)$.

It is immediate from (10.22) and (10.17) that $\int_E f\,d\mu = 0$ if $\mu(E) = 0$, and that $\int_E f\,d\mu \le \int_E g\,d\mu$ if $f \le g$ on E and both integrals exist. The familiar properties of the Lebesgue integral are shared by $\int_E f\,d\mu$; some of them are listed in the following theorem.

(10.23) Theorem

(i) $|\int_E f\,d\mu| \le \int_E |f|\,d\mu$; *furthermore, $f \in L(E;d\mu)$ if and only if $|f| \in L(E;d\mu)$.*
(ii) *If $|f| \le |g|$ a.e.(μ) in E, and if $g \in L(E;d\mu)$, then $f \in L(E;d\mu)$ and* $\int_E |f|\,d\mu \le \int_E |g|\,d\mu$.
(iii) *If $f \in L(E;d\mu)$, then f is finite a.e.(μ) in E.*
(iv) *If $f = g$ a.e.(μ) in E and if $\int_E f\,d\mu$ exists, then $\int_E g\,d\mu$ exists and* $\int_E g\,d\mu = \int_E f\,d\mu$.
(v) *If $\int_E f\,d\mu$ exists and c is a constant, then $\int_E cf\,d\mu$ exists and* $\int_E cf\,d\mu = c\int_E f\,d\mu$.
(vi) *If $f,g \in L(E;d\mu)$, then $f+g \in L(E;d\mu)$ and* $\int_E (f+g)\,d\mu = \int_E f\,d\mu + \int_E g\,d\mu$.

(vii) *If* $f \ge 0$ *and* $m \le g \le M$ *on* E, *then*

$$m\int_E f\,d\mu \le \int_E fg\,d\mu \le M\int_E f\,d\mu.$$

Proof. The proofs are similar to those for Lebesgue integrals. As examples, we will prove (iii) and (iv). For (iii), suppose that $f \in L(E;d\mu)$. Then, by (i), $|f| \in L(E;d\mu)$. Let $Z = \{x \in E : |f(x)| = +\infty\}$. Then for any positive integer k,

$$k\mu(Z) \le \int_Z |f|\,d\mu \le \int_E |f|\,d\mu.$$

Since $f \in L(E;d\mu)$, it follows that $\mu(Z) = 0$, which proves (iii).

For (iv), since both $f^+ = g^+$ and $f^- = g^-$ a.e. in E, we may assume that $f \ge 0$. Let $E_1 = \{x \in E : f(x) \ne g(x)\}$. Since $\mu(E_1) = 0$, (10.21)(ii) implies that

$$\int_E f\,d\mu = \int_{E-E_1} f\,d\mu = \int_{E-E_1} g\,d\mu = \int_E g\,d\mu,$$

as asserted.

(10.24) Theorem *If* $\{f_k\}$ *is a sequence of nonnegative measurable functions on* E, *then*

$$\int_E \left(\sum f_k\right) d\mu = \sum \int_E f_k\,d\mu.$$

Proof. Let $f = \sum_{k=1}^{\infty} f_k$. Since $f \ge \sum_{k=1}^{m} f_k$, the integral of f over E majorizes $\sum_{k=1}^{m} \int_E f_k\,d\mu$ for any m. Hence, the left side above majorizes the right. To show the opposite inequality, let $\{f_k^{(j)}\}$ be a sequence of nonnegative, simple measurable functions increasing to f_k. Let $s_j = \sum_{k=1}^{j} f_k^{(j)}$. Then s_j is nonnegative and simple, and $s_j \nearrow$. We will show that $s_j \nearrow f$. Clearly, $\lim_{j\to\infty} s_j \le f$. On the other hand, for any m,

$$\lim_{j\to\infty} s_j \ge \lim_{j\to\infty} \sum_{k=1}^{m} f_k^{(j)} = \sum_{k=1}^{m} f_k.$$

Therefore, $\lim_{j\to\infty} s_j \ge f$. It follows from (10.20) that $\int_E s_j\,d\mu \to \int_E f\,d\mu$. Since $\sum_{k=1}^{j} f_k \ge s_j$, we obtain

$$\sum_{k=1}^{\infty} \int_E f_k\,d\mu = \lim_{j\to\infty} \sum_{k=1}^{j} \int_E f_k\,d\mu \ge \lim_{j\to\infty} \int_E s_j\,d\mu = \int_E f\,d\mu.$$

This proves the desired inequality, and the theorem follows.

The next three results are essentially corollaries of (10.24).

(10.25) Theorem *If $\int_E f\,d\mu$ exists, and if $E = \bigcup E_k$ is a countable union of disjoint measurable sets E_k, then*

$$\int_E f\,d\mu = \sum \int_{E_k} f\,d\mu.$$

Proof. Suppose first that $f \geq 0$. Let $f_k = f\chi_{E_k}$ on E, so that f_k is measurable and nonnegative, and $f = \sum f_k$. By (10.24),

$$\int_E f\,d\mu = \sum \int_E f_k\,d\mu = \sum \int_{E_k} f\,d\mu.$$

For arbitrary measurable f, the existence of $\int_E f\,d\mu$ implies that of $\int_{E_k} f\,d\mu$; in fact, the integrals of f^+ and f^- over any E_k are majorized by those over E. Moreover, by the case already considered,

$$\int_E f^+\,d\mu = \sum \int_{E_k} f^+\,d\mu, \qquad \int_E f^-\,d\mu = \sum \int_{E_k} f^-\,d\mu.$$

Since at least one of these sums is finite, the conclusion follows by subtraction. [Cf. theorem (5.24).]

(10.26) Theorem *If f_k are measurable and $0 \leq f_k \nearrow f$ on E, then $\int_E f_k\,d\mu \to \int_E f\,d\mu$.*

Proof. If $\int_E f_k\,d\mu = +\infty$ for some k, the result is obvious. We may therefore assume that each $f_k \in L(d\mu)$. Write $f = f_1 + \sum_{k=2}^{\infty}(f_k - f_{k-1})$. Since each term on the right is nonnegative, we obtain from (10.24) and (10.23)(vi) that

$$\int_E f\,d\mu = \int_E f_1\,d\mu + \sum_{k=2}^{\infty}\left(\int_E f_k\,d\mu - \int_E f_{k-1}\,d\mu\right) = \lim_{k\to\infty}\int_E f_k\,d\mu.$$

(10.27) Theorem (*Monotone Convergence Theorem*) *Let $\{f_k\}$ and f be measurable functions on E.*

(i) *Suppose that $f_k \nearrow f$ a.e. (μ) on E. If there exists $\phi \in L(E;d\mu)$ such that $f_k \geq \phi$ on E for all k, then $\int_E f_k\,d\mu \to \int_E f\,d\mu$.*

(ii) *Suppose that $f_k \searrow f$ a.e. on E. If there exists $\phi \in L(E;d\mu)$ such that $f_k \leq \phi$ on E for all k, then $\int_E f_k\,d\mu \to \int_E f\,d\mu$.*

Proof. The proof of (i) follows by applying (10.26) to the functions $f_k - \phi$. The details are as in the proof of (5.32). Part (ii) follows by applying (i) to the functions $-f_k$.

(10.28) Theorem (*Uniform Convergence Theorem*) *Suppose that $f_k \in L(E;d\mu)$, $k = 1, 2, \ldots$, and that $\{f_k\}$ converges uniformly to f on E, $\mu(E) < +\infty$. Then $f \in L(E;d\mu)$ and $\int_E f_k\,d\mu \to \int_E f\,d\mu$.*

The proof is the same as for Lebesgue measure [see (5.33)], and is omitted.

Fatou's lemma and the Lebesgue dominated convergence theorem are true for abstract measures. They are stated below without proof; the proofs are like those of (5.34) and (5.36).

(10.29) Theorem (*Fatou's Lemma*) *Let $\{f_k\}$ be a sequence of measurable functions on E. If there exists $\phi \in L(E;d\mu)$ such that $f_k \geq \phi$ on E for all k, then*

$$\int_E (\liminf_{k\to\infty} f_k)\, d\mu \leq \liminf_{k\to\infty} \int_E f_k\, d\mu.$$

The case $\phi = 0$ (that is, $f_k \geq 0$) is of special importance. In this case, we also have the following useful corollary.

(10.30) Corollary *Let $\{f_k\}$ and f be nonnegative measurable functions on E such that $f_k \to f$ a.e. (μ) in E. If $\int_E f_k\, d\mu \leq M$ for all k, then $\int_E f\, d\mu \leq M$.*

(10.31) Theorem (*Lebesgue Dominated Convergence Theorem*) *Let f, $\{f_k\}$ and ϕ be measurable functions on E such that $|f_k| \leq \phi$ a.e. (μ) on E and $\phi \in L(E;d\mu)$. Then*

(i) $\int_E (\liminf_{k\to\infty} f_k)\, d\mu \leq \liminf_{k\to\infty} \int_E f_k\, d\mu \leq \limsup_{k\to\infty} \int_E f_k\, d\mu \leq \int_E (\limsup_{k\to\infty} f_k)\, d\mu$.

(ii) *If $f_k \to f$ a.e. (μ) in E, then $\int_E f_k\, d\mu \to \int_E f\, d\mu$.*

(10.32) Corollary (*Bounded Convergence Theorem*) *Let $\{f_k\}$ and f be measurable functions on E such that $f_k \to f$ a.e. in E. If $\mu(E) < +\infty$ and there is a constant M such that $|f_k| \leq M$ a.e. in E, then $\int_E f_k\, d\mu \to \int_E f\, d\mu$.*

We conclude our brief study of integration with respect to abstract measures by defining $L^p(E;d\mu) = L^p(E,\Sigma,d\mu)$, $0 < p < \infty$, to be the collection of all measurable real or complex-valued f such that $\int_E |f|^p\, d\mu < +\infty$. We set

$$\|f\|_p = \|f\|_{p,E,d\mu} = \left(\int_E |f|^p\, d\mu\right)^{1/p} \qquad (0 < p < \infty).$$

When $p = \infty$, $L^\infty(E;d\mu)$ is the collection of all measurable f such that $\|f\|_\infty < +\infty$, where

$$\|f\|_\infty = \|f\|_{\infty,E,d\mu} = \operatorname*{ess\,sup}_E |f| = \inf\{\alpha : \mu(x \in E : |f(x)| > \alpha) = 0\}.$$

Observe that l^p is $L^p(\mathscr{S},\Sigma,d\mu)$ when $\mathscr{S}$ is the set of integers, Σ is the set of all subsets of $\mathscr{S}$, and $\mu(E)$ is the number of elements of E.

For $1 \leq p \leq \infty$, Hölder's and Minkowski's inequalities hold:

$$\|fg\|_1 \leq \|f\|_p \|g\|_{p'}, \qquad \|f + g\|_p \leq \|f\|_p + \|g\|_p.$$

Moreover, L^p is a Banach space with norm $\|\cdot\|_p$. In general, L^p is not separable (see Exercise 9). However, if L^2 is separable, we can define orthogonality, linear independence, completeness, Fourier coefficients, and Fourier series as usual, obtaining Bessel's inequality and Parseval's formula, as well as the usual result relating L^2 and l^2.

3. Absolutely Continuous and Singular Set Functions and Measures

We now turn out attention from the familiar results above to some new ones arising naturally in the context of abstract measure spaces. Thus, let $(\mathscr{S},\Sigma,\mu)$ be a measure space, and let ϕ be an additive set function on Σ. If $E \in \Sigma$, then ϕ is said to be *absolutely continuous on E with respect to* μ if $\phi(A) = 0$ for every measurable $A \subset E$ with $\mu(A) = 0$. Note that this definition has a somewhat different pattern from that given for Lebesgue measure (see p. 99). However, we shall obtain in (10.34) a reformulation of the present definition in terms of the old one.

On the other hand, ϕ is said to be *singular on E with respect to* μ if there is a set $Z \subset E$ such that $\mu(Z) = 0$ and $\phi(A) = 0$ for every measurable $A \subset E - Z$. Thus, ϕ is singular if it is supported on a set of μ-measure zero, so that E splits into the union of two sets, Z and $E - Z$, one with μ-measure zero and the other with the property that ϕ is zero on each measurable subset of it.

As examples, note that if $f \in L(E;d\mu)$, then the function $\phi(A) = \int_A f\,d\mu$ is absolutely continuous on E with respect to μ. If Z is any set with $\mu(Z) = 0$ and ψ is any additive set function, then the function $\phi(A) = \psi(A \cap Z)$ is singular on E with respect to μ.

We list several simple properties of such set functions in the next theorem.

(10.33) Theorem

(i) *If ϕ is both absolutely continuous and singular on E with respect to μ, then $\phi(A) = 0$ for every measurable $A \subset E$.*

(ii) *If both ψ and ϕ are absolutely continuous (singular) on E with respect to μ, then so are $\psi + \phi$ and $c\phi$, where c is any real constant.*

(iii) *ϕ is absolutely continuous (singular) on E with respect to μ if and only if its variations $\overline{V}$ and $\underline{V}$ are, or, equivalently, if and only if its total variation V is.*

(iv) *If $\{\phi_k\}$ is a sequence of additive set functions which are absolutely continuous (singular) on E with respect to μ, and if $\phi(A) = \lim_{k\to\infty} \phi_k(A)$ exists for every measurable $A \subset E$, then ϕ is absolutely continuous (singular) on E with respect to μ.*

Proof. For part (i), suppose that ϕ is absolutely continuous and singular on E. Let Z be a subset of E with μ-measure zero such that $\phi(H) = 0$ if

H is measurable and $H \subset E - Z$. If A is any measurable subset of E, then $\phi(A) = \phi(A \cap Z) + \phi(A - Z)$. Since ϕ is absolutely continuous and $\mu(A \cap Z) = 0$, we have $\phi(A \cap Z) = 0$. Moreover, since $A - Z \subset E - Z$ and ϕ is singular, we have $\phi(A - Z) = 0$. Hence, $\phi(A) = 0$, and part (i) is proved. The proofs of parts (ii)–(iv) are left as exercises.

The next two theorems give alternate characterizations of absolutely continuous and singular set functions.

(10.34) Theorem *An additive set function ϕ is absolutely continuous on E with respect to μ if and only if given $\varepsilon > 0$, there exists $\delta > 0$ such that $|\phi(A)| < \varepsilon$ for any measurable $A \subset E$ with $\mu(A) < \delta$.*

Proof. The sufficiency of the condition is immediate since if $\mu(A) = 0$, then $\mu(A) < \delta$ for all $\delta > 0$, so that $|\phi(A)| < \varepsilon$ for all ε. Consequently, $\phi(A) = 0$. For the converse, suppose that ϕ is absolutely continuous, but that there is an $\varepsilon > 0$ for which no $\delta > 0$ gives the desired result. Then, taking $\delta = 2^{-k}$ for $k = 1, 2, \ldots$, there would exist measurable $A_k \subset E$ with $\mu(A_k) < 2^{-k}$ and $|\phi(A_k)| \geq \varepsilon$. Let $A = \limsup A_k$. Then, for any m,

$$\mu(A) \leq \mu\left(\bigcup_{k=m}^{\infty} A_k\right) \leq \sum_{k=m}^{\infty} 2^{-k},$$

so that $\mu(A) = 0$. Therefore, $\phi(A) = 0$. Assuming for the moment that $\phi \geq 0$, we obtain from (10.2) that $\phi(A) = \phi(\limsup A_k) \geq \limsup \phi(A_k) \geq \varepsilon$. This contradiction establishes the result in case $\phi \geq 0$. For the general case, the variation V of an absolutely continuous ϕ is absolutely continuous [by (10.33)(iii)] and nonnegative. Since $|\phi(A)| \leq V(A)$, the theorem follows.

(10.35) Theorem *An additive set function ϕ is singular on E with respect to μ if and only if given $\varepsilon > 0$, there is a measurable subset E_0 of E such that $\mu(E_0) < \varepsilon$ and $V(E - E_0; \phi) < \varepsilon$.*

Proof. If ϕ is singular, there exists $Z \subset E$ with $\mu(Z) = 0$ such that $V(E - Z; \phi) = 0$. Taking $E_0 = Z$, we obtain the necessity of the condition. To prove its sufficiency, choose for each $k = 1, 2, \ldots$ a measurable $E_k \subset E$ with $\mu(E_k) < 2^{-k}$ and $V(E - E_k; \phi) < 2^{-k}$. Let $Z = \limsup E_k$. Since $Z \subset \bigcup_{k=m}^{\infty} E_k$ for every m, it follows as usual that $\mu(Z) = 0$. Moreover, by (10.2),

$$\begin{aligned} V(E - Z; \phi) &= V(E - \limsup E_k; \phi) = V(\liminf (E - E_k); \phi) \\ &\leq \liminf V(E - E_k; \phi) = 0. \end{aligned}$$

Hence, ϕ is singular with respect to μ, which completes the proof.

Let ϕ be the indefinite integral of an $f \in L(E; d\mu)$: $\phi(A) = \int_A f\, d\mu$ for measurable $A \subset E$. Letting $P = \{x \in E : f(x) \geq 0\}$, we see that $\phi(A) \geq 0$ for any measurable $A \subset P$, and that $\phi(A) \leq 0$ for any measurable $A \subset E - P$.

It follows that $\overline{V}(E;\phi) = \phi(P)$ and $\underline{V}(E;\phi) = -\phi(E - P)$. Moreover, since $P = \{x \in E : f(x) = f^+(x)\}$, we have

$$\overline{V}(E;\phi) = \int_E f^+ \, d\mu, \qquad \underline{V}(E;\phi) = \int_E f^- \, d\mu.$$

For an arbitrary ϕ, we have the following basic result.

(10.36) Theorem (*Hahn Decomposition*) *Let E be a measurable set and let ϕ be an additive set function defined on the measurable subsets A of E. Then there is a measurable $P \subset E$ such that $\phi(A) \geq 0$ for $A \subset P$ and $\phi(A) \leq 0$ for $A \subset E - P$. Equivalently, $\underline{V}(P;\phi) = \overline{V}(E - P; \phi) = 0$. Hence,*

$$\overline{V}(E;\phi) = \overline{V}(P;\phi) = \phi(P),$$
$$\underline{V}(E;\phi) = \underline{V}(E - P;\phi) = -\phi(E - P).$$

Proof. Write $\overline{V}(A) = \overline{V}(A;\phi)$ and $\underline{V}(A) = \underline{V}(A;\phi)$ for $A \subset E$. For each positive integer k, choose a measurable $A_k \subset E$ such that $\phi(A_k) > \overline{V}(E) - 2^{-k}$. Then $\overline{V}(A_k) > \overline{V}(E) - 2^{-k}$. Since $\overline{V}$ is additive, $\overline{V}(E - A_k) = \overline{V}(E) - \overline{V}(A_k) < 2^{-k}$. Moreover, by the Jordan decomposition, $\overline{V}(A_k) - \underline{V}(A_k) = \phi(A_k) > \overline{V}(E) - 2^{-k}$, so that $\underline{V}(A_k) < 2^{-k}$. Let $P = \liminf A_k$. Since $\underline{V}$ is nonnegative, (10.2) implies that $\underline{V}(P) \leq \liminf \underline{V}(A_k) = 0$. Also,

$$\overline{V}(E - P) = \overline{V}(E - \liminf A_k) = \overline{V}(\limsup\,(E - A_k)) \leq \overline{V}\left(\bigcup_{k=m}^{\infty} (E - A_k)\right)$$

for any m. Therefore, for any m,

$$\overline{V}(E - P) \leq \sum_{k=m}^{\infty} \overline{V}(E - A_k) < \sum_{k=m}^{\infty} 2^{-k},$$

which gives $\overline{V}(E - P) = 0$ and completes the proof.

In the next theorem, we use the Hahn decomposition to split E into sets where ϕ is comparable to μ. In doing so, we assume that ϕ is nonnegative and μ is finite; thus, we are in fact dealing with two finite measures.

(10.37) Theorem *Let ϕ be a nonnegative additive set function defined on the measurable subsets of a measurable set E, and let μ be a measure with $\mu(E) < +\infty$. Then given $a > 0$, there is a decomposition $E = Z \cup (\bigcup_{k=1}^{\infty} E_k)$ of E into disjoint measurable sets such that*

(i) $\mu(Z) = 0$.
(ii) $a(k - 1)\mu(A) \leq \phi(A) \leq ak\mu(A)$ *for measurable* $A \subset E_k$, $k = 1, 2, \ldots$.

Proof. We may assume that $a = 1$ by considering ϕ/a. For each positive integer k, let $\psi_k(A) = \phi(A) - k\mu(A)$ for measurable $A \subset E$. Since ϕ and μ

are finite and additive, ψ_k is an additive set function. By the Hahn decomposition, there is a set $P_k \subset E$ such that $\psi_k(A) \geq 0$ if $A \subset P_k$ and $\psi_k(A) \leq 0$ if $A \subset E - P_k$. Thus, $\phi(A) \geq k\mu(A)$ if $A \subset P_k$ and $\phi(A) \leq k\mu(A)$ if $A \subset E - P_k$.

Now, let $Q_k = \bigcup_{m=k}^{\infty} P_m$ for $k = 1, 2, \ldots$, and observe that $P_k \subset Q_k$ and $Q_k \searrow$. We will show that $\phi(A) \geq k\mu(A)$ if $A \subset Q_k$, and $\phi(A) \leq k\mu(A)$ if $A \subset E - Q_k$. To see this, write

$$Q_k = P_k \cup (P_{k+1} - P_k) \cup (P_{k+2} - P_{k+1} - P_k) \cup \cdots$$

and note that the terms on the right side are disjoint. Hence, if $A \subset Q_k$, we may write $A = \bigcup_{m=k}^{\infty} A_m$, where the A_m are disjoint and $A_m \subset P_m$, by simply intersecting A with each such term of Q_k. Then

$$\phi(A) = \sum_{m=k}^{\infty} \phi(A_m) \geq \sum_{m=k}^{\infty} m\mu(A_m) \geq k \sum_{m=k}^{\infty} \mu(A_m) = k\mu(A),$$

so that $\phi(A) \geq k\mu(A)$ for $A \subset Q_k$, as claimed. On the other hand, if $A \subset E - Q_k$, then $A \subset E - P_k$, so that $\phi(A) \leq k\mu(A)$. This proves the assertion above.

We can now give the decomposition of E. Let $Z = \bigcap_{k=1}^{\infty} Q_k = \limsup P_k$, and write

$$\begin{aligned} E &= Z \cup (E - Q_1) \cup (Q_1 - Q_2) \cup (Q_2 - Q_3) \cup \cdots \\ &= Z \cup E_1 \cup E_2 \cup E_3 \cup \cdots. \end{aligned}$$

The terms in this decomposition are disjoint. If $A \subset E_1(= E - Q_1)$, then $\phi(A) \leq \mu(A)$ by what was shown above, and $\phi(A) \geq 0$ by hypothesis. For $k \geq 2$, we have $E_k = Q_{k-1} - Q_k = Q_{k-1} \cap (E - Q_k)$. Hence, if $k \geq 2$ and $A \subset E_k$, then $\phi(A) \geq (k-1)\mu(A)$ due to the fact that $A \subset Q_{k-1}$; also, $\phi(A) \leq k\mu(A)$ due to $A \subset E - Q_k$. Finally, since $Z \subset Q_k$ for all k, we have $\phi(Z) \geq k\mu(Z)$ for all k. Since ϕ is finite, it follows that $\mu(Z) = 0$, which completes the proof.

To give some idea of the significance of the last result, write $A = (A \cap Z) \cup [\bigcup_k (A \cap E_k)]$ for measurable $A \subset E$. Then

$$\phi(A) = \phi(A \cap Z) + \sum \phi(A \cap E_k).$$

The set function $\sigma(A) = \phi(A \cap Z)$ is singular with respect to μ. By (ii) of the theorem, ϕ is absolutely continuous with respect to μ on each E_k. Hence, the set function α defined by

$$\alpha(A) = \phi(A) - \sigma(A) = \sum \phi(A \cap E_k)$$

is absolutely continuous with respect to μ since if $\mu(A) = 0$, then $\mu(A \cap E_k) = 0$ and $\phi(A \cap E_k) = 0$ for all k. Note also that (ii) can be written

$$a(k-1)\int_A d\mu \le \phi(A) \le ak\int_A d\mu$$

for measurable $A \subset E_k$.

We will use these ideas to write any set function as the sum of an absolutely continuous part, which will be an indefinite integral, and a singular part. This decomposition, which is of major importance, is stated in the following theorem. We assume that the measure μ defined on the measurable subsets of E is *σ-finite*, that is, that E can be written as a countable union of measurable sets with finite μ-measure.

(10.38) Theorem (*Lebesgue Decomposition*) *Let ϕ be an additive set function on the measurable subsets of a measurable set E, and let μ be a σ-finite measure on E. Then there is a unique decomposition*

$$\phi(A) = \alpha(A) + \sigma(A) \textit{ for measurable } A \subset E,$$

where α and σ are additive set functions, α is absolutely continuous, and σ is singular with respect to μ. These functions are

$$\alpha(A) = \int_A f\,d\mu, \qquad \sigma(A) = \phi(A \cap Z)$$

for appropriate $f \in L(E;d\mu)$ and Z with $\mu(Z) = 0$. Moreover, if $\phi \ge 0$, then $f \ge 0$.

Proof. Assuming that such a decomposition exists, we will show it is unique. If $\phi = \alpha_1 + \sigma_1$ is another decomposition of ϕ into absolutely continuous and singular parts, then $\alpha - \alpha_1 = \sigma_1 - \sigma$, which (being both absolutely continuous and singular with respect to μ) must vanish identically. Hence $\alpha = \alpha_1$ and $\sigma = \sigma_1$.

To show that the decomposition exists, first assume that $\phi \ge 0$ and $\mu(E) < +\infty$. Taking $a = 2^{-m}$, $m = 1, 2, \ldots$, in (10.37), we may write E as a disjoint union $E = Z^{(m)} \cup (\bigcup_k E_k^{(m)})$, where

$$\mu(Z^{(m)}) = 0, \qquad 2^{-m}(k-1)\mu(A) \le \phi(A) \le 2^{-m}k\mu(A), \qquad A \subset E_k^{(m)}.$$

Given m, k, m', and k', let $\beta = 2^{-m}(k-1)$, $\gamma = 2^{-m}k$, $\beta' = 2^{-m'}(k'-1)$ and $\gamma' = 2^{-m'}k'$. If the intervals $[\beta,\gamma]$ and $[\beta',\gamma']$ are disjoint, we will show that the set $A = E_k^{(m)} \cap E_{k'}^{(m')}$ has μ-measure zero. In fact, we have both

$$\beta\mu(A) \le \phi(A) \le \gamma\mu(A) \quad \text{and} \quad \beta'\mu(A) \le \phi(A) \le \gamma'\mu(A).$$

If, for example, $\gamma < \beta'$, the inequalities $\beta'\mu(A) \le \phi(A) \le \gamma\mu(A)$ imply that $\mu(A) = 0$. A similar argument applies if $\gamma' < \beta$. Fixing m and k, and setting $m' = m + 1$, we see that there are at most four values of k' such that $E_{k'}^{(m+1)}$ intersects $E_k^{(m)}$ in a set of positive μ-measure: namely, $k' = 2k - 2, 2k - 1, 2k, 2k + 1$. Hence,

$$E_k^{(m)} \subset E_{2k-2}^{(m+1)} \cup E_{2k-1}^{(m+1)} \cup E_{2k}^{(m+1)} \cup E_{2k+1}^{(m+1)} \cup Y_k^{(m)},$$

where $\mu(Y_k^{(m)}) = 0$.

Let

$$Z = \Big(\bigcup_m Z^{(m)}\Big) \cup \Big(\bigcup_{k,m} Y_k^{(m)}\Big),$$

so that $\mu(Z) = 0$, and define functions $\{f_m\}_{m=1}^{\infty}$ on E by $f_m(x) = 2^{-m}(k-1)$ if $x \in E_k^{(m)} - Z$ and $f_m(x) = 0$ if $x \in Z$. Therefore, if $x \in E_k^{(m)} - Z$, $f_m(x) = 2^{-m}(k-1)$ and $f_{m+1}(x)$ takes one of the four values $2^{-m-1}j$, $j = 2k-3, 2k-2, 2k-1, 2k$. Hence, $|f_m(x) - f_{m+1}(x)| \le 2^{-m}$ if $x \in E_k^{(m)} - Z$, and so also if $x \in E$. It follows that $\{f_m\}$ converges uniformly on E to a limit f. Since $f_m \ge 0$, also $f \ge 0$.

Since E is the disjoint union $Z \cup [\bigcup_k (E_k^{(m)} - Z)]$ and ϕ is absolutely continuous on each $E_k^{(m)}$,

$$\begin{aligned}\phi(A) &= \phi(A \cap Z) + \sum_k \phi(A \cap (E_k^{(m)} - Z))\\ &= \phi(A \cap Z) + \sum_k \phi(A \cap E_k^{(m)})\end{aligned}$$

for $A \subset E$. Therefore,

$$\begin{aligned}\phi(A \cap Z) + \sum_k 2^{-m}(k-1)\mu(A \cap E_k^{(m)}) &\le \phi(A)\\ &\le \phi(A \cap Z) + \sum_k 2^{-m}k\mu(A \cap E_k^{(m)}),\end{aligned}$$

which can be rewritten

$$\phi(A \cap Z) + \int_A f_m\,d\mu \le \phi(A) \le \phi(A \cap Z) + \int_A f_m\,d\mu + 2^{-m}\mu(A).$$

Since $\mu(A)$ is finite, we obtain from the uniform convergence theorem that $\int_A f_m\,d\mu \to \int_A f\,d\mu$. Therefore,

$$\phi(A) = \phi(A \cap Z) + \int_A f\,d\mu,$$

which is the theorem in case $\phi \ge 0$ and $\mu(E) < +\infty$.

If $\phi \ge 0$ and $\mu(E) = +\infty$, then E can still be written as a disjoint union $E = \bigcup E_j$ with $\mu(E_j) < +\infty$, since E is σ-finite. Hence, there exist $Z_j \subset E_j$, $\mu(Z_j) = 0$, and nonnegative f_j on E_j such that

$$\phi(A \cap E_j) = \phi(A \cap Z_j) + \int_{A \cap E_j} f_j\,d\mu \qquad (A \subset E).$$

Letting $Z = \bigcup Z_j$ and $f = \sum f_j \chi_{E_j}$, we obtain $\mu(Z) = 0$, $f \ge 0$, and

$$\begin{aligned}\phi(A) &= \sum \phi(A \cap E_j) = \sum \phi(A \cap Z_j)\\ &\quad + \sum \int_{A \cap E_j} f_j\,d\mu = \phi(A \cap Z) + \int_A f\,d\mu.\end{aligned}$$

Of course, f is integrable since ϕ is finite. The proof is now complete if $\phi \geq 0$.

For an arbitrary ϕ, apply the decomposition to each of $\overline{V}$ and $\underline{V}$, and subtract the results. By the Jordan decomposition, we obtain $\phi(A) = \int_A f\,d\mu + \sigma(A)$, where $f \in L(E;d\mu)$ and σ is singular with respect to μ. It remains to show that there is a set Z, $\mu(Z) = 0$, such that $\sigma(A) = \phi(A \cap Z)$. Let Z be the set of μ-measure zero corresponding to σ in the definition of a singular set function. Then $\sigma(A \cap Z) = \sigma(A)$ and $\int_{A \cap Z} f\,d\mu = 0$. Hence, replacing A by $A \cap Z$ in the formula for ϕ above, we obtain $\sigma(A) = \phi(A \cap Z)$. This completes the proof. For a result concerning the uniqueness of f, see Exercise 6.

We have already noted that the indefinite integral of an integrable function is absolutely continuous. The following fundamental result gives a converse: namely, in a σ-finite space, the only absolutely continuous set functions are indefinite integrals.

(10.39) Theorem (*Radon-Nikodym*) *Let ϕ be an additive set function on the measurable subsets of a measurable E, and let μ be a σ-finite measure on E. If ϕ is absolutely continuous with respect to μ, there exists a unique $f \in L(E;d\mu)$ such that*

$$\phi(A) = \int_A f\,d\mu$$

for every measurable $A \subset E$.

Proof. The result follows from the Lebesgue decomposition. In fact, $\phi(A) = \int_A f\,d\mu + \phi(A \cap Z)$ for appropriate $f \in L(E;d\mu)$ and Z with $\mu(Z) = 0$. Since ϕ is absolutely continuous, we have $\phi(A \cap Z) = 0$, so that $\phi(A) = \int_A f\,d\mu$. For the uniqueness of f, see Exercise 6.

If ν and μ are two measures defined on the same family of measurable sets, we say that ν is *absolutely continuous with respect to* μ on a measurable set E if $\nu(A) = 0$ for every $A \subset E$ with $\mu(A) = 0$. If ν is finite, (10.34) implies that a necessary and sufficient condition for ν to be absolutely continuous with respect to μ is that given $\varepsilon > 0$, there exist $\delta > 0$ such that $\nu(A) < \varepsilon$ if $\mu(A) < \delta$. The necessity of this condition may fail if ν is not finite; see Exercise 12.

We say that ν and μ are *mutually singular* on E if E can be written as a disjoint union, $E = E_1 \cup E_2$, of two measurable sets with $\nu(E_1) = \mu(E_2) = 0$. The reader can check that the following analogue of (10.35) is valid: Two measures ν and μ are mutually singular on E if and only if given $\varepsilon > 0$, there are disjoint measurable E_1 and E_2 with $E = E_1 \cup E_2$ and $\nu(E_1) < \varepsilon$, $\mu(E_2) < \varepsilon$.

We also note that if ν and μ are mutually singular on E and if $g \in L(E;d\nu)$, then the set function $\int_A g\, d\nu$ is singular with respect to μ. To see this, write $E = E_1 \cup E_2$, where E_1 and E_2 are disjoint with $\nu(E_1) = \mu(E_2) = 0$. Setting $Z = E_2$, we have $\mu(Z) = 0$ and $\nu(A) = 0$ for every measurable $A \subset E - Z = E_1$. Hence, $\int_A g\, d\nu = 0$ for such A, which proves the assertion.

We have the following analogue of the Lebesgue decomposition.

(10.40) Theorem *Let ν and μ be two σ-finite measures defined on the measurable subsets of a measurable E. Then there is a unique, nonnegative measurable f on E and a unique measure σ on the measurable subsets of E such that σ and μ are mutually singular on E and $\nu(A) = \int_A f\, d\mu + \sigma(A)$ for every measurable $A \subset E$. Moreover,*

$$\int_A g\, d\nu = \int_A gf\, d\mu + \int_A g\, d\sigma$$

whenever $\int_A g\, d\nu$ exists.

Before giving the proof, we add several remarks. First, $\int_A f\, d\mu$ is an absolutely continuous measure with respect to μ since $f \geq 0$. Next, if $E = Z \cup (E - Z)$, where $\mu(Z) = \sigma(E - Z) = 0$, then σ has the form

$$\sigma(A) = \nu(A \cap Z),$$

as can be seen by replacing A by $A \cap Z$ in the decomposition of ν. Note also that if $\nu(E)$ is finite, then $f \in L(E;d\mu)$. Finally, if $g \in L(E;d\nu)$, then $g \in L(E;d\sigma)$. In this case, the second formula in the theorem implies that $fg \in L(E;d\mu)$ and expresses the Lebesgue decomposition of $\int_A g\, d\nu$ with respect to μ.

Proof. If $\nu(E) < +\infty$, the Lebesgue decomposition implies that $\nu(A) = \int_A f\, d\mu + \sigma(A)$ for measurable $A \subset E$, where $f \geq 0$, $f \in L(E;d\mu)$, and σ and μ are mutually singular. If $\nu(E) = +\infty$, then since ν is σ-finite, we have $E = \bigcup E_j$ with E_j disjoint and $\nu(E_j) < +\infty$. Choose $Z_j \subset E_j$ and f_j on E_j such that $\mu(Z_j) = 0$, $f_j \geq 0$ and

$$\nu(A \cap E_j) = \int_{A \cap E_j} f_j\, d\mu + \nu(A \cap Z_j) \text{ for measurable } A \subset E.$$

Let $Z = \bigcup Z_j$ and $f = \sum f_j \chi_{E_j}$. Then $\mu(Z) = 0$, and adding over j, we have

$$\nu(A) = \int_A f\, d\mu + \nu(A \cap Z) = \int_A f\, d\mu + \sigma(A),$$

as claimed. The proof of the uniqueness of f and σ is left as an exercise.

If g is the characteristic function χ_B of a measurable set B, the formula in question, namely, $\int_A g\, d\nu = \int_A gf\, d\mu + \int_A g\, d\sigma$, reduces to $\nu(A \cap B) = \int_{A \cap B} f\, d\mu + \sigma(A \cap B)$. Hence, the formula is valid for any simple measurable g, and therefore, by the monotone convergence theorem, for any measurable $g \geq 0$. Now let g be any measurable function for which $\int_A g\, d\nu$ exists. Then at least one of $\int_A g^+ \, d\nu$ and $\int_A g^- \, d\nu$ is finite, and the formula for g follows by subtracting those for g^+ and g^-.

(10.41) Corollary *Let ν and μ be two σ-finite measures defined on the measurable subsets of a measurable E.*

(i) *Then ν is absolutely continuous with respect to μ on E if and only if there is a nonnegative measurable f such that $\nu(A) = \int_A f \, d\mu$ for every measurable $A \subset E$. In this case,*

$$\int_A g \, d\nu = \int_A gf \, d\mu$$

for any measurable g and $A \subset E$ for which $\int_A g \, d\nu$ exists.

(ii) *Let $g \in L(E;d\nu)$. Then $\int_A g \, d\nu = \int_A gf \, d\mu$ for some nonnegative f and all measurable $A \subset E$ if and only if $\int_A g \, d\nu$ is absolutely continuous with respect to μ.*

Proof. Part (i) follows from (10.40) since $\sigma \equiv 0$ if and only if ν is absolutely continuous with respect to μ. Let $\nu(A) = \int_A f \, d\mu + \sigma(A)$ be the decomposition given by (10.40). Part (ii) follows from the fact that the formula $\int_A g \, d\nu = \int_A gf \, d\mu + \int_A g \, d\sigma$ is the Lebesgue decomposition of $\int_A g \, d\nu$.

4. The Dual Space of L^p

If B is a Banach space (or, more generally, a normed linear space) over the real numbers, a real-valued *linear functional* l on B is by definition a real-valued function $l(f)$, $f \in B$, which satisfies

$$l(f_1 + f_2) = l(f_1) + l(f_2), \qquad l(\alpha f) = \alpha l(f), \; -\infty < \alpha < +\infty.$$

A linear functional l is said to be *bounded* if there is a constant c such that $|l(f)| \le c\|f\|$ for all $f \in B$. A bounded linear functional l is continuous with respect to the norm in B, by which we mean that if $\|f - f_k\| \to 0$ as $k \to \infty$, then $l(f_k) \to l(f)$, since $|l(f) - l(f_k)| = |l(f - f_k)| \le c\|f - f_k\| \to 0$.

The *norm* $\|l\|$ of a bounded linear functional l is defined as

(10.42) $$\|l\| = \sup_{\|f\| \le 1} |l(f)|.$$

Since $f/\|f\|$ has norm 1 for any $f \ne 0$, and since l is linear, we have $\|l\| = \sup |l(f)|/\|f\|$.

The collection of all bounded linear functionals on B is called the *dual space* B' of B. We shall consider the case when $B = L^p = L^p(E;d\mu)$, and for simplicity restrict our attention to real-valued functions. Our goal is to show that if $1 \le p < \infty$ and μ is σ-finite, then the dual space $(L^p)'$ of L^p can be identified in a natural way with $L^{p'}$. The main tool in doing so is the Radon-Nikodym theorem. The first result is the following.

(10.43) Theorem *Let* $1 \le p \le \infty$, $1/p + 1/p' = 1$. *If* $g \in L^{p'}(E;d\mu)$, *then the formula*

$$l(f) = \int_E fg\,d\mu$$

defines a bounded linear functional $l \in [L^p(E;d\mu)]'$. *Moreover,* $\|l\| \le \|g\|_{p'}$.

Proof. This follows immediately from Hölder's inequality and the linear properties of the integral. We have

$$|l(f)| = \left|\int_E fg\,d\mu\right| \le \|g\|_{p'}\|f\|_p,$$

so that $\|l\| \le \|g\|_{p'}$.

The theorem shows that with each $g \in L^{p'}$ we can associate a bounded linear functional, $l(f) = \int_E fg\,d\mu$, on L^p. The correspondence between g and l is unique (see Exercise 6) and defines an embedding of $L^{p'}$ in $(L^p)'$. We now give the characterization of $(L^p)'$, $1 \le p < \infty$.

(10.44) Theorem *Let* $1 \le p < \infty$, $1/p + 1/p' = 1$, *and let* μ *be* σ*-finite. If* $l \in [L^p(E;d\mu)]'$, *there is a unique* $g \in L^{p'}(E;d\mu)$ *such that*

$$l(f) = \int_E fg\,d\mu.$$

Moreover, $\|l\| = \|g\|_{p'}$, *so that the correspondence between* l *and* g *defines an isometry between* $(L^p)'$ *and* $L^{p'}$.

Proof. Suppose first that $\mu(E) < +\infty$. Let $l \in (L^p)'$ and write $\|l\| = c$. Define a set function ϕ on the measurable sets $A \subset E$ by

$$\phi(A) = l(\chi_A).$$

Note that ϕ is finite; in fact, $|\phi(A)| \le c\|\chi_A\|_p = c\mu(A)^{1/p}$. Clearly, ϕ is finitely additive. To show that it is countably additive, suppose that $A = \bigcup_{k=1}^\infty A_k$, A_k disjoint. Write $A = (\bigcup_{k=1}^m A_k) \cup (\bigcup_{k=m+1}^\infty A_k) = A' \cup A''$. Then

$$\phi(A) = \phi(A') + \phi(A'') = \sum_{k=1}^m \phi(A_k) + \phi(A'').$$

Since $|\phi(A'')| \le c\mu(A'')^{1/p}$, $\phi(A'')$ tends to zero as $m \to \infty$. Hence, $\phi(A) = \sum_{k=1}^\infty \phi(A_k)$, which shows that ϕ is countably additive. The fact that $|\phi(A)| \le c\mu(A)^{1/p}$ implies that ϕ is absolutely continuous with respect to μ.

By the Radon-Nikodym theorem, there is a $g \in L^1(E;d\mu)$ such that $\phi(A) = \int_A g\,d\mu$ for measurable $A \subset E$. This means that $l(\chi_A) = \int_E \chi_A g\,d\mu$, so that $l(f) = \int_E fg\,d\mu$ for any simple measurable f. To show the same formula holds for any $f \in L^p$, we first claim that $g \in L^{p'}$ and $\|g\|_{p'} \le c$. If $p > 1$, choose simple functions h_k with $0 \le h_k \nearrow |g|^{p'}$. Let $\{g_k\}$ be the simple functions defined by

$$g_k = h_k^{1/p} \operatorname{sign} g.$$

Then $\|g_k\|_p = \|h_k\|_1^{1/p}$, and

$$\int_E g_k g \, d\mu = l(g_k) \le c\|g_k\|_p = c\|h_k\|_1^{1/p}.$$

Since $g_k g = h_k^{1/p}|g| \ge h_k^{1/p+1/p'} = h_k$, we obtain $\|h_k\|_1 \le c\|h_k\|_1^{1/p}$. We may assume that $\|h_k\|_1 \ne 0$ for large k. (Otherwise, g would be zero a.e., and there would be nothing to prove.) Hence, dividing both sides of the last inequality by $\|h_k\|_1^{1/p}$, we have $\|h_k\|_1^{1/p'} \le c$, so that $\|g\|_{p'} \le c$ by the monotone convergence theorem. This proves the claim when $p > 1$. The case $p = 1$ is left as an exercise.

To show that $l(f) = \int_E fg \, d\mu$ for any $f \in L^p$, choose simple f_k converging to f in L^p norm (see Exercise 8). Then $l(f_k) \to l(f)$, and $\int_E f_k g \, d\mu \to \int_E fg \, d\mu$ by Hölder's inequality:

$$\left| \int_E f_k g \, d\mu - \int_E fg \, d\mu \right| \le \int_E |f_k - f||g| \, d\mu \le \|f_k - f\|_p \|g\|_{p'}.$$

The fact that the formula holds for f_k thus implies that it holds for f by passing to the limit.

To complete the proof for the case $\mu(E) < +\infty$, it remains to show that $\|g\|_{p'} = c$ and that the correspondence between l and g is unique. However, we already know that $\|g\|_{p'} \le c$, and the opposite inequality follows from (10.43). For the uniqueness of the correspondence, see Exercise 6.

If $\mu(E) = +\infty$, then since μ is σ-finite, there exist $E_j \nearrow E$ with $\mu(E_j) < +\infty$. Let $l \in [L^p(E)]'$. Since the restriction of l to $L^p(E_j)$ is a bounded linear functional, there is a unique $g_j \in L^{p'}(E_j)$, $\|g_j\|_{p',E_j} \le \|l\|$, such that

$$l(f) = \int_{E_j} fg_j \, d\mu$$

for every f in L^p which vanishes outside E_j. For such f, the fact that $E_j \subset E_{j+1}$ also gives

$$l(f) = \int_{E_{j+1}} fg_{j+1} \, d\mu = \int_{E_j} fg_{j+1} \, d\mu.$$

Therefore, $g_{j+1} = g_j$ a.e. in E_j. We may assume that $g_{j+1} = g_j$ everywhere in E_j. Define $g(x)$ by $g(x) = g_j(x)$ if $x \in E_j$. Then g is measurable and it follows that $\|g\|_{p'} \le \|l\|$. If $f \in L^p(E)$, then

$$l(f\chi_{E_j}) = \int_{E_j} fg_j \, d\mu = \int_{E_j} fg \, d\mu.$$

Since $f\chi_{E_j}$ converges in L^p to f and $\int_{E_j} fg\, d\mu \to \int_E fg\, d\mu$ (note that $fg \in L^1$ by Hölder's inequality), we obtain $l(f) = \int_E fg\, d\mu$ in the limit. Therefore, by (10.43), $\|l\| \le \|g\|_{p'}$, so that $\|l\| = \|g\|_{p'}$, and the proof is complete.

We remark that not every bounded linear functional on L^∞ can be represented $l(f) = \int fg\, d\mu$ for some $g \in L^1$; an example is indicated in Exercise 18.

5. Relative Differentiation of Measures

Lebesgue's differentiation theorem (7.11) states that if f is locally integrable in $\mathbf{R}^n$, then

$$\lim_{h\to 0} \frac{1}{|Q_{\mathbf{x}}(h)|} \int_{Q_{\mathbf{x}}(h)} f(\mathbf{y})\, d\mathbf{y} = f(\mathbf{x}) \text{ a.e.},$$

where $Q_{\mathbf{x}}(h)$ is the cube with center $\mathbf{x}$ and edge length h. We will now study an analogue of this result for other measures on $\mathbf{R}^n$. Specifically, if μ and ν are two σ-finite measures on the Borel subsets of $\mathbf{R}^n$, we will study the existence of

$$\lim_{h\to 0} \frac{\nu(Q_{\mathbf{x}}(h))}{\mu(Q_{\mathbf{x}}(h))},$$

and its relation to the Lebesgue decomposition of ν with respect to μ.

We will follow the method used to prove Lebesgue's differentiation theorem. To do this, we must find a replacement for Vitali's lemma: the simple form of Vitali's lemma (see (7.4)) relies heavily on the fact that expanding a cube concentrically by a factor (say 5) only enlarges its Lebesgue measure proportionally, whereas no such relation may hold for general measures. In order to bypass this difficulty, we shall present a covering lemma which is purely geometric in nature, that is, which makes no mention of measure.

We consider only cubes whose edges are parallel to the coordinate axes, and write $Q = Q_{\mathbf{x}}$ for those with center $\mathbf{x}$. We say that a family K of cubes has *bounded overlaps* if there is a constant c such that every $\mathbf{x} \in \mathbf{R}^n$ belongs to at most c cubes from K. Thus, K has bounded overlaps if and only if

$$\sum_{Q\in K} \chi_Q(\mathbf{x}) \le c \qquad (\mathbf{x} \in \mathbf{R}^n).$$

(10.45) Theorem (***Besicovitch Covering Lemma***) *Let E be a bounded subset of $\mathbf{R}^n$, and let K be a family of cubes covering E which contains a cube $Q_{\mathbf{x}}$ with center $\mathbf{x}$ for each $\mathbf{x} \in E$. Then there exist points $\{\mathbf{x}_k\}$ in E such that*

(i) $E \subset \bigcup Q_{\mathbf{x}_k}$.
(ii) $\{Q_{\mathbf{x}_k}\}$ *has bounded overlaps.*

Moreover, the constant c for which $\sum \chi_{Q_{\mathbf{x}_k}} \le c$ can be chosen to depend only on n.

In order to prove this, it will be convenient to first prove the following lemma.

(10.46) Lemma *Let $\{Q_k\}_{k=1}^{\infty}$ be a sequence of cubes with centers $\{\mathbf{x}_k\}$ such that if $j < k$, then $\mathbf{x}_k \notin Q_j$ and $|Q_k| \le 2|Q_j|$. Then $\{Q_k\}$ has bounded overlaps, and the constant c for which $\sum \chi_{Q_k} \le c$ can be chosen to depend only on n.*

Proof. We will consider only $n = 2$; the case $n > 2$ is similar and left as an exercise. Let Q_{k_m}, $m = 1, 2, \ldots$, be those Q_k which contain the origin and whose centers are in the first quadrant, and let h_m denote the edge length of Q_{k_m}. Then Q_{k_1} covers at least the region

$$A = \{(x,y) : 0 \le x \le \tfrac{1}{2}h_1,\ 0 \le y \le \tfrac{1}{2}h_1\}.$$

Hence, no Q_{k_m} can have its center outside the set $\{(x,y) : 0 \le x \le h_1, 0 \le y \le h_1\}$; otherwise, we would have $h_m > 2h_1$ for some m, so that $|Q_{k_m}| > 4|Q_{k_1}|$, a contradiction. Therefore, the center of each Q_{k_m}, $m \ge 2$, must lie in one of the regions A, B, C, or D indicated below:

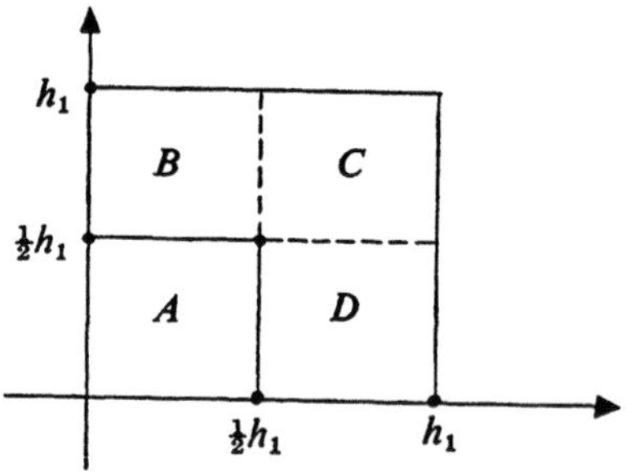

The center cannot be in A since that would contradict the assumption that the center of Q_{k_m}, $m \ge 2$, does not lie in Q_{k_1}. If it lies in B, then since Q_{k_m} contains 0, it covers B, and so there is at most one Q_{k_m} with center in B. Similar arguments hold for C and D. Applying the same reasoning to each quadrant, we see that there are at most 16 cubes in $\{Q_k\}$ which contain the origin. By translation, the same holds at any point of the plane.

Proof of Besicovitch's lemma. Let

$$\alpha_1 = \sup\{|Q_{\mathbf{x}}| : \mathbf{x} \in E\}.$$

If $\alpha_1 = +\infty$, there are arbitrarily large $Q_{\mathbf{x}}$, and since E is bounded, we simply choose one which contains E. If $\alpha_1 < +\infty$, write $E_1 = E$ and choose

$\mathbf{x}_1 \in E_1$ with $|Q_{\mathbf{x}_1}| > \alpha_1/2$. Let

$$E_2 = E_1 - Q_{\mathbf{x}_1}, \qquad \alpha_2 = \sup\{|Q_{\mathbf{x}}| : \mathbf{x} \in E_2\}.$$

If $\alpha_2 \neq 0$, choose $\mathbf{x}_2 \in E_2$ with $|Q_{\mathbf{x}_2}| > \alpha_2/2$. Proceed in this way, obtaining at the kth stage

$$E_k = E_{k-1} - Q_{\mathbf{x}_{k-1}} = E - \bigcup_{j=1}^{k-1} Q_{\mathbf{x}_j}, \qquad \alpha_k = \sup\{|Q_{\mathbf{x}}| : \mathbf{x} \in E_k\},$$
$$\mathbf{x}_k \in E_k, \qquad |Q_{\mathbf{x}_k}| > \alpha_k/2.$$

The process can be continued as long as $\alpha_k > 0$.

Since $\mathbf{x}_k \in E_k$, we have $\mathbf{x}_k \in E_j$ for all $j \leq k$. Therefore, $|Q_{\mathbf{x}_k}| \leq \alpha_k \leq \alpha_j < 2|Q_{\mathbf{x}_j}|$ if $j \leq k$. It follows that $\{Q_{\mathbf{x}_k}\}$ satisfies the hypothesis of (10.46), and so has bounded overlaps. It remains only to show that $E \subset \bigcup Q_{\mathbf{x}_k}$.

If some $\alpha_{k_0} = 0$, then E_{k_0} is empty, and E is contained in the union of the $Q_{\mathbf{x}_k}$, $k \leq k_0 - 1$. If no $\alpha_k = 0$, there are infinitely many $Q_{\mathbf{x}_k}$. Since $\alpha_k \searrow$ and $\alpha_k/2 < |Q_{\mathbf{x}_k}| \leq \alpha_k$, it follows that either $|Q_{\mathbf{x}_k}| \to 0$ or that there exists $\delta > 0$ such that $|Q_{\mathbf{x}_k}| \geq \delta$ for all k. The second possibility cannot arise; otherwise, E would not be bounded since $\mathbf{x}_k \in E$ but $\mathbf{x}_k$ is not in any $Q_{\mathbf{x}_j}$ with $j \leq k$. Hence, $|Q_{\mathbf{x}_k}| \to 0$, or equivalently, $\alpha_k \to 0$. If $\mathbf{x} \in E - \bigcup Q_{\mathbf{x}_k}$, then $\mathbf{x} \in E_k$ for all k. Therefore, $|Q_{\mathbf{x}}| \leq \alpha_k$ for all k, which means that $|Q_{\mathbf{x}}| = 0$. This shows that $E - \bigcup Q_{\mathbf{x}_k}$ is actually empty, and completes the proof.

A Borel measure μ on $\mathbf{R}^n$ (i.e., a measure on the Borel subsets of $\mathbf{R}^n$) is called *regular* if

$$\mu(E) = \inf\{\mu(G) : G \supset E,\ G \text{ open}\}$$

for every Borel set E. From now on, we will consider two regular Borel measures μ and ν on $\mathbf{R}^n$ which are finite on the bounded Borel sets. If $Q_{\mathbf{x}}(h)$ denotes the cube with center $\mathbf{x}$ and edge length h, we will also assume that

(i) $\mu(Q_{\mathbf{x}}(h)) > 0$ for $\mathbf{x} \in \mathbf{R}^n$, $h > 0$;
(ii) sets of the form

$$\left\{\mathbf{x} : \sup_{h>0} \frac{\nu(Q_{\mathbf{x}}(h))}{\mu(Q_{\mathbf{x}}(h))} > \alpha\right\}, \ \left\{\mathbf{x} : \limsup_{h\to 0} \frac{\nu(Q_{\mathbf{x}}(h))}{\mu(Q_{\mathbf{x}}(h))} > \alpha\right\}, \text{ etc.,}$$

are measurable.

Assumption (ii) is made for simplicity and is not necessary (see Exercise 17 of Chapter 11). Also, the assumption that μ and ν are regular is redundant (see (11.24) and the remarks following it).

We will use the fact that the class of continuous functions with compact support is dense in $L(d\mu)$; i.e., given $f \in L(d\mu)$, there exist continuous g_k with compact support such that $\int |f - g_k|\, d\mu \to 0$. (See Exercise 27.)

Note that when μ is Lebesgue measure and $\nu(E) = \int_E |f|\, d\mathbf{x}$ for a locally integrable f, then $\sup_{h>0} \nu(Q_{\mathbf{x}}(h))/\mu(Q_{\mathbf{x}}(h))$ is the Hardy-Littlewood maximal function of f [see (7.5)]. In the lemma below, we estimate the size of this expression for general μ and ν. The results we will prove on differentiation are corollaries of this estimate.

(10.47) Lemma *Let μ and ν satisfy the stated conditions. Then there is a constant c depending only on n such that*

(a) $\mu\{\mathbf{x} \in \mathbf{R}^n : \sup_{h>0} [\nu(Q_{\mathbf{x}}(h))/\mu(Q_{\mathbf{x}}(h))] > \alpha\} \le (c/\alpha)\nu(\mathbf{R}^n)$.
(b) $\mu\{\mathbf{x} \in E : \limsup_{h\to 0} [\nu(Q_{\mathbf{x}}(h))/\mu(Q_{\mathbf{x}}(h))] > \alpha\} \le (c/\alpha)\nu(E)$

for any Borel set $E \subset \mathbf{R}^n$ and any $\alpha > 0$.

Proof. (a) Fix $\alpha > 0$, and let

$$S = \left\{\mathbf{x} \in \mathbf{R}^n : \sup_{h>0} \frac{\nu(Q_{\mathbf{x}}(h))}{\mu(Q_{\mathbf{x}}(h))} > \alpha\right\}.$$

If B is any bounded Borel set and $\mathbf{x} \in S \cap B$, there is a cube $Q_{\mathbf{x}}$ with center $\mathbf{x}$ such that $\nu(Q_{\mathbf{x}})/\mu(Q_{\mathbf{x}}) > \alpha$. Using Besicovitch's lemma, select $\{Q_{\mathbf{x}_k}\}$ and c such that $\nu(Q_{\mathbf{x}_k}) > \alpha\mu(Q_{\mathbf{x}_k})$, $S \cap B \subset \bigcup Q_{\mathbf{x}_k}$ and $\sum \chi_{Q_{\mathbf{x}_k}} \le c$. We then have

$$\mu(S \cap B) \le \mu\left(\bigcup Q_{\mathbf{x}_k}\right) \le \sum \mu(Q_{\mathbf{x}_k}) < \frac{1}{\alpha}\sum \nu(Q_{\mathbf{x}_k}),$$

$$\sum \nu(Q_{\mathbf{x}_k}) = \sum \int_{\bigcup Q_{\mathbf{x}_k}} \chi_{Q_{\mathbf{x}_k}}\, d\nu \le c\int_{\bigcup Q_{\mathbf{x}_k}} d\nu = c\nu\left(\bigcup Q_{\mathbf{x}_k}\right).$$

Therefore, $\mu(S \cap B) \le c\nu(\bigcup Q_{\mathbf{x}_k})/\alpha$, so that $\mu(S \cap B) \le c\nu(\mathbf{R}^n)/\alpha$. Letting $B \nearrow \mathbf{R}^n$, we obtain $\mu(S) \le c\nu(\mathbf{R}^n)/\alpha$, as desired.

(b) Fix $\alpha > 0$, and let

$$T = \left\{\mathbf{x} \in E : \limsup_{h\to 0} \frac{\nu(Q_{\mathbf{x}}(h))}{\mu(Q_{\mathbf{x}}(h))} > \alpha\right\}.$$

If $\nu(E) = +\infty$, there is nothing to prove. Otherwise, choose an open set $G \supset E$ with $\nu(G) < \nu(E) + \varepsilon$, and let B be a bounded Borel set. If $\mathbf{x} \in T \cap B$, there is a cube $Q_{\mathbf{x}}$ such that $Q_{\mathbf{x}} \subset G$ and $\nu(Q_{\mathbf{x}})/\mu(Q_{\mathbf{x}}) > \alpha$. As above, there exists $\{Q_{\mathbf{x}_k}\}$, $Q_{\mathbf{x}_k} \subset G$, such that $\mu(T \cap B) \le c\nu(\bigcup Q_{\mathbf{x}_k})/\alpha$. Therefore, $\mu(T \cap B) \le c\nu(G)/\alpha \le c[\nu(E) + \varepsilon]/\alpha$. The result now follows by first letting $\varepsilon \to 0$ and then letting $B \nearrow \mathbf{R}^n$.

The first result about differentiation of measures is the following.

(10.48) Theorem *Let ν and μ satisfy the stated conditions. If ν and μ are mutually singular, then*

$$\lim_{h\to 0} \frac{\nu(Q_{\mathbf{x}}(h))}{\mu(Q_{\mathbf{x}}(h))} = 0 \text{ a.e. } (\mu).$$

Proof. Since ν and μ are mutually singular, there is a set Z with $\nu(\mathbf{R}^n - Z) = \mu(Z) = 0$. Let $E = \mathbf{R}^n - Z$, and consider the sets

$$T_\alpha = \left\{\mathbf{x} \in E : \limsup_{h\to 0} \frac{\nu(Q_\mathbf{x}(h))}{\mu(Q_\mathbf{x}(h))} > \alpha\right\}, \qquad (\alpha > 0),$$

$$T = \left\{\mathbf{x} \in E : \limsup_{h\to 0} \frac{\nu(Q_\mathbf{x}(h))}{\mu(Q_\mathbf{x}(h))} > 0\right\}.$$

By (10.47)(b), we have $\mu(T_\alpha) \le c\nu(E)/\alpha = 0$. Since T is the union of the T_{α_k} for any sequence $\alpha_k \to 0$, it also has μ-measure zero, and the result follows.

(10.49) Theorem *Let μ satisfy the stated conditions, and let f be a Borel measurable function which is integrable $(d\mu)$ over every bounded Borel set in $\mathbf{R}^n$. Then*

$$\lim_{h\to 0} \frac{1}{\mu(Q_\mathbf{x}(h))} \int_{Q_\mathbf{x}(h)} f\,d\mu = f(\mathbf{x}) \text{ a.e. } (\mu).$$

Proof. Assume first that $f \in L(\mathbf{R}^n; d\mu)$. For any integrable g, we have

$$\left|\frac{1}{\mu(Q_\mathbf{x}(h))} \int_{Q_\mathbf{x}(h)} f\,d\mu - f(\mathbf{x})\right| \le \frac{1}{\mu(Q_\mathbf{x}(h))} \int_{Q_\mathbf{x}(h)} |f - g|\,d\mu + \left|\frac{1}{\mu(Q_\mathbf{x}(h))} \int_{Q_\mathbf{x}(h)} g\,d\mu - f(\mathbf{x})\right|.$$

If g is also continuous, the last term on the right converges to $|g(\mathbf{x}) - f(\mathbf{x})|$ as $h \to 0$. Hence, letting $L(\mathbf{x})$ denote the limsup as $h \to 0$ of the term on the left, we obtain

$$L(\mathbf{x}) \le \sup_{h>0} \frac{1}{\mu(Q_\mathbf{x}(h))} \int_{Q_\mathbf{x}(h)} |f - g|\,d\mu + |g(\mathbf{x}) - f(\mathbf{x})|.$$

Therefore, the set S_ε where $L(\mathbf{x}) > \varepsilon$, $\varepsilon > 0$, is contained in the union of the two sets where the corresponding terms on the right side of the last inequality exceed $\frac{1}{2}\varepsilon$. From (10.47) and Tchebyshev's inequality, we obtain

$$\mu(S_\varepsilon) \le c(\tfrac{1}{2}\varepsilon)^{-1} \int_{\mathbf{R}^n} |f - g|\,d\mu + (\tfrac{1}{2}\varepsilon)^{-1} \int_{\mathbf{R}^n} |f - g|\,d\mu.$$

As noted earlier, g can be chosen such that $\int_{\mathbf{R}^n} |f - g|\,d\mu$ is arbitrarily small. Hence, $\mu(S_\varepsilon) = 0$ for every $\varepsilon > 0$, and the result follows.

The case when $f \notin L(\mathbf{R}^n; d\mu)$ is left as an exercise [cf. theorem (7.11)].

Combining the last two theorems, we obtain the main result.

(10.50) Corollary *Let ν and μ satisfy the stated conditions. If $\nu(E) =$*

$\int_E f\,d\mu + \sigma(E)$ is the decomposition of v into parts which are absolutely continuous and singular with respect to μ, then

$$\lim_{h\to 0}\frac{\nu(Q_{\mathbf{x}}(h))}{\mu(Q_{\mathbf{x}}(h))} = f(\mathbf{x}) \text{ a.e. } (\mu).$$

Exercises

1. Prove theorem (10.13).
2. A measure space $(\mathscr{S},\Sigma,\mu)$ is said to be *complete* if Σ contains all subsets of sets with measure zero; i.e., $(\mathscr{S},\Sigma,\mu)$ is complete if $Y \in \Sigma$ whenever $Y \subset Z$, $Z \in \Sigma$, and $\mu(Z) = 0$. In this case, show that if f is measurable and $g = f$ a.e., then g is also measurable. Is this true if $(\mathscr{S},\Sigma,\mu)$ is not complete?
3. Prove Egorov's theorem (10.14).
4. If $(\mathscr{S},\Sigma,\mu)$ is a measure space, and if f and $\{f_k\}$ are measurable and finite a.e. in a measurable set E, then $\{f_k\}$ is said to *converge in μ-measure* on E to limit f if

$$\lim_{k\to\infty} \mu\{x \in E : |f(x) - f_k(x)| > \varepsilon\} = 0,\ \varepsilon > 0.$$

   Formulate and prove analogues of theorems (4.21)–(4.23).
5. Complete the proof of lemma (10.18).
6. (a) If $f_1,f_2 \in L(d\mu)$ and $\int_E f_1\,d\mu = \int_E f_2\,d\mu$ for all measurable E, show that $f_1 = f_2$ a.e. (μ).
   (b) Prove the uniqueness of f and σ in (10.40).
   (c) Let μ be σ-finite, and let $f_1,f_2 \in L^{p'}(d\mu)$, $1/p + 1/p' = 1$, $1 \le p \le \infty$. If $\int f_1 g\,d\mu = \int f_2 g\,d\mu$ for all $g \in L^p(d\mu)$, show that $f_1 = f_2$ a.e.
7. Prove the integral convergence results in (10.27)–(10.31).
8. Show that for $1 \le p < \infty$, the class of simple functions vanishing outside sets of finite measure is dense in $L^p(d\mu)$.
9. The *symmetric difference* of two sets E_1 and E_2 is defined as

$$E_1 \,\Delta\, E_2 = (E_1 - E_2) \cup (E_2 - E_1).$$

   Let $(\mathscr{S},\Sigma,\mu)$ be a measure space, and identify E_1 and E_2 if $\mu(E_1 \,\Delta\, E_2) = 0$. Show that Σ is a metric space with distance $d(E_1,E_2) = \mu(E_1 \,\Delta E_2)$, and that if μ is finite $L^p(\mathscr{S},\Sigma,\mu)$ is separable if and only if Σ is, $1 \leqslant p < \infty$.
10. If ϕ is a set function whose Jordan decomposition is $\phi = \overline{V} - \underline{V}$, define

$$\int_E f\,d\phi = \int_E f\,d\overline{V} - \int_E f\,d\underline{V},$$

   provided not both integrals on the right are infinite with the same sign. If V is the total variation of ϕ on E, and if $|f| \le M$, prove that $|\int_E f\,d\phi| \le MV$.
11. Prove parts (ii)–(iv) of (10.33).

12. Give an example of a pair of measures ν and μ such that ν is absolutely continuous with respect to μ, but given $\varepsilon > 0$, there is no $\delta > 0$ such that $\nu(A) < \varepsilon$ for every A with $\mu(A) < \delta$. [Thus, the analogue for measures of (10.34) may fail.]

    Prove the analogue of (10.35) for mutually singular measures ν and μ.

13. Show that the set P of the Hahn decomposition is unique up to null sets. [By a null set for ϕ, we mean a set N such that $\phi(A) = 0$ for every measurable $A \subset N$.]

14. Complete the proof of (10.44) for $p = 1$.

15. (Converse of Hölder's inequality) If μ is σ-finite and $1 \le p \le \infty$, show that

$$||f||_p = \sup |\textstyle\int fg \, d\mu|,$$

where the supremum is taken over all bounded g which vanish outside sets of finite measure, and for which $||g||_{p'} \le 1$ and $\int fg \, d\mu$ exists. [If $1 < p \le \infty$ and $||f||_p < +\infty$, this can be deduced from (10.44).]

16. Consider a convolution operator $Tf(\mathbf{x}) = \int_{\mathbf{R}^n} f(\mathbf{y})K(\mathbf{x} - \mathbf{y}) \, d\mathbf{y}$ with $K \ge 0$. If $||Tf||_p \le M||f||_p$, $1 \le p \le \infty$, for $f \in L^p(\mathbf{R}^n, d\mathbf{x})$, show that $||Tf||_{p'} \le M||f||_{p'}$, $1/p + 1/p' = 1$. [Use Exercise 15 to write $||Tf||_{p'} = \sup_{||g||_p \le 1} |\int_{\mathbf{R}^n} (Tf)g \, d\mathbf{x}|$, and note that

$$\int_{\mathbf{R}^n} (Tf)(\mathbf{x})g(\mathbf{x}) \, d\mathbf{x} = \int_{\mathbf{R}^n} (T\tilde{g})(-\mathbf{y})f(\mathbf{y}) \, d\mathbf{y},$$

where $\tilde{g}(\mathbf{x}) = g(-\mathbf{x})$.]

17. Let μ be σ-finite and define $\mathscr{L}^p(d\mu)$ to be the class of *complex*-valued f with $\int |f|^p \, d\mu < +\infty$. Let l be a *complex*-valued bounded linear functional on $\mathscr{L}^p(d\mu)$. If $1 \le p < \infty$, show that there is a function $g \in \mathscr{L}^{p'}(d\mu)$ such that $l(f) = \int fg \, d\mu$. (Here, as usual, we define $\int h \, d\mu = \int h_1 \, d\mu + i \int h_2 \, d\mu$ if $h = h_1 + ih_2$ with h_1 and h_2 real-valued.) (HINT: Reduce to the real case.)

18. Give an example to show that $(L^\infty)' \neq L^1$. (Consider $L^\infty[-1,1]$ with Lebesgue measure, and let $\mathscr{C}$ be the subspace of continuous functions on $[-1,1]$ with the sup norm. Define $l(f) = f(0)$ for $f \in \mathscr{C}$. Then l is a bounded linear functional on $\mathscr{C}$, so by the Hahn-Banach theorem (see, e.g., p. 62 of N. Dunford and J. T. Schwartz, *Linear Operators*, Part 1, 4th printing, Interscience, New York, 1967.), l has an extension $\bar{l} \in (L^\infty)'$. If there were a function $g \in L^1[-1,1]$ such that $\bar{l}(f) = \int_{-1}^{1} fg \, dx$ for $f \in L^\infty[-1,1]$, then we would have $f(0) = \int_{-1}^{1} fg \, dx$ for $f \in \mathscr{C}$. Show that this implies that $g = 0$ a.e., so that $l \equiv 0$. The functional l is called the "δ-function.")

19. Complete the proof of (10.49).

20. Under the hypothesis of (10.49), prove that

$$\lim_{h \to 0} \frac{1}{\mu(Q_{\mathbf{x}}(h))} \int_{Q_{\mathbf{x}}(h)} |f(\mathbf{y}) - f(\mathbf{x})| \, d\mu(\mathbf{y}) = 0 \text{ a.e. } (\mu).$$

21. Derive an analogue of the Besicovitch covering lemma (10.45) for the case of two dimensions (x,y) when the squares $Q_{(x,y)}$ are replaced by rectangles $R_{(x,y)}(h)$

whose x and y dimensions are h and h^2, respectively. Use this result to prove that under the hypothesis of (10.49),

$$\frac{1}{\mu(R_{(x,y)}(h))}\int_{R_{(x,y)}(h)} f\,d\mu \to f(x,y) \text{ a.e. } (\mu) \text{ as } h \to 0.$$

22. Let μ be a finite measure on A, let f be measurable and bounded on A, and let ϕ be convex in an interval containing the range of f. Prove that

$$\phi\left(\frac{\int_A f\,d\mu}{\int_A d\mu}\right) \leq \frac{\int_A \phi(f)\,d\mu}{\int_A d\mu}.$$

[This is Jensen's inequality for measures. See (7.44).]

23. A sequence $\{\phi_k\}$ of set functions is said to be *uniformly absolutely continuous* with respect to a measure μ if given $\varepsilon > 0$, there exists $\delta > 0$ such that if E satisfies $\mu(E) < \delta$, then $|\phi_k(E)| < \varepsilon$ for all k. If $\{f_k\}$ is a sequence of integrable functions on a finite measure space $(\mathscr{S},\Sigma,\mu)$ which converges pointwise a.e. to an integrable f, show that $f_k \to f$ in $L(d\mu)$ norm if and only if the indefinite integrals of the f_k are uniformly absolutely continuous with respect to μ.

24. Let $(\mathscr{S},\Sigma,\mu)$ be a σ-finite measure space, and let f be Σ-measurable and integrable over $\mathscr{S}$. Let Σ_0 be a σ-algebra satisfying $\Sigma_0 \subset \Sigma$. Of course, f may not be Σ_0-measurable. Show that there is a unique function f_0 which is Σ_0-measurable such that $\int fg\,d\mu = \int f_0 g\,d\mu$ for every Σ_0-measurable g for which the integrals are finite. The function f_0 is called the *conditional expectation* of f with respect to Σ_0, denoted $f_0 = E(f|\Sigma_0)$. [Apply the Radon-Nikodym theorem to the set function $\phi(E) = \int_E f\,d\mu$, $E \in \Sigma_0$.]

25. Using the notation of the preceding exercise, prove the following:
(a) $E(af + bg|\Sigma_0) = aE(f|\Sigma_0) + bE(g|\Sigma_0)$, a,b constants.
(b) $E(f|\Sigma_0) \geq 0$ if $f \geq 0$.
(c) $E(fg|\Sigma_0) = gE(f|\Sigma_0)$ if g is Σ_0-measurable.
(d) If $\Sigma_1 \subset \Sigma_0 \subset \Sigma$, then $E(f|\Sigma_1) = E(E(f|\Sigma_0)|\Sigma_1)$.

26. (*Hardy's inequality*) Let $f \geq 0$ on $(0,\infty)$, $1 \leq p < \infty$, $d\mu(x) = x^\alpha\,dx$ and $d\nu(x) = x^{\alpha+p}\,dx$ on $(0,\infty)$. Prove there exists a constant c independent of f such that

(i) $\displaystyle\int_0^\infty \left(\int_0^x f(t)\,dt\right)^p d\mu(x) \leq c\int_0^\infty f^p(x)\,d\nu(x) \qquad (\alpha < -1)$

(ii) $\displaystyle\int_0^\infty \left(\int_x^\infty f(t)\,dt\right)^p d\mu(x) \leq c\int_0^\infty f^p(x)\,d\nu(x) \qquad (\alpha > -1).$

[For (i), $(\int_0^x f(t)\,dt)^p \leq cx^{p-\eta-1}\int_0^x f(t)t^\eta dt$ by Hölder's inequality, provided $p - \eta - 1 > 0$. Multiply both sides by x^α, integrate over $(0,\infty)$, change the order of integration, and observe that an appropriate η exists since $\alpha < -1$.]

27. If μ is a regular Borel measure on $\mathbf{R}^n$, show that the class of continuous functions with compact support is dense in $L^p(d\mu)$, $1 \leq p < \infty$. [By Exercise 8, it is enough to approximate χ_E, where E is a Borel set with finite measure. Given $\varepsilon > 0$, there exist open G and closed F with $F \subset E \subset G$ and $\mu(G - F) < \varepsilon$. Now use Urysohn's lemma: if F_1 and F_2 are disjoint closed sets in $\mathbf{R}^n$, there is a continuous f on $\mathbf{R}^n$ with $0 \leq f \leq 1$, $f = 1$ on F_1, $f = 0$ on F_2.]

Chapter 11

Outer Measure; Measure

1. Constructing Measures from Outer Measures

A function $\Gamma = \Gamma(A)$ which is defined for every subset A of a space $\mathscr{S}$ is called an *outer measure* if it satisfies the following:

(i) $\Gamma(A) \geq 0$, $\Gamma(\emptyset) = 0$.
(ii) $\Gamma(A_1) \leq \Gamma(A_2)$ if $A_1 \subset A_2$.
(iii) $\Gamma(\bigcup A_k) \leq \sum \Gamma(A_k)$ for any countable collection of sets $\{A_k\}$.

For example, ordinary Lebesgue outer measure is an outer measure on the subsets of $\mathbf{R}^n$. Some other concrete examples will be constructed later in the chapter.

As with Lebesgue outer measure, it is possible to use any outer measure to introduce a class of measurable sets and a corresponding measure. In doing so, we base the definition of measurability on Carathéodory's theorem (3.30). Thus, given an outer measure Γ, we say that a subset E of $\mathscr{S}$ is Γ-*measurable*, or simply *measurable*, if

(11.1) $$\Gamma(A) = \Gamma(A \cap E) + \Gamma(A - E)$$

for *every* $A \subset \mathscr{S}$. Equivalently, E is measurable if and only if

$$\Gamma(A_1 \cup A_2) = \Gamma(A_1) + \Gamma(A_2) \text{ whenever } A_1 \subset E,\ A_2 \subset \mathscr{S} - E.$$

As a simple example, we will show that any set Z with $\Gamma(Z) = 0$ is measurable. In fact, for such Z and any $A \subset \mathscr{S}$, property (ii) gives

$$\Gamma(A \cap Z) + \Gamma(A - Z) \leq \Gamma(Z) + \Gamma(A) = \Gamma(A).$$

But by (iii), the opposite inequality $\Gamma(A) \leq \Gamma(A \cap Z) + \Gamma(A - Z)$ is always true, and the measurability of Z follows.

If E is a measurable set, then $\Gamma(E)$ is called its Γ-*measure*, or simply its *measure*. The terminology is justified by the following theorem.

(11.2) Theorem *Let Γ be an outer measure on the subsets of $\mathscr{S}$.*

(i) *The family of Γ-measurable subsets of $\mathscr{S}$ forms a σ-algebra.*

(ii) *If $\{E_k\}$ is a countable collection of disjoint measurable sets, then $\Gamma(\bigcup E_k) = \sum \Gamma(E_k)$. More generally, for any A, measurable or not,*

$$\Gamma(A \cap \bigcup E_k) = \sum \Gamma(A \cap E_k) \text{ and } \Gamma(A) = \sum \Gamma(A \cap E_k) + \Gamma(A - \bigcup E_k).$$

Proof. Let $\{E_k\}$ be a collection of disjoint measurable sets, let $H = \bigcup_{k=1}^{\infty} E_k$ and $H_j = \bigcup_{k=1}^{j} E_k$, $j = 1, 2, \ldots$. We first claim that

$$\Gamma(A) = \sum_{k=1}^{j} \Gamma(A \cap E_k) + \Gamma(A - H_j).$$

The proof will be by induction on j. If $j = 1$, the formula follows from the measurability of E_1. Assuming that the formula holds for $j - 1$, we have

$$\begin{aligned}\Gamma(A) &= \Gamma(A \cap E_j) + \Gamma(A - E_j)\\ &= \Gamma(A \cap E_j) + \sum_{k=1}^{j-1} \Gamma((A - E_j) \cap E_k) + \Gamma((A - E_j) - H_{j-1}).\end{aligned}$$

Since the E_k are disjoint, $(A - E_j) \cap E_k = A \cap E_k$ for $k \le j - 1$. Hence, since $(A - E_j) - H_{j-1} = A - H_j$, we obtain $\Gamma(A) = \sum_{k=1}^{j} \Gamma(A \cap E_k) + \Gamma(A - H_j)$, as required. This proves the claim.

Since $H_j \subset H$, we have $\Gamma(A - H_j) \ge \Gamma(A - H)$. Using this fact in the formula above and letting $j \to \infty$, it follows that

$$\Gamma(A) \ge \sum_{k=1}^{\infty} \Gamma(A \cap E_k) + \Gamma(A - H) \ge \Gamma(A \cap H) + \Gamma(A - H).$$

However, we also have $\Gamma(A) \le \Gamma(A \cap H) + \Gamma(A - H)$. Therefore, H is measurable and $\Gamma(A) = \sum_{k=1}^{\infty} \Gamma(A \cap E_k) + \Gamma(A - H)$. Replacing A by $A \cap H$ in this equation, we obtain $\Gamma(A \cap H) = \sum_{k=1}^{\infty} \Gamma(A \cap E_k)$, and the proof of (ii) is complete.

Note we have also shown that a countable union of disjoint measurable sets is measurable. To prove (i), it remains to show that a countable union of arbitrary measurable sets is measurable, and that the complement of a measurable set is measurable. To prove these facts, we shall use the following lemma.

(11.3) Lemma *If E_1 and E_2 are measurable, then so is $E_1 - E_2$.*

Proof. We will show that $\Gamma(A \cup B) = \Gamma(A) + \Gamma(B)$ whenever $A \subset E_1 - E_2$ and $B \subset C(E_1 - E_2)$. Since $B = (B \cap E_2) \cup (B - E_2)$, we have $A \cup B = [A \cup (B - E_2)] \cup [B \cap E_2]$. Hence, since $A \cup (B - E_2) \subset CE_2$ and $B \cap E_2 \subset E_2$, it follows from the measurability of E_2 that $\Gamma(A \cup B) = \Gamma(A \cup (B - E_2)) + \Gamma(B \cap E_2)$. However, $A \subset E_1$ and $B - E_2 \subset$

$C(E_1 - E_2) - E_2 \subset CE_1$. Therefore, since E_1 is measurable, $\Gamma(A \cup (B - E_2)) = \Gamma(A) + \Gamma(B - E_2)$. Combining equalities and using the measurability of E_2, we obtain

$$\Gamma(A \cup B) = \Gamma(A) + \Gamma(B - E_2) + \Gamma(B \cap E_2) = \Gamma(A) + \Gamma(B),$$

which proves the lemma.

Returning now to the proof of (11.2), we see from the lemma and the fact that the whole space $\mathscr{S}$ is measurable that the complement of a measurable set is measurable. Moreover, since $E_1 \cup E_2 = C(CE_1 - E_2)$, it follows that $E_1 \cup E_2$ is measurable if E_1 and E_2 are. Therefore, any finite union of measurable sets is measurable. Now, let $\{E_k\}$ be a countable collection of measurable sets. If $H_j = \bigcup_{k=1}^{j} E_k$, then

$$\bigcup_{k=1}^{\infty} E_k = H_1 \cup \left[\bigcup_{j=1}^{\infty} (H_{j+1} - H_j)\right],$$

and since the H_j are measurable and increasing, the terms on the right are measurable and disjoint. Thus, by the case already proved, it follows that $\bigcup_{k=1}^{\infty} E_k$ is measurable. This completes the proof of the theorem.

According to (11.2), an outer measure Γ is a measure on the σ-algebra of Γ-measurable sets, and so enjoys the usual properties of measures. We also have the following result.

(11.4) Corollary *Let Γ be an outer measure on $\mathscr{S}$, let $\{E_k\}$ be a collection of measurable sets, and let A be any set.*

(i) *If $E_k \nearrow$, then $\Gamma(A \cap \lim E_k) = \lim_{k\to\infty} \Gamma(A \cap E_k)$; if $E_k \searrow$ and if $\Gamma(A \cap E_{k_0})$ is finite for some k_0, then $\Gamma(A \cap \lim E_k) = \lim_{k\to\infty} \Gamma(A \cap E_k)$.*

(ii) *$\Gamma(A \cap \liminf E_k) \le \liminf_{k\to\infty} \Gamma(A \cap E_k)$; if $\Gamma(A \cap \bigcup_{k=k_0}^{\infty} E_k)$ is finite for some k_0, then $\Gamma(A \cap \limsup E_k) \ge \limsup_{k\to\infty} \Gamma(A \cap E_k)$.*

Proof. We will prove the first statements in (i) and (ii); the proofs of the second statements are left as exercises. Let $E_k \nearrow$ and be measurable. To prove the first part of (i), we may assume that $\Gamma(A \cap E_k)$ is finite for each k; otherwise, the result is clear. The sets $E_1, E_2 - E_1, \ldots, E_{k+1} - E_k, \ldots$, are disjoint and measurable. Since

$$\lim E_k = \bigcup_{k=1}^{\infty} E_k = E_1 \cup \left[\bigcup_{k=1}^{\infty} (E_{k+1} - E_k)\right],$$

it follows from (11.2) that

$$\Gamma(A \cap \lim E_k) = \Gamma(A \cap E_1) + \sum_{k=1}^{\infty} \Gamma(A \cap (E_{k+1} - E_k)).$$

Moreover, since E_k and $E_{k+1} - E_k$ are disjoint and measurable with finite measures, we have $\Gamma(A \cap (E_{k+1} - E_k)) = \Gamma(A \cap E_{k+1}) - \Gamma(A \cap E_k)$. Therefore,

$$\Gamma(A \cap \lim E_k) = \Gamma(A \cap E_1) + \sum_{k=1}^{\infty} [\Gamma(A \cap E_{k+1}) - \Gamma(A \cap E_k)]$$
$$= \lim_{k \to \infty} \Gamma(A \cap E_{k+1}),$$

which proves the first part of (i).

For the first part of (ii), let $\{E_k\}$ be measurable and define $X_j = \bigcap_{k=j}^{\infty} E_k$, $j = 1, 2, \ldots$. Then $X_j \nearrow \liminf E_k$, so that by (i), $\Gamma(A \cap \liminf E_k) = \lim_{j\to\infty} \Gamma(A \cap X_j)$. But since $A \cap X_j \subset A \cap E_j$, we have $\lim_{j\to\infty} \Gamma(A \cap X_j) \le \liminf_{j\to\infty} \Gamma(A \cap E_j)$, and the result follows.

2. Metric Outer Measures

Now let us introduce a new assumption concerning the underlying space $\mathscr{S}$: namely, that it is a metric space with metric d. The distance between two sets A_1 and A_2 is then defined by

$$d(A_1,A_2) = \inf \{d(x,y) : x \in A_1, y \in A_2\},$$

as in Euclidean space (see p. 5). An outer measure Γ on $\mathscr{S}$ is called a *metric outer measure*, or *an outer measure in the sense of Carathéodory*, if

$$\Gamma(A_1 \cup A_2) = \Gamma(A_1) + \Gamma(A_2) \text{ whenever } d(A_1,A_2) > 0.$$

For example, by (3.16) of Chapter 3, Lebesgue outer measure satisfies this condition.

Since $\mathscr{S}$ is a metric space, it has the topology induced by its metric. Thus, a set G in $\mathscr{S}$ is said to be open if for every $x \in G$, there is a $\delta > 0$ such that the "ball" $\{y : d(x,y) < \delta\}$ lies in G. A closed set is by definition the complement of an open set, and $\mathscr{B}$ denotes the σ-algebra of *Borel subsets* of $\mathscr{S}$; i.e., $\mathscr{B}$ is the smallest σ-algebra containing all the open (closed) subsets of $\mathscr{S}$.

(11.5) Theorem *Let Γ be a metric outer measure on a metric space $\mathscr{S}$. Then every Borel subset of $\mathscr{S}$ is Γ-measurable.*

Proof. Since the collection of Γ-measurable sets is a σ-algebra, it is enough to prove that every closed set is Γ-measurable. To prove this, we will use the following fact.

(11.6) Lemma *Let Γ be a metric outer measure on a space $\mathscr{S}$ with metric d. Let A be any set contained in an open set G, and let $A_k = \{x \in A : d(x,CG) \ge (1/k)\}$, $k = 1, 2, \ldots$. Then $\lim_{k\to\infty} \Gamma(A_k) = \Gamma(A)$.*

Proof. Since G is open, we have $A_k \nearrow A$. Clearly, $\lim_{k\to\infty} \Gamma(A_k) \le \Gamma(A)$.

To prove the opposite inequality, let $D_k = A_{k+1} - A_k$, $k = 1, 2, \ldots$. Then $d(D_{k+1}, A_k) \geq [(1/k) - (1/(k+1))] > 0$ and

$$A = A_k \cup D_k \cup D_{k+1} \cup \cdots, \qquad \Gamma(A) \leq \Gamma(A_k) + \Gamma(D_k) + \Gamma(D_{k+1}) + \cdots.$$

If $\sum \Gamma(D_j) < +\infty$, then $\sum_{j \geq k} \Gamma(D_j)$ tends to zero as $k \to \infty$, and it follows that $\Gamma(A) \leq \lim_{k\to\infty} \Gamma(A_k)$, as desired.

If $\sum \Gamma(D_j) = +\infty$, then at least one of $\sum \Gamma(D_{2j})$ and $\sum \Gamma(D_{2j+1})$ is infinite. We can therefore choose N so that $\Gamma(D_N) + \Gamma(D_{N-2}) + \Gamma(D_{N-4}) + \cdots$ is arbitrarily large. However, the fact that $D_{k-1} \subset A_k$ implies that the distance between D_{k+1} and D_{k-1} is positive. Therefore,

$$\Gamma(D_N \cup D_{N-2} \cup D_{N-4} \cup \cdots) = \Gamma(D_N) + \Gamma(D_{N-2}) + \Gamma(D_{N-4}) + \cdots.$$

Since A_{N+1} contains $D_N \cup D_{N-2} \cup D_{N-4} \cup \cdots$, it follows that $\lim \Gamma(A_k) = +\infty$, and the lemma is proved.

Proof of (11.5). Let F be any closed set. It is enough to show that $\Gamma(A \cup B) = \Gamma(A) + \Gamma(B)$ for $A \subset CF$, $B \subset F$. If $A_k = \{x \in A : d(x,F) \geq (1/k)\}$, then $d(A_k,B) \geq (1/k)$, so that $\Gamma(A_k \cup B) = \Gamma(A_k) + \Gamma(B)$. Therefore, $\Gamma(A \cup B) \geq \Gamma(A_k) + \Gamma(B)$. Letting $k \to \infty$, it follows from the lemma that $\Gamma(A \cup B) \geq \Gamma(A) + \Gamma(B)$. Since the opposite inequality is also true, the theorem is proved.

If $\mathscr{S}$ is a metric space, the notions of upper and lower semicontinuity of functions can be defined just as in $\mathbf{R}^n$. For example, a function f defined near a point x_0 is said to be upper semicontinuous at x_0 if

$$\limsup_{x \to x_0} f(x) \leq f(x_0).$$

Here, of course, the notation $x \to x_0$ means that $d(x,x_0) \to 0$. The results of Theorem (4.14) are valid for metric spaces; for example f is usc at every point of $\mathscr{S}$ if and only if $\{f \geq a\}$ is closed for every a. We thus obtain the following fact.

(11.7) Corollary *Let Γ be a metric outer measure on $\mathscr{S}$. Then every semicontinuous function on $\mathscr{S}$ is Γ-measurable.*

Proof. Suppose for example that f is upper semicontinuous on $\mathscr{S}$. Then $\{x : f(x) \geq a\}$ is closed for every a, and so is Γ-measurable by (11.5). Hence, f is Γ-measurable. If f is lower semicontinuous, then $-f$ is upper semicontinuous, and the corollary follows.

3. Lebesgue-Stieltjes Measure

In this section and the next, we will consider two specific examples of Carathéodory outer measures. The first of these, known as Lebesgue-Stieltjes outer measure, elucidates the connection between measures and monotone functions. The situation is relatively simple for measures on $\mathbf{R}^1$ and monotone functions of a single variable, and we shall restrict our attention to this case. Extensions to higher dimensions are possible, but more complicated.

To construct Lebesgue-Stieltjes outer measure, consider any fixed function f which is finite and monotone increasing on $(-\infty,+\infty)$. For each half-open finite interval of the form $(a,b]$, let

$$\lambda(a,b] = \lambda_f((a,b]) = f(b) - f(a).$$

Note that $\lambda \geq 0$ since f is increasing. If A is a nonempty subset of $\mathbf{R}^1$, let

$$\Lambda^*(A) = \Lambda_f^*(A) = \inf \sum \lambda(a_k,b_k],$$

where the inf is taken over all countable collections $\{(a_k,b_k]\}$ such that $A \subset \bigcup (a_k,b_k]$. Further, define $\Lambda^*(\emptyset) = 0$.

(11.8) Theorem *Λ^* is a Carathéodory outer measure on $\mathbf{R}^1$.*

Proof. We have $\Lambda^* \geq 0$ and $\Lambda^*(\emptyset) = 0$. First, we will show that if $A_1 \subset A_2$, then $\Lambda^*(A_1) \leq \Lambda^*(A_2)$. This is obvious if either $A_1 = \emptyset$ or $\Lambda^*(A_2) = +\infty$. In any other case, choose $\{(a_k,b_k]\}$ such that $A_2 \subset \bigcup (a_k,b_k]$ and $\sum \lambda(a_k,b_k] < \Lambda^*(A_2) + \varepsilon$. Then $A_1 \subset \bigcup (a_k,b_k]$, so that $\Lambda^*(A_1) \leq \sum \lambda(a_k,b_k]$. Therefore, $\Lambda^*(A_1) < \Lambda^*(A_2) + \varepsilon$, and the result follows by letting $\varepsilon \to 0$.

To show that Λ^* is subadditive, let $\{A_j\}_{j=1}^{\infty}$ be a collection of nonempty subsets of $\mathbf{R}^1$ and let $A = \bigcup A_j$. We may assume that $\Lambda^*(A_j) < +\infty$ for each j. Choose $\{(a_k^j,b_k^j]\}$ such that

$$A_j \subset \bigcup_k (a_k^j,b_k^j], \qquad \sum_k \lambda(a_k^j,b_k^j] < \Lambda^*(A_j) + \varepsilon 2^{-j}.$$

Since $A \subset \bigcup_{j,k} (a_k^j,b_k^j]$, we have

$$\Lambda^*(A) \leq \sum_{j,k} \lambda(a_k^j,b_k^j] < \sum_j \Lambda^*(A_j) + \varepsilon.$$

It follows that $\Lambda^*(A) \leq \sum \Lambda^*(A_j)$, and therefore that Λ^* is an outer measure.

To show that Λ^* is a Carathéodory outer measure, observe that if $a = a_0 < a_1 < \cdots < a_N = b$, then

$$\lambda(a,b] = f(b) - f(a) = \sum_{k=1}^{N} [f(a_k) - f(a_{k-1})] = \sum_{k=1}^{N} \lambda(a_{k-1},a_k].$$

It follows that in defining Λ^*, we can always work with arbitrarily short intervals $(a_k,b_k]$. Hence, if A_1 and A_2 satisfy $d(A_1,A_2) > 0$, then given $\varepsilon > 0$, we can choose $\{(a_k,b_k]\}$ such that each $(a_k,b_k]$ has length less than $d(A_1,A_2)$ and

$$A_1 \cup A_2 \subset \bigcup (a_k,b_k], \qquad \sum \lambda(a_k,b_k] \leq \Lambda^*(A_1 \cup A_2) + \varepsilon.$$

Thus, $\{(a_k,b_k]\}$ splits into two coverings, one of A_1 and the other of A_2. Therefore, $\Lambda^*(A_1) + \Lambda^*(A_2) \le \sum \lambda(a_k,b_k]$, so that since ε is arbitrary, $\Lambda^*(A_1) + \Lambda^*(A_2) \le \Lambda^*(A_1 \cup A_2)$. But the opposite inequality is always true, which completes the proof.

Λ_f^* is called the *Lebesgue-Stieltjes outer measure corresponding to* f, and its restriction to those sets which are Λ_f^*-measurable is called the *Lebesgue-Stieltjes measure corresponding to* f, and denoted Λ_f or simply Λ.

We leave it as an exercise to show that the Lebesgue-Stieltjes outer measure Λ_x^* corresponding to $f(x) = x$ coincides with ordinary Lebesgue outer measure. Hence, by Carathéodory's theorem (3.30), a set if Λ_x^*-measurable if and only if it is Lebesgue measurable.

An outer measure Γ defined on the subsets of a set $\mathscr{S}$ is said to be *regular* if for every $A \subset \mathscr{S}$, there is a Γ-measurable set E such that $A \subset E$ and $\Gamma(A) = \Gamma(E)$. The next theorem shows that any Lebesgue-Stieltjes outer measure is regular; in fact, it shows that any set in $\mathbf{R}^1$ can be included in a Borel set with the same Lebesgue-Stieltjes outer measure. Of course, Borel sets are Λ^*-measurable by (11.5).

(11.9) Theorem *Let Λ^* be a Lebesgue-Stieltjes outer measure. If A is a subset of $\mathbf{R}^1$, there is a Borel set B containing A such that $\Lambda^*(A) = \Lambda(B)$.*

Proof. Given $j = 1, 2, \ldots$, choose $\{(a_k,b_k]\}$ such that

$$A \subset \bigcup_k (a_k^j,b_k^j], \qquad \sum_k \lambda(a_k^j,b_k^j] \le \Lambda^*(A) + \frac{1}{j}.$$

Let $B_j = \bigcup_k (a_k^j,b_k^j]$ and $B = \bigcap B_j$. Then $A \subset B$ and B is a Borel set. Moreover,

$$\Lambda(B_j) \le \sum_k \lambda(a_k^j,b_k^j] \le \Lambda^*(A) + \frac{1}{j}.$$

Since $B \subset B_j$, it follows that $\Lambda(B) \le \Lambda^*(A) + (1/j)$, so that $\Lambda(B) \le \Lambda^*(A)$. But the opposite inequality is also true since $A \subset B$, and the theorem follows.

If μ is a finite Borel measure on $\mathbf{R}^1$, define

$$f_\mu(x) = \mu((-\infty,x]), \quad -\infty < x < +\infty.$$

Note that f_μ is monotone increasing and that $\mu((a,b]) = f_\mu(b) - f_\mu(a)$. It is natural to ask if the Lebesgue-Stieltjes measure induced by f_μ agrees with μ as a Borel measure. An affirmative answer would mean that every finite Borel measure is a Lebesgue-Stieltjes measure. We shall see later [(11.22)] that this is actually the case, and that the continuity from the right of f_μ (see Exercise 2) plays a role.

(11.10) Theorem *If f is an increasing function which is continuous from the right, then its Lebesgue-Stieltjes measure Λ satisfies*

$$\Lambda((a,b]) = f(b) - f(a).$$

In particular, $\Lambda(\{a\}) = f(a) - f(a-)$.

Proof. Since $(a,b]$ covers itself, we always have $\Lambda((a,b]) \le f(b) - f(a)$. To show the opposite inequality, suppose that $(a,b] \subset \bigcup (a_k,b_k]$. Given $\varepsilon > 0$, use the right continuity of f to choose $\{b_k'\}$ with

$$b_k < b_k', \qquad f(b_k) > f(b_k') - \varepsilon 2^{-k}.$$

If a' satisfies $a < a' < b$, then $[a',b]$ is covered by the (a_k,b_k'), and therefore, there is a finite N such that $[a',b] \subset \bigcup_{k=1}^{N} (a_k,b_k')$. By discarding any unnecessary (a_k,b_k') and reindexing the rest, we may assume that $a_{k+1} < b_k'$ for $k = 1, \ldots, N-1$. Also, $a_1 < a'$ and $b < b_N'$, so that $f(a_1) \le f(a')$ and $f(b) \le f(b_N')$. We have

$$\begin{aligned}\sum_k \lambda(a_k,b_k] \ge \sum_{k=1}^{N} \lambda(a_k,b_k] &= \sum_{k=1}^{N} [f(b_k) - f(a_k)] \\ &= f(b_N) - f(a_1) + \sum_{k=1}^{N-1} [f(b_k) - f(a_{k+1})].\end{aligned}$$

Now,

$$\begin{aligned}f(b_N) - f(a_1) &= [f(b_N) - f(b_N')] + [f(b_N') - f(a_1)] \\ &\ge -\varepsilon + [f(b) - f(a')].\end{aligned}$$

Also, since $f(b_k') - f(a_{k+1}) \ge 0$ for $k = 1, \ldots, N-1$,

$$\begin{aligned}\sum_{k=1}^{N-1} [f(b_k) - f(a_{k+1})] &= \sum_{k=1}^{N-1} [f(b_k) - f(b_k')] + \sum_{k=1}^{N-1} [f(b_k') - f(a_{k+1})] \\ &\ge \sum_{k=1}^{\infty} (-\varepsilon 2^{-k}) \ge -\varepsilon.\end{aligned}$$

Combining estimates, we obtain

$$\sum_k \lambda(a_k,b_k] \ge -2\varepsilon + [f(b) - f(a')].$$

Letting $\varepsilon \to 0$ and $a' \to a$, we have $\sum_k \lambda(a_k,b_k] \ge f(b) - f(a)$. Hence, $\Lambda((a,b]) \ge f(b) - f(a)$, and the first statement of the theorem follows. The second statement is proved by applying the first to the intervals $(a - (1/k), a]$, $k = 1, 2, \ldots$, which decrease to $\{a\}$.

Let g be a Borel measurable function defined on $\mathbf{R}^1$, and let Λ_f be a Lebesgue-Stieltjes measure. Then the integral $\int g \, d\Lambda_f$ is called the *Lebesgue-Stieltjes integral of g with respect to* Λ_f.† The next theorem gives a relation between $\int g \, d\Lambda_f$ and the usual Riemann-Stieltjes integral $\int g \, df$.

†In some other texts, any integral $\int f \, d\mu$ of the kind considered in Chapter 10 is called a Lebesgue-Stieltjes integral. We shall use the terminology only when μ is a Lebesgue-Stieltjes measure.

(11.11) Theorem *Let f be an increasing function which is right continuous on $[a,b]$, and let g be a bounded Borel measurable function on $[a,b]$. If the Riemann-Stieltjes integral $\int_a^b g\,df$ exists, then*

$$\int_{(a,b]} g\,d\Lambda_f = \int_a^b g\,df.$$

Proof. Let $\Gamma = \{x_j\}$ be a partition of $[a,b]$, and let m_j and M_j be the inf and sup of g in $[x_{j-1},x_j]$, respectively. Let

$$L_\Gamma = \sum m_j[f(x_j) - f(x_{j-1})], \qquad U_\Gamma = \sum M_j[f(x_j) - f(x_{j-1})]$$

denote the corresponding lower and upper Riemann-Stieltjes sums. Define functions g_1 and g_2 by setting $g_1 = m_j$ in $(x_{j-1},x_j]$ and $g_2 = M_j$ in $(x_{j-1},x_j]$. Since f is right continuous, it follows from (11.10) that

$$\int_{(a,b]} g_1\,d\Lambda = L_\Gamma, \qquad \int_{(a,b]} g_2\,d\Lambda = U_\Gamma.$$

Therefore, since $g_1 \le g \le g_2$, we obtain $L_\Gamma \le \int_{(a,b]} g\,d\Lambda \le U_\Gamma$. However, as $|\Gamma| \to 0$, both L_Γ and U_Γ converge to $\int_a^b g\,df$ by (2.29). This completes the proof.

We remark in passing that a right continuous function f of bounded variation can be written $f = f_1 - f_2$, where f_1 and f_2 are right continuous, bounded, and increasing. If Λ_1 and Λ_2 are the Lebesgue-Stieltjes measures corresponding to f_1 and f_2, consider the Borel set function $\Phi = \Lambda_1 - \Lambda_2$, and define

$$\int g\,d\Phi = \int g\,d\Lambda_1 - \int g\,d\Lambda_2.$$

If $\int_a^b g\,df_1$ and $\int_a^b g\,df_2$ exist and are finite, it then follows from (11.11) and (2.16) that

$$\int_{(a,b]} g\,d\Phi = \int_a^b g\,df.$$

4. Hausdorff Measure

Our second example of a Carathéodory outer measure is Hausdorff outer measure. To define it, fix $\alpha > 0$, and let A be any subset of $\mathbf{R}^n$. Given $\varepsilon > 0$, let

$$H_\alpha^{(\varepsilon)}(A) = \inf \sum_k \delta(A_k)^\alpha,$$

where $\delta(A_k)$ denotes the diameter of A_k, and the inf is taken over all countable collections $\{A_k\}$ such that $A \subset \bigcup A_k$ and $\delta(A_k) < \varepsilon$ for all k. We may always assume that the A_k in a given covering are disjoint and that $A = \bigcup A_k$.

If $\varepsilon' < \varepsilon$, each covering of A by sets with diameters less than ε' is also such a cover for ε. Hence, as ε decreases, the collection of coverings decreases, and consequently $H_\alpha^{(\varepsilon)}(A)$ increases. Define

$$H_\alpha(A) = \lim_{\varepsilon \to 0} H_\alpha^{(\varepsilon)}(A).$$

(11.12) Theorem *For $\alpha > 0$, H_α is a Carathéodory outer measure on $\mathbf{R}^n$.*

Proof. Clearly, $H_\alpha \geq 0$ and $H_\alpha(\emptyset) = 0$. If $A_1 \subset A_2$, then any covering of A_2 is also one of A_1, so that $H_\alpha^{(\varepsilon)}(A_1) \leq H_\alpha^{(\varepsilon)}(A_2)$. Letting $\varepsilon \to 0$, we obtain $H_\alpha(A_1) \leq H_\alpha(A_2)$. To show that H_α is subadditive, let $A = \bigcup A_k$, and choose a cover of A_k for each k. The union of these is a cover of A, and it is easy to show that $H_\alpha^{(\varepsilon)}(A) \leq \sum H_\alpha^{(\varepsilon)}(A_k) \leq \sum H_\alpha(A_k)$. Letting $\varepsilon \to 0$, we get $H_\alpha(A) \leq \sum H_\alpha(A_k)$. The details of this argument and the proof that H_α is a Carathéodory outer measure are left as an exercise.

H_α is called *Hausdorff outer measure of dimension* α on $\mathbf{R}^n$, and the corresponding measure is called *Hausdorff measure of dimension* α, and also denoted H_α. It has the following basic property.

(11.13) Theorem

(i) *If $H_\alpha(A) < +\infty$, then $H_\beta(A) = 0$ for $\beta > \alpha$.*
(ii) *If $H_\alpha(A) > 0$, then $H_\beta(A) = +\infty$ for $\beta < \alpha$.*

Proof. Statements (i) and (ii) are equivalent. To prove (i), let $A = \bigcup A_k$, $\delta(A_k) < \varepsilon$. If $\beta > \alpha$, then

$$H_\beta^{(\varepsilon)}(A) \leq \sum \delta(A_k)^\beta < \varepsilon^{\beta-\alpha} \sum \delta(A_k)^\alpha.$$

Therefore, $H_\beta^{(\varepsilon)}(A) \leq \varepsilon^{\beta-\alpha} H_\alpha^{(\varepsilon)}(A)$. Letting $\varepsilon \to 0$, we obtain $H_\beta(A) = 0$ if $H_\alpha(A) < +\infty$.

The next theorem shows that Hausdorff outer measure is regular, by showing that any set in $\mathbf{R}^n$ can be included in a Borel set with the same Hausdorff outer measure.

(11.14) Theorem *Given $A \subset \mathbf{R}^n$ and $\alpha > 0$, there is a set B of type G_δ containing A such that $H_\alpha(A) = H_\alpha(B)$.*

Proof. Given $\varepsilon > 0$, choose $\{A_k\}$ such that $A = \bigcup A_k$, $\delta(A_k) < \varepsilon/2$ and

$$\sum \delta(A_k)^\alpha \leq H_\alpha^{(\varepsilon/2)}(A) + \varepsilon \leq H_\alpha(A) + \varepsilon.$$

Enclose A_k in an open set G_k with $\delta(G_k) \leq (1 + \varepsilon)\delta(A_k)$; this can be done by letting $G_k = \{\mathbf{x} : d(\mathbf{x}, A_k) < \varepsilon\delta(A_k)/2\}$. Let $G = \bigcup G_k$. Then G is open and $A \subset G$. Since $\delta(G_k) < (1 + \varepsilon)\varepsilon/2 < \varepsilon$ for $0 < \varepsilon < 1$, we have

$$\begin{aligned} H_\alpha^{(\varepsilon)}(G) &\leq \sum \delta(G_k)^\alpha \leq (1 + \varepsilon)^\alpha \sum \delta(A_k)^\alpha \\ &\leq (1 + \varepsilon)^\alpha [H_\alpha(A) + \varepsilon]. \end{aligned}$$

Now let $\varepsilon \to 0$ through a sequence $\{\varepsilon_j\}$, and let G_j be the corresponding open sets. If $B = \bigcap G_j$, then B is of type G_δ and $A \subset B$. Also, since $B \subset G_j$ for each j, we have

$$H_\alpha^{(\varepsilon)}(B) \le (1 + \varepsilon)^\alpha [H_\alpha(A) + \varepsilon]$$

for $\varepsilon = \varepsilon_j$. Letting $j \to \infty$, we obtain $H_\alpha(B) \le H_\alpha(A)$. Since the opposite inequality is clearly true, the result follows.

If A is a subset of $\mathbf{R}^1$ and $A = \bigcup A_k$ with $\delta(A_k) < \varepsilon$, then $\delta(A_k) = |I_k|$, where I_k is the smallest interval containing A_k. Hence, in the one-dimensional case,

$$H_\alpha^{(\varepsilon)}(A) = \inf \sum |I_k|^\alpha \qquad (n = 1),$$

where the I_k's are intervals with lengths less than ε such that $A \subset \bigcup I_k$. If $\alpha = 1$, it follows that $H_1(A)$ is the usual Lebesgue outer measure of A. In $\mathbf{R}^n$, $n > 1$, H_n is not the same as Lebesgue outer measure (see Exercise 10). Nevertheless, there is a simple relation between the two, which is a corollary of the next lemma.

Let

$$H_\alpha'^{(\varepsilon)}(A) = \inf \sum \delta(Q_k)^\alpha,$$

where $\{Q_k\}$ is any collection of *cubes* with edges parallel to the axes such that $A \subset \bigcup Q_k$ and $\delta(Q_k) < \varepsilon$. Also, let

$$H_\alpha'(A) = \lim_{\varepsilon \to 0} H_\alpha'^{(\varepsilon)}(A).$$

Thus, H_α' is defined in the same way that H_α is, except that cubes are used instead of arbitrary sets.

(11.15) Lemma *There is a constant c depending only on n and α such that*

$$H_\alpha(A) \le H_\alpha'(A) \le cH_\alpha(A), \quad A \subset \mathbf{R}^n.$$

Proof. Since every covering of A by cubes is a covering of A, we obtain $H_\alpha(A) \le H_\alpha'(A)$. Any set with diameter δ, say, is contained in a cube with edge length 2δ, and so with diameter $2\sqrt{n}\delta$. Now let $A = \bigcup A_k$, $\delta(A_k) < \varepsilon$. Select cubes $Q_k \supset A_k$ with $\delta(Q_k) = 2\sqrt{n}\delta(A_k)$. Then

$$\sum \delta(A_k)^\alpha = (2\sqrt{n})^{-\alpha} \sum \delta(Q_k)^\alpha \ge (2\sqrt{n})^{-\alpha} H_\alpha'^{(2\sqrt{n}\,\varepsilon)}(A).$$

Therefore, $H_\alpha^{(\varepsilon)}(A) \ge (2\sqrt{n})^{-\alpha} H_\alpha'^{(2\sqrt{n}\varepsilon)}(A)$. Letting $\varepsilon \to 0$, we obtain $H_\alpha(A) \ge (2\sqrt{n})^{-\alpha} H_\alpha'(A)$, which completes the proof.

(11.16) Theorem

(i) *There are positive constants c_1 and c_2 depending only on the dimension n such that $c_1 H_n(A) \le |A|_e \le c_2 H_n(A)$ for $A \subset \mathbf{R}^n$.*

(ii) *If* $\alpha > n$, *then* $H_\alpha(A) = 0$ *for every* $A \subset \mathbf{R}^n$.

Proof. For a cube Q, $\delta(Q)^n$ is proportional to the volume of Q: $\delta(Q)^n = n^{n/2}|Q|$. Hence, due to the additivity of volume,

$$H_n'(A) = \inf \sum \delta(Q_k)^n, \; A \subset \mathbf{R}^n,$$

where $\{Q_k\}$ is any collection of cubes covering A, without restriction on the size of the diameters. Therefore, $H_n'(A) = n^{n/2}|A|_e$, $A \subset \mathbf{R}^n$. Part (i) now follows from the fact that H_n' and H_n are comparable [(11.15)].

For part (ii), if $H_n(A)$ is finite, then $H_\alpha(A) = 0$ for $\alpha > n$ by (11.13). If $H_n(A) = +\infty$, write $A = \bigcup (A \cap Q_j)$, where the Q_j are disjoint (partly open) cubes. Since $|A \cap Q_j|_e$ is finite, so is $H_n(A \cap Q_j)$. Hence, $H_\alpha(A \cap Q_j) = 0$ for $\alpha > n$. Therefore, $H_\alpha(A) \le \sum H_\alpha(A \cap Q_j) = 0$.

It is natural to ask if $H_\alpha(A)$ is comparable to the expression

$$\inf \sum \delta(A_k)^\alpha,$$

where the inf is taken over all coverings $\{A_k\}$ of A, without any requirement on the size of the diameters. It is not difficult to see that the answer in general is *no*. In fact, this expression is finite for any bounded A, as is easily seen from covering A by itself. On the other hand, it is clear from (11.13) and (11.16) that if $|A|_e > 0$, then $H_\alpha(A) = +\infty$ for $\alpha < n$.

However, in case $\alpha = n$, $H_\alpha(A)$ is comparable to the expression above. To see this, it is enough by (11.15) to prove that $H_n'(A)$ is comparable to it. But since $H_n'(A) = \inf \sum \delta(Q_k)^n$, where $\{Q_k\}$ is any collection of cubes covering A, this follows from the fact that $\inf \sum \delta(Q_k)^n$ and $\inf \sum \delta(A_k)^n$ are comparable [cf. (11.15)].

We remark in passing that Hausdorff measure is particularly useful in measuring sets with Lebesgue measure zero since these may have positive Hausdorff measure for some $\alpha < n$. For example, it can be shown that the Cantor set in [0,1] has Hausdorff measure of dimension log 2/log 3 equal to 1.

5. The Carathéodory-Hahn Extension Theorem

In this section, we will settle the question which arose in the discussion preceding theorem (11.10). We recall the situation. Let μ be a finite Borel measure on $\mathbf{R}^1$, and let Λ be the Lebesgue-Stieltjes measure induced by the function $f(x) = \mu((-\infty,x])$. Since f is continuous from the right, we have by (11.10) that

$$\Lambda((a,b]) = \mu((a,b]) \qquad [= f(b) - f(a)].$$

The point in question is whether this implies that μ and Λ agree on every Borel set in $\mathbf{R}^1$. More generally, we may ask if two Borel measures can be finite and equal on every $(a,b]$ without being identical. It is worthwhile to consider this question in a still more general context, which we now present.

Let $\mathscr{S}$ be a fixed set. By an *algebra* $\mathscr{A}$ of subsets of $\mathscr{S}$, we mean a non-empty collection of subsets of $\mathscr{S}$ which is closed under the operations of taking complements and *finite* unions; that is, $\mathscr{A}$ is an algebra if it satisfies the following:

(i) If $A \in \mathscr{A}$, then $CA(= \mathscr{S} - A) \in \mathscr{A}$.
(ii) If $A_1, \ldots, A_N \in \mathscr{A}$, then $\bigcup_{k=1}^{N} A_k \in \mathscr{A}$.

What distinguishes an algebra from a σ-algebra is that an algebra is only closed under *finite* unions. It follows from the definition that an algebra is also closed under finite intersections and differences (relative complements), and that both the empty set $\varnothing$ and the whole space $\mathscr{S}$ belong to it.

The collection of finite intervals $(a,b]$ on the line is clearly not an algebra. However, we can generate an algebra from it by adjoining $\varnothing$, $\mathbf{R}^1$, and all intervals of the form $(-\infty,a]$ and $(b,+\infty)$, as well as all possible finite disjoint unions of these and the intervals $(a,b]$. This algebra will be called the *algebra generated by the intervals* $(a,b]$.

By a *measure* λ *on an algebra* $\mathscr{A}$, we mean a function λ which is defined on the elements of $\mathscr{A}$ and which satisfies

(i) $\lambda(A) \geq 0$, $\lambda(\varnothing) = 0$.
(ii) $\lambda(\bigcup_{k=1}^{\infty} A_k) = \sum_{k=1}^{\infty} \lambda(A_k)$ whenever $\{A_k\}$ is a countable collection of disjoint sets in $\mathscr{A}$ whose union also belongs to $\mathscr{A}$.

A measure λ on $\mathscr{A}$ is called *σ-finite* (with respect to $\mathscr{A}$) if $\mathscr{S}$ can be written $\mathscr{S} = \bigcup S_k$ with $S_k \in \mathscr{A}$ and $\lambda(S_k) < +\infty$. For example, any Lebesgue-Stieltjes measure Λ on the line is a σ-finite measure on the algebra generated by the intervals $(a,b]$.

Using the ideas behind the construction of Lebesgue-Stieltjes outer measure, we can construct an outer measure λ^* from λ. Thus, let λ be a measure on an algebra $\mathscr{A}$ of subsets of $\mathscr{S}$. For any subset A of $\mathscr{S}$, define

(11.17) $$\lambda^*(A) = \inf \sum \lambda(A_k),$$

where the infimum is taken over all countable collections $\{A_k\}$ such that $A \subset \bigcup A_k$, $A_k \in \mathscr{A}$. It is always possible to find such a covering of A since $\mathscr{S}$ itself belongs to $\mathscr{A}$. The fact that $\mathscr{A}$ is an algebra allows us to assume without loss of generality that the sets A_k covering A are disjoint.

(11.18) Theorem *Let λ be a measure on an algebra $\mathscr{A}$, and let λ^* be defined by* (11.17). *Then λ^* is an outer measure.*

The proof is similar to the first part of the proof of (11.8), and is left as an exercise.

(11.19) Theorem *Let λ be a measure on an algebra $\mathscr{A}$, and let λ^* be the corresponding outer measure. If $A \in \mathscr{A}$, then $\lambda^*(A) = \lambda(A)$ and A is measurable with respect to λ^*.*

Proof. Let $A \in \mathscr{A}$. Clearly, $\lambda^*(A) \le \lambda(A)$. On the other hand, given disjoint $A_k \in \mathscr{A}$ with $A \subset \bigcup A_k$, let $A'_k = A_k \cap A$. Then $A'_k \in \mathscr{A}$ and A is the disjoint union of the A'_k. Hence, $\lambda(A) = \sum \lambda(A'_k)$. Since $A'_k \subset A_k$, it follows that $\lambda(A) \le \sum \lambda(A_k)$. Therefore, $\lambda(A) \le \lambda^*(A)$, and the proof of the first part of the theorem is complete.

For the second part, let $A \in \mathscr{A}$. To show that A is measurable, we must show that

$$\lambda^*(E) = \lambda^*(E \cap A) + \lambda^*(E - A) \text{ for every } E \subset \mathscr{S}.$$

Since λ^* is subadditive, the right side majorizes the left. To show the opposite inequality, we may assume that $\lambda^*(E)$ is finite. Given $\varepsilon > 0$, choose $\{E_k\}$ such that $E_k \in \mathscr{A}$, $E \subset \bigcup E_k$ and $\sum \lambda(E_k) < \lambda^*(E) + \varepsilon$. Since E_k and A are in $\mathscr{A}$ and $(E_k \cap A) \cup (E_k - A) = E_k$, we have $\lambda(E_k \cap A) + \lambda(E_k - A) = \lambda(E_k)$. Hence,

$$\sum \lambda(E_k \cap A) + \sum \lambda(E_k - A) < \lambda^*(E) + \varepsilon.$$

Therefore, since $E \cap A \subset \bigcup (E_k \cap A)$ and $E - A \subset \bigcup (E_k - A)$, it follows from the definition of λ^* that

$$\lambda^*(E \cap A) + \lambda^*(E - A) < \lambda^*(E) + \varepsilon.$$

Letting $\varepsilon \to 0$, we obtain the desired inequality, which completes the proof.

Let λ be a measure on an algebra $\mathscr{A}$, and let μ be a measure on a σ-algebra Σ which contains $\mathscr{A}$. Then μ is said to be an *extension of* λ *to* Σ if $\mu(A) = \lambda(A)$ for every $A \in \mathscr{A}$. If λ^* is the outer measure generated by λ and if $\mathscr{A}^*$ denotes the σ-algebra of λ^*-measurable sets, it follows from the last theorem that λ^* is an extension of λ to $\mathscr{A}^*$. This proves the first part of the following theorem, which is the main result of this section.

(11.20) Theorem (*Carathéodory-Hahn Extension Theorem*) *Let λ be a measure on an algebra $\mathscr{A}$, let λ^* be the corresponding outer measure, and let $\mathscr{A}^*$ be the σ-algebra of λ^*-measurable sets. Then*

(i) *the restriction of λ^* to $\mathscr{A}^*$ is an extension of λ;*
(ii) *if λ is σ-finite with respect to $\mathscr{A}$, and if Σ is any σ-algebra with $\mathscr{A} \subset \Sigma \subset \mathscr{A}^*$, then λ^* is the only measure on Σ which is an extension of λ.*

Proof. As we have already observed, (i) follows from (11.19). To prove (ii), which states the uniqueness of the extension, let μ be any measure on Σ, $\mathscr{A} \subset \Sigma \subset \mathscr{A}^*$, which agrees with λ on $\mathscr{A}$. Given a set $E \in \Sigma$, consider any countable collection $\{A_k\}$ such that $E \subset \bigcup A_k$ and each $A_k \in \mathscr{A}$. Then

$$\mu(E) \le \mu(\bigcup A_k) \le \sum \mu(A_k) = \sum \lambda(A_k).$$

Therefore, by the definition of λ^*, we have $\mu(E) \le \lambda^*(E)$. To show that equality holds, first suppose that there exists a set $A \in \mathscr{A}$ with $E \subset A$ and $\lambda(A) < +\infty$. Applying what has just been proved to $A - E$ (which belongs to Σ), we obtain $\mu(A - E) \leq \lambda^*(A - E)$. However,

$$\mu(E) + \mu(A - E) = \mu(A) = \lambda^*(A) = \lambda^*(E) + \lambda^*(A - E).$$

Since all these terms are finite (due to the fact that $\lambda(A)$ is finite), it follows that $\mu(E) = \lambda^*(E)$ in this case.

In the general case, since λ is σ-finite, there exist disjoint $S_k \in \mathscr{A}$ such that $\mathscr{S} = \bigcup S_k$ and $\lambda(S_k) < +\infty$. We may apply the result above to each $E \cap S_k$ (which is a subset of S_k), obtaining $\mu(E \cap S_k) = \lambda^*(E \cap S_k)$. By adding over k, we obtain $\mu(E) = \lambda^*(E)$, which completes the proof.

As a corollary, we can answer the questions raised at the beginning of this section.

(11.21) Corollary *Let μ and v be two Borel measures on $\mathbf{R}^1$ which are finite and equal on every half-open interval $(a,b]$, $-\infty < a < b < +\infty$. Then $\mu(B) = v(B)$ for every Borel set $B \subset \mathbf{R}^1$.*

Proof. Such μ and v must agree on the algebra generated by the $(a,b]$, and are σ-finite with respect to this algebra. Since the smallest σ-algebra containing all $(a,b]$ is the Borel σ-algebra, it follows from (11.20) that μ and v are identical.

Although we have confined ourselves to $n = 1$, we note that analogous results can be formulated in higher dimensions. See Exercise 18.

Now let μ be any finite Borel measure on $\mathbf{R}^1$, and define f_μ by $f_\mu(x) = \mu((-\infty,x])$. Clearly, $\mu((a,b]) = f_\mu(b) - f_\mu(a)$. Moreover, if Λ denotes the Lebesgue-Stieltjes measure constructed from f_μ, then since f_μ is continuous from the right, it follows from (11.10) that $\Lambda((a,b]) = f_\mu(b) - f_\mu(a)$. Therefore, by the corollary above, μ and Λ are identical as Borel measures, and we easily obtain

(11.22) Corollary *The class of finite Borel measures on $\mathbf{R}^1$ is identical with the class of Lebesgue-Stieltjes measures induced by bounded increasing functions that are continuous from the right.*

See also Exercise 4.

Let μ be a Borel measure on $\mathbf{R}^1$ which is finite on every $(a,b]$, $-\infty < a < b < +\infty$ (equivalently, μ is finite on every bounded Borel set). Consider the restriction of μ to the algebra $\mathscr{A}$ generated by the $(a,b]$, and let μ^* be the corresponding outer measure. The smallest σ-algebra containing $\mathscr{A}$ is the Borel sets $\mathscr{B}$. Thus, $\mathscr{A} \subset \mathscr{B} \subset \mathscr{A}^*$. Since μ and μ^* are measures on $\mathscr{B}$ which agree on $\mathscr{A}$, it follows that $\mu = \mu^*$ on $\mathscr{B}$. In particular, if $B \in \mathscr{B}$, we see from the definition of μ^* [see (11.17)] that

$$\mu(B) = \inf \{\sum \mu(A_k) : B \subset \bigcup A_k,\ A_k \in \mathscr{A}\}.$$

Each $A_k \in \mathscr{A}$ is a countable union of disjoint $(a,b]$. Hence, we obtain the formula

(11.23) $$\mu(B) = \inf \{\textstyle\sum \mu(a_k,b_k] : B \subset \bigcup (a_k,b_k]\}, \qquad B \in \mathscr{B}.$$

We recall (see p. 187) that a Borel measure μ is said to be regular if

$$\mu(B) = \inf \{\mu(G) : B \subset G,\ G \text{ open}\}, \qquad B \in \mathscr{B}.$$

(11.24) Theorem *If μ is a Borel measure on $\mathbf{R}^1$ which is finite on every bounded Borel set, then μ is regular.*

Proof. This is a corollary of (11.23). Given a Borel set B and $\varepsilon > 0$, find a cover $\{(a_k,b_k]\}$ of B such that

$$\sum \mu(a_k,b_k] \le \mu(B) + \varepsilon.$$

Since μ is finite on intervals, it follows from (10.11) that $\mu(a,b] = \lim_{\varepsilon \to 0+} \mu(a, b + \varepsilon)$. Hence, by slightly enlarging each $(a_k,b_k]$, we see that there is an open set G, $G = \bigcup (a_k, b_k + \varepsilon_k)$ for sufficiently small ε_k, containing B such that

$$\mu(G) \le \sum \mu(a_k, b_k + \varepsilon_k) \le \mu(B) + 2\varepsilon.$$

This completes the proof.

In this theorem, as in (11.21), we have limited ourselves to $n = 1$. For $n > 1$, see Exercise 18. In particular, note that the assumption on p. 187, Chapter 10, concerning the regularity of μ and ν is redundant.

Exercises

1. Prove the second statements in both parts of (11.4).
2. Let μ be a finite Borel measure on $\mathbf{R}^1$, and define $f_\mu(x) = \mu((-\infty,x])$, $-\infty < x < +\infty$. Show that f_μ is monotone increasing, $\mu((a,b]) = f_\mu(b) - f_\mu(a)$, f_μ is continuous from the right, and $\lim_{x \to -\infty} f_\mu(x) = 0$.
3. Let f be monotone increasing on $\mathbf{R}^1$.
   (a) Show that Λ_f is finite if and only if f is bounded.
   (b) Let f be bounded and right continuous, let $\mu = \Lambda_f$ and let $\bar{f}$ denote the function f_μ defined in Exercise 2. Show that f and $\bar{f}$ differ by a constant.
   Thus, if we make the additional assumption that $\lim_{x \to -\infty} f(x) = 0$, then $f = \bar{f}$.
4. If we identify two functions on $\mathbf{R}^1$ which differ by a constant, prove that there is a one-to-one correspondence between the class of finite Borel measures on $\mathbf{R}^1$ and the class of bounded increasing functions that are continuous from the right.
5. Let f be monotone increasing and right continuous on $\mathbf{R}^1$.
   (a) Show that Λ_f is absolutely continuous with respect to Lebesgue measure if and only if f is absolutely continuous. (By absolutely continuous on $\mathbf{R}^1$, we mean absolutely continuous on every compact interval.

(b) If Λ_f is absolutely continuous with respect to Lebesgue measure, show that its Radon-Nikodym derivative equals (df/dx).

6. Prove that the Lebesgue-Stieltjes outer measure constructed from $f(x) = x$ is the same as Lebesgue outer measure.

7. If f is monotone increasing and right continuous on $\mathbf{R}^1$, show that $\Lambda_f^*(A) = \mathring{\Lambda}_f^*(A)$, where $\mathring{\Lambda}_f^*$ is defined in the same way as Λ_f^* except that we use *open* intervals (a_k,b_k).

8. If f is monotone increasing and right continuous, derive formulas for $\Lambda_f([a,b])$ and $\Lambda_f((a,b))$.

9. Complete the proof of theorem (11.12).

10. Show that in $\mathbf{R}^n$, $n > 1$, Hausdorff outer measure H_n is not identical to Lebesgue outer measure. (For example, let $n = 2$, and write $A = \bigcup A_k$, $\delta(A_k) < \varepsilon$. Enclose A_k in a circle C_k with the same diameter, and show that $\sum \delta(A_k)^2 \geq (4/\pi)|A|_e$. Thus, $H_2^{(\varepsilon)}(A) \geq (4/\pi)|A|_e$.)

11. If A is a subset of $\mathbf{R}^n$, define the *Hausdorff dimension* of A as follows: If $H_\alpha(A) = 0$ for all $\alpha > 0$, let $\dim A = 0$; otherwise, let

$$\dim A = \sup \{\alpha : H_\alpha(A) = +\infty\}.$$

(a) Show that $H_\alpha(A) = 0$ if $\alpha > \dim A$, and that $H_\alpha(A) = +\infty$ if $\alpha < \dim A$. Show that in $\mathbf{R}^n$ we have $\dim A \leq n$.

(b) If $\dim A_k = d$ for each A_k in a countable collection $\{A_k\}$, show that $\dim (\bigcup A_k) = d$. Hence, show that every countable set has Hausdorff dimension 0.

12. Let Γ be an outer measure on $\mathscr{S}$, and let Γ' denote Γ restricted to the Γ-measurable sets. Since Γ' is a measure on an algebra, it induces an outer measure Γ^*. Show that $\Gamma^*(A) \geq \Gamma(A)$ for $A \subset \mathscr{S}$, and that equality holds for a given A if and only if there is a Γ-measurable set E such that $A \subset E$ and $\Gamma(E) = \Gamma(A)$. Thus, $\Gamma = \Gamma^*$ if Γ is regular.

13. Let λ be a measure on an algebra $\mathscr{A}$, and let λ^* be the corresponding outer measure. Given A, show that there is a set H of the form $\bigcap_k \bigcup_j A_{k,j}$ such that $A_{k,j} \in \mathscr{A}$, $A \subset H$ and $\lambda^*(A) = \lambda^*(H)$. Thus, every outer measure which is induced by a measure on an algebra is regular.

14. Prove theorem (11.18).

15. (a) Show that the intersection of a family of algebras is an algebra.
(b) A collection $\mathscr{C}$ of subsets of $\mathscr{S}$ is called a *subalgebra* if it is closed under finite intersections and if the complement of any set in $\mathscr{C}$ is the union of a finite number of disjoint sets in $\mathscr{C}$. Give an example of a subalgebra. Show that a subalgebra $\mathscr{C}$ generates an algebra by adding $\varnothing$, $\mathscr{S}$, and all finite disjoint unions of sets of $\mathscr{C}$.

16. If μ is a finite Borel measure on $\mathbf{R}^1$, show that $\mu(B) = \sup \mu(F)$ for every Borel set B, where F denotes any closed subset of B.

17. Show that the conclusions of theorems (10.48) and (10.49) of Chapter 10, and therefore also the conclusion of (10.50), remain true without the assumption (ii) stated before (10.47). [Show that without this assumption, the conclu-

sions of (10.47) are true with μ replaced by μ^*; for example,

$$\mu^*\left\{\mathbf{x} \in E : \sup_{h>0} \frac{\nu(Q_{\mathbf{x}}(h))}{\mu(Q_{\mathbf{x}}(h))} > \alpha\right\} \leq c\,\frac{\nu(\mathbf{R}^n)}{\alpha}.]$$

18. Derive analogues of (11.21) and (11.24) in $\mathbf{R}^n$, $n > 1$. (Use partly open n-dimensional intervals in place of the intervals $(a,b]$.)

Chapter 12

A Few Facts from Harmonic Analysis

1. Trigonometric Fourier Series

Lebesgue measure and integration have been decisive in the development of many branches of analysis and are applied there in ever greater degree. But, conversely, some of the applications have had considerable impact on the theory of integration. In this chapter, we will consider one topic where this interdependence has been particularly fruitful: *harmonic analysis* (see p. 214). We begin by describing some elementary notions and facts; the concept of an orthogonal system, and in particular of the trigonometric system, is basic here.

The notion of an orthogonal system, defined generally on a subset E of positive measure in $\mathbf{R}^n$, was introduced in Chapter 8, and we refer the reader to that place for the definitions and properties of general orthogonal systems, restating only a few facts here.

A system of complex-valued functions $\{\phi_\alpha(x)\}$, all belonging to $L^2(E)$, is called *orthogonal* over E if

$$\langle \phi_\alpha, \phi_\beta \rangle = \int_E \phi_\alpha \bar{\phi}_\beta \begin{cases} = 0 & \alpha \neq \beta. \\ > 0 & \alpha = \beta. \end{cases}$$

The second condition means that $\phi_\alpha \not\equiv 0$. If $\langle \phi_\alpha, \phi_\alpha \rangle = 1$ for all α, the orthogonal system is called *normal*, or *orthonormal*. If $\{\phi_\alpha\}$ is orthogonal, the system $\{\phi_\alpha / \|\phi_\alpha\|_2\}$ is orthonormal. Thus, by merely multiplying the functions of an orthogonal system by suitable constants, we can *normalize* it, and formulas for orthonormal systems can be easily and automatically extended to general, not necessarily normal, orthogonal systems. On the other hand, certain important orthogonal systems very naturally appear, often for historical reasons, in a nonnormalized form, and because of this, it may be desirable not to insist on the normality of the system under consideration. Let us, therefore, briefly restate the definitions in this somewhat more general set-up.

Since orthogonal systems are countable, we may index them by integers. Let $\phi_1(\mathbf{x}), \phi_2(\mathbf{x}), \ldots$ be an orthogonal system on $E \subset \mathbf{R}^n$. Thus,

$$\int_E \phi_k \bar{\phi}_l = \begin{cases} 0 & k \neq l. \\ \lambda_k > 0 & k = l. \end{cases}$$

Given any (complex-valued) $f \in L^2(E)$, we call the numbers

$$c_k = \frac{1}{\lambda_k} \int_E f \bar{\phi}_k$$

the *Fourier coefficients* of f, and the series $S[f] = \sum c_k \phi_k(\mathbf{x})$ the *Fourier series* of f, with respect to $\{\phi_k\}$. As before, we write

$$f \sim \sum c_k \phi_k(\mathbf{x}).$$

If we set

$$\psi_k = \lambda_k^{-1/2} \phi_k, \qquad d_k = \lambda_k^{1/2} c_k = \int_E f \bar{\psi}_k, \tag{12.1}$$

then $\{\psi_k\}$ is orthonormal over E, and $\{d_k\}$ is the sequence of Fourier coefficients of f with respect to $\{\psi_k\}$. Clearly,

$$d_k \psi_k = c_k \phi_k. \tag{12.2}$$

This set of formulas enables us to rewrite relations for orthonormal systems in forms valid for general orthogonal systems. Thus, Bessel's inequality $\sum |d_k|^2 \le \int_E |f|^2$ and Parseval's formula $\sum |d_k|^2 = \int_E |f|^2$ take the forms

$$\sum \lambda_k |c_k|^2 \le \int_E |f|^2, \qquad \sum \lambda_k |c_k|^2 = \int_E |f|^2. \tag{12.3}$$

The notion of completeness of an orthogonal system ("the vanishing of all the Fourier coefficients implies the vanishing of the function") remains unchanged in the general case, and as in the case of normalized systems, the validity of the second formula (12.3) is a necessary and sufficient condition for the completeness of $\{\phi_k\}$.

Let s_n denote the nth partial sum of $S[f]$. As a corollary of the corresponding result for orthonormal systems, we see that the equation $\sum \lambda_k |c_k|^2 = \int_E |f|^2$ is equivalent to

$$\int_E |f - s_n|^2 \to 0.$$

Thus, if an orthogonal system is complete, the Fourier series of every $f \in L^2(E)$ converges to f, convergence being understood in the metric L^2. Of course, this says nothing about the pointwise convergence of $\sum c_k \phi_k(\mathbf{x})$. On the other hand, it holds for any rearrangement of the terms of $\sum c_k \phi_k$ since the orthogonality and completeness of a system are not affected by a permutation of the functions within the system.

We shall now consider a special orthogonal system, the *trigonometric system*. This name is given to the system of functions

$$e^{ikx} = \cos kx + i \sin kx \qquad (k = 0, \pm 1, \pm 2, \ldots).$$

These functions are all periodic, with period 2π, and it is immediate that they form an orthogonal system over any interval $Q = (a, a + 2\pi)$ of length 2π, since if k and m are distinct integers, then

$$\int_Q e^{ikx}\overline{e^{imx}}\,dx = \left[\frac{e^{i(k-m)x}}{k-m}\right]_a^{a+2\pi} = 0.$$

The system is not orthonormal since $\lambda_k = \int_Q |e^{ikx}|^2\,dx = 2\pi$ for all k. Thus, with any $f \in L(Q)$, we may associate its Fourier coefficients

$$\text{(12.4)} \quad c_k = \frac{1}{2\pi}\int_Q f(t)\overline{e^{ikt}}\,dt = \frac{1}{2\pi}\int_Q f(t)e^{-ikt}\,dt \qquad (k = 0, \pm 1, \pm 2, \ldots),$$

and its Fourier series

$$\text{(12.5)} \qquad f \sim \sum_{-\infty}^{+\infty} c_k e^{ikx}.$$

In what follows, this series will be designated by $S[f]$, and its coefficients by $c_k[f]$.

Observe that if two functions ϕ and ψ are orthogonal over a set E and if $\int_E |\phi|^2 = \int_E |\psi|^2$, then the pair $\phi \pm \psi$ is also orthogonal over E, as seen from the equation

$$\int_E (\phi + \psi)(\bar{\phi} - \bar{\psi}) = \int_E |\phi|^2 - \int_E |\psi|^2 = 0.$$

Applying this to the pairs $e^{\pm ikx}$ $(k = 1, 2, \ldots)$, we see that the functions

$$\text{(12.6)} \qquad \tfrac{1}{2}, \ldots, \frac{e^{ikx} + e^{-ikx}}{2}, \frac{e^{ikx} - e^{-ikx}}{2i}, \ldots \quad (k = 1, 2, \ldots),$$

or what is the same thing, the functions

$$\text{(12.7)} \qquad \tfrac{1}{2}, \cos x, \sin x, \ldots, \cos kx, \sin kx, \ldots$$

are orthogonal over any interval Q of length 2π. Using the form (12.6), we find that the numbers for (12.7) are

$$\tfrac{1}{2}\pi, \pi, \pi, \ldots.$$

Thus, any $f \in L(Q)$ can be developed into a new Fourier series

$$\text{(12.8)} \qquad f \sim \tfrac{1}{2}a_0 + \sum_{k=1}^{\infty} (a_k \cos kx + b_k \sin kx),$$

where

$$a_0 = (\tfrac{1}{2}\pi)^{-1}\int_Q \tfrac{1}{2}f = \frac{1}{\pi}\int_Q f, \tag{12.9}$$
$$a_k = \frac{1}{\pi}\int_Q f(t)\cos kt\,dt, \qquad b_k = \frac{1}{\pi}\int_Q f(t)\sin kt\,dt.$$

The numbers a_k and b_k are easily expressible in terms of the coefficients c_k of (12.4):

$$\tfrac{1}{2}a_0 = c_0, \qquad a_k = c_k + c_{-k}, \qquad b_k = i(c_k - c_{-k}). \tag{12.10}$$

Hence,

$$\sum_{k=-n}^{n} c_k e^{ikx} = c_0 + \sum_{k=1}^{n} c_k e^{ikx} + \sum_{k=1}^{n} c_{-k}e^{-ikx}$$
$$= \tfrac{1}{2}a_0 + \sum_{k=1}^{n}(a_k\cos kx + b_k\sin kx),$$

and the nth partial sum of the series in (12.8) turns out to be the nth *symmetric* partial sum of $\sum c_k e^{ikx}$. The numbers $a_k = a_k[f]$ and $b_k = b_k[f]$ are called the *Fourier cosine* and *sine coefficients* of f, respectively.

To sum up, we may consider the trigonometric system in two forms. One consists of the functions e^{ikx} ($k = 0, \pm 1, \pm 2, \ldots$), and the Fourier series has the form $\sum_{-\infty}^{+\infty} c_k e^{ikx}$, where the c_k are given by (12.4). The other consists of the functions (12.7), the Fourier series is (12.8), and the coefficients a_k and b_k are given by (12.9). The partial sums of (12.8) are the symmetric partial sums of (12.5). In both cases, the terms of the Fourier series are *harmonic oscillations*, and for this reason, the study of $S[f]$ is called the *harmonic analysis* of f.

Each form of the trigonometric system has its advantages. For example, if f is real-valued, then the numbers a_k and b_k are real, while the c_k have the property $c_{-k} = \bar{c}_k$. Note also that if $Q = (-\pi, \pi)$ and f is an *even function*, i.e., if $f(-x) = f(x)$, then

$$a_k = \frac{2}{\pi}\int_0^{\pi} f(t)\cos kt\,dt, \qquad b_k = 0,$$

and if f is an *odd function*, i.e., if $f(-x) = -f(x)$, then

$$a_k = 0, \qquad b_k = \frac{2}{\pi}\int_0^{\pi} f(t)\sin kt\,dt.$$

Thus, if f is even, (12.8) reduces to the cosine series $\tfrac{1}{2}a_0 + \sum_{k=1}^{\infty} a_k\cos kx$, and if f is odd, to the sine series $\sum_{k=1}^{\infty} b_k \sin kx$.

Since the terms of a (trigonometric) Fourier series are periodic with period 2π, if we expect to represent a function f by its Fourier series, we may assume from the start that f is defined everywhere (or almost everywhere) on the real axis and is periodic with period 2π. This amounts to considering the function as defined on the circumference of the unit circle. We do not distinguish between points which are congruent mod 2π. By an integrable function, we shall mean a function integrable over a period. Similarly for L^p and other classes of functions. In what follows, *periodic* will mean of period 2π.

In the preceding chapters, we proved a number of theorems about functions in $L^p(\mathbf{R}^n)$, and in particular, in $L^p(\mathbf{R}^1)$. Usually these results have analogues for periodic functions, where integrals over $\mathbf{R}^1$ are replaced by integrals over a period. The proofs are usually in essence identical with those for $\mathbf{R}^1$ (or are merely corollaries of the results for $\mathbf{R}^1$), and may be left as exercises.

We would like to stress one point. The definition of a general orthogonal system presupposes that the functions in the system are of class L^2. This makes it possible to define Fourier coefficients for any $f \in L^2$. If f is not in L^2, it may be impossible to define its Fourier coefficients with respect to certain orthogonal systems. The situation is different for special orthogonal systems. For example, in the case of the functions $\{e^{ikx}\}$, which are bounded, the coefficients c_k are defined for any f which is merely integrable over Q, and in particular, for any $f \in L^p(Q)$, $1 \le p \le \infty$. Thus, the trigonometric system is richer in properties than general orthogonal systems.

Some simple developments are important for the general theory of Fourier series. We consider two here and refer the reader to Exercise 5 for others.

Example (a). Let f be periodic and equal to $\frac{1}{2}(\pi - x)$ for $0 < x < 2\pi$, with $f(0) = f(2\pi) = 0$. Since f is odd, its Fourier series is a sine series, and integration by parts gives

$$b_k = \frac{2}{\pi}\int_0^{\pi} \frac{1}{2}(\pi - x)\sin kx\,dx = \frac{1}{\pi}\left(\frac{\pi}{k} - \frac{1}{k}\int_0^{\pi}\cos kx\,dx\right) = \frac{1}{k}.$$

Thus,

$$f \sim \sum_{k=1}^{\infty} \frac{\sin kx}{k} = \frac{1}{2}\sum_{-\infty}^{+\infty}{}' \frac{e^{ikx}}{ik},$$

where Σ' denotes $\sum_{k\neq 0}$.

Example (b). Let f be periodic and equal in $(-\pi,\pi)$ to the characteristic function of the interval $(-h,h)$, $0 < h < \pi$. Then f is even, and if $k \neq 0$, its cosine coefficient is

$$a_k = \frac{2}{\pi}\int_0^{h} \cos kx\,dx = \frac{2}{\pi}\frac{\sin kh}{k}.$$

Since $a_0 = 2h/\pi$, we obtain

$$f \sim \frac{2h}{\pi}\left\{\frac{1}{2} + \sum_{k=1}^{\infty} \frac{\sin kh}{kh} \cos kx\right\}$$
$$= \frac{h}{\pi}\left\{1 + \sum_{-\infty}^{+\infty}{}' \frac{\sin kh}{kh} e^{ikx}\right\}.$$

Series of the form

$$\sum_{-\infty}^{+\infty} c_k e^{ikx}, \qquad \tfrac{1}{2}a_0 + \sum_{k=1}^{\infty} (a_k \cos kx + b_k \sin kx),$$

whether they are Fourier series or not, are called *trigonometric* series. In defining the convergence of $\sum_{-\infty}^{+\infty} c_k e^{ikx}$, we usually consider the limit, ordinary or generalized, of the symmetric partial sums $\sum_{-n}^{+n}$, and it is immediate that

$$\sum_{-n}^{n} c_k e^{ikx} = \tfrac{1}{2}a_0 + \sum_{k=1}^{n} (a_k \cos kx + b_k \sin kx),$$

where

$$\tfrac{1}{2}a_0 = c_0, \qquad a_k = c_k + c_{-k}, \qquad b_k = i(c_k - c_{-k}).$$

A finite sum $T = \sum_{-n}^{n} c_k e^{ikx}$ is called a *trigonometric polynomial of order n*, and if $|c_{-n}| + |c_n| \neq 0$, T is *strictly of order n*. If T is of order n and vanishes at more than $2n$ distinct points (i.e., distinct mod 2π), then it vanishes identically, i.e., all the c_k are 0. For $Te^{inx} = \sum_{-n}^{n} c_k e^{i(k+n)x}$ is a (power) polynomial in $z = e^{ix}$ of degree $\leq 2n$, and if it has more than $2n$ zeros, then it vanishes identically.

If the numbers a_k and b_k are real, the trigonometric series

$$S = \frac{1}{2}a_0 + \sum_{k=1}^{\infty} (a_k \cos kx + b_k \sin kx)$$

is the real part of the power series

$$\frac{1}{2}a_0 + \sum_{k=1}^{\infty} (a_k - ib_k)z^k$$

on the unit circle $z = e^{ix}$. The imaginary part is then the series

(12.11) $$\sum_{k=1}^{\infty} (a_k \sin kx - b_k \cos kx)$$

(with vanishing constant term). If S is written in the complex form $\sum_{-\infty}^{+\infty} c_k e^{ikx}$, it is easy to see that (12.11) is

(12.12) $$\sum_{-\infty}^{+\infty} (-i \operatorname{sign} k) c_k e^{ikx}$$

(where, by convention, sign $0 = 0$). The series (12.11), or (12.12), is said to be *conjugate* to S. A series conjugate to a trigonometric series S is denoted by $\tilde{S}$. If S has constant term 0, then

$$\tilde{\tilde{S}} = -S.$$

It is natural to study the properties of $\tilde{S}[f]$ simultaneously with those of $S[f]$.

One more remark. Properties of functions in $L^p(\mathbf{R}^n)$, and in particular in $L^p(\mathbf{R}^1)$, are important for the theory of *Fourier integrals*, which for non-periodic functions play the same role as Fourier series in the periodic case. Though the two theories run largely parallel, we shall limit ourselves to Fourier series since our primary aim is to show the role that Lebesgue integration plays in problems of representability of functions, and both the results and techniques of Fourier series are sufficiently indicative of the situation.

2. Theorems about Fourier Coefficients

(12.13) Theorem *If a periodic f is the indefinite integral of its derivative f' (i.e., if f is periodic and absolutely continuous), and if $f \sim \sum c_k e^{ikx}$, then*

$$f' \sim \sum_{-\infty}^{+\infty} c_k(ik)e^{ikx}.$$

In symbols,

$$S[f'] = S'[f],$$

where $S'[f]$ denotes the result of the termwise differentiation of $S[f]$.

Proof. It is clear that f' is also periodic and that its constant term equals

$$(2\pi)^{-1}\int_0^{2\pi} f'\,dx = (2\pi)^{-1}[f(2\pi) - f(0)] = 0.$$

If $k \neq 0$, integrating by parts and observing that the integrated term is zero, we have

$$(2\pi)^{-1}\int_0^{2\pi} f'e^{-ikx}\,dx = (2\pi)^{-1}ik\int_0^{2\pi} fe^{-ikx}\,dx = ikc_k,$$

which proves the theorem.

By repeated application of this result, we see that if a periodic f is the mth indefinite integral of an integrable function $f^{(m)}$, then

$$S[f^{(m)}] = S^{(m)}[f] = \sum (ik)^m c_k e^{ikx}.$$

(12.14) Theorem *If f is periodic, $f \sim \sum c_k e^{ikx}$, and if F is the indefinite* *integral of f, then* $F(x) - c_0 x$ *is periodic and*

$$F(x) - c_0 x \sim C_0 + \sum{}' \frac{c_k}{ik} e^{ikx},$$

where C_0 *is a suitable constant (depending on the choice of the arbitrary constant of integration in F), and* Σ' *denotes* $\sum_{k \neq 0}$.

Proof. Let $G(x) = F(x) - c_0 x$. The periodicity of G follows from the equation

$$G(x + 2\pi) - G(x) = \int_x^{x+2\pi} f\,dt - c_0(2\pi) = 2\pi c_0 - 2\pi c_0 = 0.$$

Since G is also absolutely continuous, (12.13) gives

$$S'[G] = S[G'] = S[f - c_0] = \sum_{k \neq 0} c_k e^{ikx}.$$

Hence, $S[G]$ is obtained by termwise integration of $\sum' c_k e^{ikx}$, which leads to the result, C_0 being the constant term of $S[F - c_0 x]$.

For the trigonometric system, we certainly have *Bessel's inequality* (see (12.3))

$$\sum_{-\infty}^{+\infty} |c_k|^2 \leq \frac{1}{2\pi} \int_0^{2\pi} |f|^2\,dx,$$

but actually, as we shall see, we also have *Parseval's formula*

(12.15) $$\sum_{-\infty}^{+\infty} |c_k|^2 = \frac{1}{2\pi} \int_0^{2\pi} |f|^2\,dx$$

for every $f \in L^2$. We know (see (8.31)) that this is a corollary of the following

(12.16) Theorem *The trigonometric system is complete. More precisely, if all the Fourier coefficients of an integrable f are zero, then* $f = 0$ *a.e.*

Proof. Assume first that f is continuous and real-valued, with all $c_k = 0$. If $f \not\equiv 0$, then $|f|$ attains a nonzero maximum M at some point x_0. Suppose, e.g., that $f(x_0) = M > 0$. Let $\delta > 0$ be so small that $f(x) > \frac{1}{2}M$ in the interval $I = (x_0 - \delta, x_0 + \delta)$. Consider the trigonometric polynomial

$$t(x) = 1 + \cos(x - x_0) - \cos\delta.$$

It is strictly greater than 1 inside I and does not exceed 1 in absolute value elsewhere. The hypothesis that all the Fourier coefficients of f are 0 implies that $\int_{-\pi}^{\pi} fT\,dx = 0$ for any trigonometric polynomial T, and in particular,

$$\int_{-\pi}^{\pi} f t^N\,dx = 0 \qquad (N = 1, 2, \ldots).$$

We claim that this is impossible for N large enough. The absolute value of the part of the last integral extended over the complement of I is $\leq 2\pi \cdot M \cdot 1^N = 2\pi M$. If I' is the middle half of I, then $t(x) \geq \theta > 1$ in I', so that

$$\int_I ft^N\,dx \geq \int_{I'} ft^N\,dx \geq \tfrac{1}{2}M \cdot |I'|\theta^N \to +\infty.$$

Collecting results, we see that $\int_{-\pi}^{\pi} ft^N\,dx \to +\infty$; this contradiction shows that $f \equiv 0$.

If f is continuous but not real-valued, the hypothesis $\int_0^{2\pi} fe^{-ikx}\,dx = 0$ for all k implies that $\int_0^{2\pi} \bar{f}e^{-ikx}\,dx = 0$ for all k. By adding and subtracting the last two equations, we see that both the real and imaginary parts of f have all their Fourier coefficients equal to 0, and so vanish identically.

Finally, if f is merely integrable, the hypothesis $c_0 = 0$ implies that the function $F(x) = \int_0^x f\,dt$ is periodic, and by (12.14), for a suitable C_0, the Fourier coefficients of the continuous function $F - C_0$ are all 0. Hence, $F - C_0 \equiv 0$, F is a constant, and $f = F' = 0$ a.e. This completes the proof of the theorem. Another proof is given on p. 241.

An immediate corollary is the following

(12.17) Theorem *Parseval's formula* (12.15) *holds for any* $f \in L^2$.

Parseval's formula can be written in more general forms, which are, however, corollaries of (12.15). Thus, besides $f \sim \sum c_k e^{ikx} \in L^2$, consider another function $g \sim \sum d_k e^{ikx} \in L^2$. Then, by an argument like that on p. 140,

(12.18)
$$\frac{1}{2\pi}\int_0^{2\pi} f\bar{g}\,dx = \sum_{-\infty}^{+\infty} c_k \bar{d}_k.$$

This reduces to (12.15) in the special case $f = g$.

The completeness of the trigonometric system also gives the following two theorems.

(12.19) Theorem *If the Fourier series of a continuous f converges uniformly, then the sum of the series is f.*

For let g be the sum of the uniformly convergent series $S[f]$. The Fourier coefficients of g can be obtained by multiplying $S[f]$ by e^{-ikx} and integrating the result termwise. Thus, $c_k[g] = c_k[f]$ for all k, so that $f \equiv g$.

(12.20) Theorem *If a periodic f is the integral of a function in L^2, then $S[f]$ converges absolutely and uniformly. In particular, the Fourier series of a continuously differentiable function converges uniformly to the function.*

For let f be the integral of $g \in L^2$, $g \sim \sum c_k e^{ikx}$ ($c_0 = 0$). Then $S[f] = C_0 + \sum' C_k e^{ikx}$, $C_k = c_k/ik$, $k \neq 0$. We have $\sum |c_k|^2 < +\infty$ by Bessel's inequality, so that $\sum |C_k| < +\infty$ by Schwarz's inequality. This completes the proof.

The theorem which follows is of basic importance.

(12.21) Theorem (***Riemann-Lebesgue***) *The Fourier coefficients c_k of any integrable f tend to 0 as $k \to \pm\infty$. Hence, also $a_k, b_k \to 0$ as $k \to +\infty$.*

Proof. First, we note the obvious but important inequality

$$|c_k[f]| \le \frac{1}{2\pi}\int_0^{2\pi} |f|\, dx.$$

We will give two proofs of the theorem.

(a) (See also Exercise 15 of Chapter 8.) If $f \in L^2$, then $c_k \to 0$ as a corollary of Bessel's inequality (p. 218). If $f \in L$, write $f = g + h$, where $g \in L^2$ and $\int_0^{2\pi} |h| < \varepsilon$. (This decomposition can be made in various ways: we may, e.g., take M large enough and define h to be f wherever $|f| \ge M$ and 0 elsewhere; clearly, $|g| \le M$, and so $g \in L^2$.) Then

$$c_k[f] = c_k[g] + c_k[h].$$

Since $c_k[g] \to 0$ and $|c_k[h]| \le (2\pi)^{-1} \int_0^{2\pi} |h| < \varepsilon/2\pi$ for all k, the relation $c_k[f] \to 0$ follows.

(b) Observe that

$$c_k[f] = \frac{1}{2\pi}\int_0^{2\pi} f(x)e^{-ikx}\, dx = \frac{1}{2\pi}\int_{-(\pi/k)}^{2\pi-(\pi/k)} f\left(x + \frac{\pi}{k}\right)e^{-ik(x+(\pi/k))}\, dx$$

$$= -\frac{1}{2\pi}\int_0^{2\pi} f\left(x + \frac{\pi}{k}\right)e^{-ikx}\, dx.$$

Taking the semi-sum of the first and third integrals, we obtain

$$c_k[f] = \frac{1}{2\pi}\int_0^{2\pi} \frac{1}{2}\left[f(x) - f\left(x + \frac{\pi}{k}\right)\right]e^{-ikx}\, dx,$$

$$|c_k[f]| \le \frac{1}{2\pi}\int_0^{2\pi} \frac{1}{2}\left|f(x) - f\left(x + \frac{\pi}{k}\right)\right| dx.$$

However, we know that the last integral tends to 0 as $k \to \pm\infty$ [cf. theorem (8.19); the analogous result for periodic functions is left to the reader]. This completes the proof.

Given any continuous periodic f, the expression

$$\sup_{x,h;|h|<\delta} |f(x + h) - f(x)| \qquad (\delta > 0)$$

is called the *modulus of continuity* of f, and denoted by $\omega(\delta)$ or $\omega(\delta, f)$ (cf., Chapter 1, Exercise 17). If f is in L^p, $1 \le p < \infty$, the expression

$$\sup_{|h|<\delta}\left[\frac{1}{2\pi}\int_0^{2\pi}|f(x+h)-f(x)|^p\,dx\right]^{1/p}$$

is called the *pth modulus of continuity* of f, and denoted $\omega_p(\delta)$. Clearly, $\omega_p(\delta) \le \omega(\delta)$ and, as is easily seen from Hölder's inequality,

$$\omega_p(\delta) \le \omega_q(\delta) \quad \text{if} \quad p \le q.$$

We know that if $f \in L^p$, then $\omega_p(\delta;f) \to 0$ with δ [theorem (8.19)]. The last inequality in proof (b) above gives

(12.22) $$|c_k[f]| \le \frac{1}{2}\omega\left(\frac{\pi}{|k|},f\right), \qquad |c_k[f]| \le \frac{1}{2}\omega_1\left(\frac{\pi}{|k|},f\right).$$

These two inequalities contain the Riemann-Lebesgue theorem in a sharp form since they *quantitatively* estimate the magnitude of the Fourier coefficients of f in terms of various moduli of continuity.

The estimates (12.22) are also useful for families of functions. The following special case deserves a separate mention. A continuous periodic f is said to satisfy a *Lipschitz condition of order* α, $0 < \alpha \le 1$, if $\omega(\delta;f) = O(\delta^\alpha)$. or equivalently, if there is a finite constant M independent of x, h such that

$$|f(x+h)-f(x)| \le M|h|^\alpha.$$

(12.23) Theorem *If f satisfies a Lipschitz condition of order α, $0 < \alpha < 1$, then*

$$|c_k[f]| = O(|k|^{-\alpha}).$$

If $\alpha = 1$, the stronger estimate

$$|c_k[f]| = o\left(\frac{1}{|k|}\right)$$

is valid.

Proof. The first part follows from the first inequality (12.22). If $\alpha = 1$, then f is absolutely continuous [see (7.26)] and so equals the indefinite integral of its derivative f'. Since f' is bounded (and so is in L^2), its Fourier coefficients tend to zero. Hence, the coefficients of f are $o(1/|k|)$ by (12.14).

(12.24) Theorem *If a periodic f is of bounded variation over a period, then* $|c_k[f]| = O(1/|k|)$. More precisely,

$$|c_k| \le V/2\pi|k|,$$

where V is the total variation of f over a closed interval of periodicity.

Proof. Integrating by parts and taking account of the periodicity of f, we have

$$2\pi c_k[f] = \int_{-\pi}^{\pi} e^{-ikx} f(x)\, dx = \frac{1}{ik} \int_{-\pi}^{\pi} e^{-ikx}\, df(x),$$

where the last integral is a Riemann-Stieltjes integral. Hence,

$$2\pi |c_k[f]| \leq |k|^{-1} \int_{-\pi}^{\pi} |df(x)| = |k|^{-1} V.$$

It must be stressed that V is the total variation over a *closed* interval of periodicity.

3. Convergence of $S[f]$ and $\tilde{S}[f]$

We shall now briefly discuss the problem of pointwise convergence of $S[f]$, treating side-by-side the parallel problem for $\tilde{S}[f]$. Among many existing results, we will consider only the simplest. Without loss of generality, we may restrict our attention to real-valued f.

We begin by computing the partial sums of $S[f]$ and $\tilde{S}[f]$. If a_k and b_k denote the cosine and sine coefficients of f, then the nth partial sum of $S[f]$ is

$$\begin{aligned} s_n(x) &= \frac{1}{2} a_0 + \sum_{k=1}^{n} (a_k \cos kx + b_k \sin kx) \\ &= \frac{1}{2\pi} \int_{-\pi}^{\pi} f(t)\, dt + \sum_{k=1}^{n} \left\{ \cos kx \cdot \frac{1}{\pi} \int_{-\pi}^{\pi} f(t) \cos kt\, dt + \sin kx \right. \\ &\qquad \left. \cdot \frac{1}{\pi} \int_{-\pi}^{\pi} f(t) \sin kt\, dt \right\} \\ &= \frac{1}{\pi} \int_{-\pi}^{\pi} f(t) \left[\frac{1}{2} + \sum_{k=1}^{n} \cos k(x-t) \right] dt = \frac{1}{\pi} \int_{-\pi}^{\pi} f(t) D_n(x-t)\, dt, \end{aligned}$$

say, where

$$D_n(t) = \frac{1}{2} + \sum_{k=1}^{n} \cos kt.$$

The trigonometric polynomial D_n is called the nth *Dirichlet kernel.* Similarly, the nth partial sum of $\tilde{S}[f]$ is

$$\begin{aligned} \tilde{s}_n(x) &= \sum_{k=1}^{n} (a_k \sin kx - b_k \cos kx) = \frac{1}{\pi} \int_{-\pi}^{\pi} f(t) \left\{ \sum_{k=1}^{n} \sin k(x-t) \right\} dt \\ &= \frac{1}{\pi} \int_{-\pi}^{\pi} f(t) \tilde{D}_n(x-t)\, dt, \end{aligned}$$

where

$$\tilde{D}_n(t) = \sum_{k=1}^{n} \sin kt$$

is the nth *conjugate Dirichlet kernel*. Notice that D_n and $\tilde{D}_n$ are even and odd functions of t, respectively, and that

$$\text{(12.25)} \qquad \frac{1}{\pi}\int_{-\pi}^{\pi} D_n(t)\,dt = 1, \qquad \frac{1}{\pi}\int_{-\pi}^{\pi} \tilde{D}_n(t)\,dt = 0.$$

Moreover, D_n and $\tilde{D}_n$ are respectively the real and imaginary parts of

$$\frac{1}{2} + \sum_{k=1}^{n} e^{ikt} = \frac{1}{2} + \sum_{k=1}^{n} z^k = \frac{1}{2} + \frac{z^n - z}{z - 1} \qquad (z = e^{it}),$$

and an elementary computation gives

$$\text{(12.26)} \quad D_n(t) = \frac{\sin(n+\frac{1}{2})t}{2\sin\frac{1}{2}t}, \qquad \tilde{D}_n(t) = \frac{\cos\frac{1}{2}t - \cos(n+\frac{1}{2})t}{2\sin\frac{1}{2}t}.$$

A quicker, though somewhat artificial, method of obtaining the first formula is to multiply $D_n(t)$ termwise by $2\sin\frac{1}{2}t$, replace the products $2\sin\frac{1}{2}t\cos kt$ by $\sin(k+\frac{1}{2})t - \sin(k-\frac{1}{2})t$, and make use of cancellation of terms. Likewise for $\tilde{D}_n(t)$.

Given a function f and a fixed point x, let us consider the expressions

$$\phi_x(t) = \tfrac{1}{2}[f(x+t) + f(x-t)], \qquad \psi_x(t) = \tfrac{1}{2}[f(x+t) - f(x-t)]$$

as functions of t. They are called the *even* and *odd parts of* f at the point x, respectively. Clearly,

$$f(x+t) = \phi_x(t) + \psi_x(t).$$

It turns out that the behaviors of $\phi_x(t)$ and $\psi_x(t)$ near $t = 0$ are decisive for the behaviors of $S[f]$ and $\tilde{S}[f]$, as the case may be, at the point x.

Returning to the formula for $s_n(x)$ and making use of the even character of $D_n(t)$, we can write

$$\begin{aligned} s_n(x) &= \frac{1}{\pi}\int_{-\pi}^{\pi} f(t)D_n(x-t)\,dt = \frac{1}{\pi}\int_{-\pi}^{\pi} f(t)D_n(t-x)\,dt \\ &= \frac{1}{\pi}\int_{-\pi}^{\pi} f(x+t)D_n(t)\,dt = \frac{2}{\pi}\int_{0}^{\pi} \frac{1}{2}[f(x+t) + f(x-t)]D_n(t)\,dt \\ &= \frac{2}{\pi}\int_{0}^{\pi} \phi_x(t)D_n(t)\,dt. \end{aligned}$$

The first formula (12.25) immediately gives

$$s_n(x) - f(x) = \frac{2}{\pi}\int_0^\pi [\phi_x(t) - f(x)]D_n(t)\,dt$$
$$= \frac{2}{\pi}\int_0^\pi [\phi_x(t) - f(x)]\frac{\sin(n+\frac{1}{2})t}{2\sin\frac{1}{2}t}\,dt.$$

It will be convenient to modify this formula slightly by replacing n by $n - 1$ and taking the semi-sum of the two formulas. Writing

(12.27) $$s_n^{\#}(x) = \tfrac{1}{2}[s_n(x) + s_{n-1}(x)] = s_n(x) - \tfrac{1}{2}(a_n \cos nx + b_n \sin nx),$$

we obtain

$$s_n^{\#}(x) - f(x) = \frac{2}{\pi}\int_0^\pi [\phi_x(t) - f(x)]\frac{\sin nt}{2\tan\frac{1}{2}t}\,dt.$$

The right side here is the nth Fourier sine coefficient of the odd function

$$[\phi_x(t) - f(x)]\tfrac{1}{2}\cot\tfrac{1}{2}t,$$

and if this function happens to be integrable near $t = 0$, the Riemann-Lebesgue theorem immediately gives $s_n^{\#}(x) - f(x) \to 0$. Hence, making use of the fact that $a_n, b_n \to 0$, we obtain from (12.27) the following basic result.

(12.28) Theorem (*Dini's Test*) *If the integral*

$$\int_0^\pi |\phi_x(t) - f(x)|\tfrac{1}{2}\cot\tfrac{1}{2}t\,dt$$

is finite, then $S[f]$ converges at the point x to the value $f(x)$.

Since only small values of t matter here, and since for small t we have $\frac{1}{2}\cot\frac{1}{2}t \simeq t^{-1}$, Dini's condition can be restated in the form

$$\int_0^\pi \frac{|\phi_x(t) - f(x)|}{t}\,dt < +\infty,$$

or what is the same thing,

(12.29) $$\int_0^\pi \frac{|f(x+t) + f(x-t) - 2f(x)|}{t}\,dt < +\infty.$$

The following special case is useful. Suppose that f has a jump discontinuity at x, so that the one-sided limits $f(x+)$, $f(x-)$ exist. Since changing f at a single point does not affect $S[f]$, we may assume that

$$f(x) = \tfrac{1}{2}[f(x+) + f(x-)],$$

in which case we say that f has a *regular discontinuity* at x. Condition (12.29) is then certainly satisfied if both

$$\int_0^\pi \frac{|f(x+t) - f(x+)|}{t}\,dt < +\infty, \qquad \int_0^\pi \frac{|f(x-t) - f(x-)|}{t}\,dt < +\infty.$$

Thus, a corollary of Dini's test is that *if both* $f(x+)$ and $f(x-)$ *exist, and if both of the last two integrals are finite, then* $S[f]$ *converges at the point* x *to the value*

$$\frac{f(x+) + f(x-)}{2}.$$

There is a result analogous to (12.28) for $\tilde{S}[f]$, and we will be brief here. Using the formula for $\tilde{s}_n$ and the odd character of $\tilde{D}_n$, we have

$$\tilde{s}_n(x) = -\frac{1}{\pi}\int_{-\pi}^{\pi} f(x+t)\tilde{D}_n(t)\,dt = -\frac{2}{\pi}\int_0^\pi \psi_x(t)\frac{\cos\frac{1}{2}t - \cos(n+\frac{1}{2})t}{2\sin\frac{1}{2}t}\,dt,$$

$$\tilde{s}_n^{\#}(x) = \frac{\tilde{s}_n(x) + \tilde{s}_{n-1}(x)}{2} = -\frac{2}{\pi}\int_0^\pi \psi_x(t)\tfrac{1}{2}\cot\tfrac{1}{2}t\,dt + \frac{2}{\pi}\int_0^\pi \psi_x(t)\frac{\cos nt}{2\sin\frac{1}{2}t}\,dt,$$

provided that

(12.30) $$\int_0^\pi |\psi_x(t)|\frac{dt}{t} < +\infty.$$

Under this hypothesis, the last term in the preceding equation tends to zero by the Riemann-Lebesgue theorem, and we obtain

(12.31) Theorem *Under the hypothesis* (12.30), *the series* $\tilde{S}[f]$ *converges at the point* x *to the sum*

$$-\frac{2}{\pi}\int_0^\pi \psi_x(t)\tfrac{1}{2}\cot\tfrac{1}{2}t\,dt = -\frac{1}{\pi}\int_0^\pi \frac{f(x+t) - f(x-t)}{2\tan\frac{1}{2}t}\,dt.$$

We denote the last integral, if it exists, by $\tilde{f}(x)$:

(12.32) $$\tilde{f}(x) = -\frac{1}{\pi}\int_0^\pi \frac{f(x+t) - f(x-t)}{2\tan\frac{1}{2}t}\,dt.$$

This function is called the *conjugate function* of f, and is intimately connected with the behavior of $\tilde{S}[f]$. We shall study the existence and properties of $\tilde{f}$ in detail later.

Observe that condition (12.30) is of a nature completely different from (12.29); the former precludes the possibility that f may have a jump at x. See Exercise 15.

The proofs of (12.28) and (12.31) are based on the Riemann-Lebesgue theorem and give convergence results only at individual points. They cannot give uniform convergence in an interval without additional and rather strong assumptions. We consider one such assumption which, though very restrictive, leads to an important result.

(12.33) Theorem *If $f = 0$ in an interval (a,b), then $S[f]$ and $\tilde{S}[f]$ converge uniformly in every smaller interval $(a + \varepsilon, b - \varepsilon)$. The sum of $S[f]$ is 0.*

Proof. The pointwise convergence in (a,b) is a corollary of (12.28) and (12.31), and it is only the question of uniformity which requires additional comment. We will consider only $S[f]$; the argument for $\tilde{S}[f]$ is similar. Fix $\varepsilon > 0$. From (12.27), we deduce that

$$s_n^*(x_0) = \frac{s_n(x_0) + s_{n-1}(x_0)}{2} = \frac{1}{\pi}\int_{-\pi}^{\pi} f(x_0 + t)\frac{\sin nt}{2\tan\frac{1}{2}t}\,dt$$

$$= \frac{1}{\pi}\int_{-\pi}^{\pi} f(x_0 + t)\chi(t)\sin nt\,dt, \qquad x_0 \in (a - \varepsilon, b + \varepsilon),$$

where $\chi(t)$ is periodic, equals $\frac{1}{2}\cot\frac{1}{2}t$ for $\varepsilon \le |t| \le \pi$ and is arbitrary for $|t| < \varepsilon$. Suppose χ is defined so that it is continuous everywhere. Write $f(x_0 + t)\chi(t) = g_{x_0}(t)$, treating t as the variable and x_0 as a parameter, and consider the modulus of continuity of $g_{x_0}(t)$ in the metric L^1. If we show that $\omega_1(\delta, g_{x_0})$ tends to 0 with δ uniformly for $x_0 \in (a + \varepsilon, b - \varepsilon)$, then the Fourier coefficients of $g_{x_0}(t)$ will also tend to 0 uniformly for such x_0 [see the second formula (12.22)], and the theorem will follow. Now, for $h > 0$,

$$\int_0^{2\pi} |g_{x_0}(t + h) - g_{x_0}(t)|\,dt = \int_0^{2\pi} |f(x_0 + t + h)\chi(t + h) - f(x_0 + t)\chi(t)|\,dt$$

$$\le \int_0^{2\pi} |f(x_0 + t + h) - f(x_0 + t)||\chi(t + h)|\,dt$$

$$+ \int_0^{2\pi} |f(x_0 + t)||\chi(t + h) - \chi(t)|\,dt.$$

The last integral clearly tends to 0 with h, uniformly in x_0, since $\max_t |\chi(t + h) - \chi(t)| \to 0$ as $h \to 0$. If $M = \max|\chi|$, the preceding integral is majorized by

$$M\int_0^{2\pi} |f(x_0 + t + h) - f(x_0 + t)|\,dt = M\int_0^{2\pi} |f(t + h) - f(t)|\,dt,$$

a quantity independent of x_0 and tending to 0. This completes the proof.

Two trigonometric series T_1 and T_2 are said to be *equiconvergent* at a point x_0 if their difference $T_1 - T_2$ converges to 0 at x_0. If $T_1 - T_2$ merely converges, but not necessarily to 0, then T_1 and T_2 are said to be *equiconvergent in the wider sense* at x_0. Each of two equiconvergent series may be individually divergent, but the character of divergence is so similar that divergence cancels out in $T_1 - T_2$.

(12.34) Theorem *Let f_1 and f_2 be two periodic functions which are equal in an* *interval (a,b). Then $S[f_1]$ and $S[f_2]$ are uniformly equiconvergent in every subinterval $(a + \varepsilon, b - \varepsilon)$; $\tilde{S}[f_1]$ and $\tilde{S}[f_2]$ are uniformly equiconvergent in the wider sense in every $(a + \varepsilon, b - \varepsilon)$.*

This is a corollary of (12.33), since, for example, $S[f_1] - S[f_2] = S[f]$ where $f\,(= f_1 - f_2)$ vanishes in (a,b).

Thus, if we change the values of f in an *arbitrary way* outside an interval (a,b), we do not affect the behavior of $S[f]$ in $(a + \varepsilon, b - \varepsilon)$. Likewise for $\tilde{S}[f]$, although in this case, if the series converges, the value of the sum may change. Therefore, the convergence or divergence of $S[f]$ and $\tilde{S}[f]$ at a point x_0 is a *local* property, i.e., depends only on the behavior of f near x_0.

4. Divergence of Fourier Series

(12.35) Theorem *There exists a continuous periodic f such that $S[f]$ diverges (more specifically, the partial sums of $S[f]$ are unbounded) at some point.*

Proof. Let $1 \le m < n$ and consider the polynomials

$$Q_{m,n}(x) = \frac{\cos mx}{n} + \frac{\cos (m+1)x}{n-1} + \cdots + \frac{\cos (m+n-1)x}{1} - \frac{\cos (m+n+1)x}{1} - \frac{\cos (m+n+2)x}{2} - \cdots - \frac{\cos (m+2n)x}{n}.$$

We will show that all these polynomials are uniformly bounded, but that their partial sums are not. To prove the first statement, we need the fact that the partial sums of the series

$$\sum_{k=1}^{\infty} \frac{\sin kx}{k},$$

which we considered on p. 215, are uniformly bounded. This is an elementary fact which can be proved in many ways [see, e.g., (12.50(c))], but here we take it for granted. Thus, since

$$Q_{m,n}(x) = \sum_{k=1}^{n} \frac{\cos (m+n-k)x - \cos (m+n+k)x}{k} = 2 \sin (m+n)x \sum_{k=1}^{n} \frac{\sin kx}{k},$$

we obtain $|Q_{m,n}(x)| \le C$, where C is independent of m and n. On the other hand, when $x = 0$, the partial sum

$$Q^{\#}_{m,n}(x) = \frac{\cos mx}{n} + \cdots + \frac{\cos (m+n-1)x}{1}$$

has the value $1 + (1/2) + \cdots + (1/n)$, which is of order $\log n$.

Now select integers m_k and n_k such that

$$m_k + 2n_k < m_{k+1} \qquad (k = 1, 2, \ldots),$$

and take a series of positive numbers α_k such that $\sum \alpha_k < +\infty$, $\alpha_k \log n_k \to +\infty$. (We will make the construction in a moment.) The series

$$\sum_{k=1}^{\infty} \alpha_k Q_{m_k,n_k}(x)$$

then converges uniformly to a continuous function f. In view of the inequality above relating m_k and m_{k+1}, the polynomials Q_{m_k,n_k} do not overlap. Hence, the last series can be written as a single trigonometric series, whose coefficients (because of uniform convergence) are the Fourier coefficients of f. Thus, this series, unbracketed, is $S[f]$. But $S[f]$ has unbounded partial sums at $x = 0$ since a single block of terms, namely, $\alpha_k Q^{\#}_{m_k,n_k}(x)$, is of order $\alpha_k \log n_k$ at $x = 0$.

It is easy to verify that if we set

$$m_k = 5^{k^3}, \qquad n_k = 2m_k = 2(5^{k^3}), \qquad \alpha_k = 1/k^2,$$

then all the conditions required above are fulfilled. This completes the proof.

We leave it to the reader to check that if we choose

$$m_k = 5^{k^2}, \qquad n_k = 2m_k, \qquad \alpha_k = 1/k^2$$

in the construction above, we get a continuous f whose partial sums are divergent but bounded at $x = 0$.

Theorem (12.35) asserts that the partial sums of $S[f]$ can be unbounded even if f is continuous. It is of interest to know "how unbounded" they can be. From the formula

$$s_n(x,f) = \frac{1}{\pi}\int_{-\pi}^{\pi} f(x+t)D_n(t)\,dt,$$

we see that if $|f| \le 1$, then

$$|s_n(x,f)| \le \frac{1}{\pi}\int_{-\pi}^{\pi} |D_n(t)|\,dt = \frac{2}{\pi}\int_0^{\pi} |D_n(t)|\,dt,$$

uniformly in x. The right side here is called the nth *Lebesgue constant*, and will be denoted by L_n. Note that L_n is actually the value of $s_n(0,f)$ for a specific f, namely, $f(t) = \operatorname{sign} D_n(t)$.

(12.36) Theorem *We have*

$$L_n = \frac{2}{\pi}\int_0^{\pi} |D_n(t)|\,dt \simeq \frac{4}{\pi^2}\log n \qquad \text{as} \quad n \to \infty.$$

Proof. Write

$$L_n = \frac{2}{\pi}\int_0^\pi |D_n(t)|\,dt = \frac{2}{\pi}\int_0^\pi \left|\sin\left(n+\frac{1}{2}\right)t\right| \frac{1}{2\sin\frac{1}{2}t}\,dt$$
$$= \frac{2}{\pi}\int_0^\pi \left|\sin\left(n+\frac{1}{2}\right)t\right|\frac{dt}{t} + \frac{2}{\pi}\int_0^\pi \left|\sin\left(n+\frac{1}{2}\right)t\right|\left(\frac{1}{2\sin\frac{1}{2}t} - \frac{1}{t}\right)dt.$$

Since the expression in brackets is bounded for $0 < t \le \pi$, and $|\sin(n+\frac{1}{2})t| \le 1$, the last integral is majorized by an absolute constant. The change of variable $(n+\frac{1}{2})t = u$ shows that the preceding term equals

$$\frac{2}{\pi}\int_0^{(n+(1/2))\pi} \frac{|\sin u|}{u}\,du.$$

We may disregard the parts of this integral extended over $(0,\pi)$ and $(n\pi, (n+\frac{1}{2})\pi)$, since the integrand is bounded. In view of the periodicity of $\sin u$, what remains can be written

$$\frac{2}{\pi}\int_\pi^{n\pi} \frac{|\sin u|}{u}\,du = \frac{2}{\pi}\int_0^\pi \sin u\left(\sum_{k=1}^{n-1}\frac{1}{u+k\pi}\right)du.$$

For $0 \le u \le \pi$, the sum in brackets is contained between $\pi^{-1}\sum_{k=2}^{n}(1/k)$ and $\pi^{-1}\sum_{k=1}^{n-1}(1/k)$, and so is strictly of order $\pi^{-1}\log n$. If we now note that $\int_0^\pi \sin u\,du = 2$, and collect estimates, we obtain $L_n \simeq (4/\pi^2)\log n$.

(12.37) Theorem *If f is integrable, then at each point x_0 of continuity of f,*

$$s_n(x_0,f) = o(\log n).$$

The estimate is uniform over every closed interval of continuity of f.

Proof. We will prove only the first statement, leaving the second to the reader. Suppose, as we may, that $x_0 = 0$, $f(x_0) = 0$. Because of our results about localization [see (12.34)], we may assume that f vanishes outside an arbitrarily small fixed interval $(-\delta,\delta)$. Then

$$|s_n(0)| = \left|\frac{1}{\pi}\int_{-\delta}^{\delta} f(t)D_n(t)\,dt\right| \le \sup_{|t|\le\delta}|f(t)|\cdot\int_{-\pi}^{\pi}|D_n(t)|\,dt.$$

Since the sup here is small with δ and the integral is of order $\log n$, the assertion follows.

5. Summability of Sequences and Series

Theorem (12.35) shows that even continuous functions, when developed into Fourier series, may not be representable by those series in terms of pointwise convergence. The situation can be remedied by considering *generalized* sums of the series. This topic is vast and basic for analysis, and we will study only a few facts important for the theory of Fourier series.

Consider a fixed doubly infinite matrix of numbers (real or complex):

$$(\mathscr{M}) \qquad \begin{matrix} \alpha_{00} & \alpha_{01} & \cdots & \alpha_{0n} & \cdots \\ \alpha_{10} & \alpha_{11} & \cdots & \alpha_{1n} & \cdots \\ & \vdots & & & \\ \alpha_{m0} & \alpha_{m1} & \cdots & \alpha_{mn} & \cdots \\ & \vdots & & & \end{matrix}$$

Given an infinite sequence of numbers $s_0, s_1, \ldots, s_n, \ldots$, we transform it into a sequence $\sigma_0, \sigma_1, \ldots, \sigma_m, \ldots$ by means of the formulas

$$\sigma_m = \alpha_{m0}s_0 + \alpha_{m1}s_1 + \cdots + \alpha_{mn}s_n + \cdots \qquad (m = 0, 1, 2, \ldots),$$

assuming that the series defining σ_m converges for each m. We may ask what conditions on $(\mathscr{M})$ will guarantee that whenever $\{s_n\}$ converges to a finite limit s, $\lim \sigma_m$ also exists and equals s. An answer is given by the following theorem.

(12.38) Theorem *Suppose that $(\mathscr{M})$ satisfies the following three conditions:*

(i) $\sum_n |\alpha_{mn}| \le A$ *(for all m, with A independent of m).*
(ii) $\lim_{m\to\infty} (\sum_n \alpha_{mn}) = 1$.
(iii) $\lim_{m\to\infty} \alpha_{mn} = 0$ *for each n.*

Then for any sequence $\{s_n\}$ converging to a finite limit s, $\lim \sigma_m$ exists and equals s.

Proof. First of all, since $\{s_n\}$ is bounded, (i) implies that σ_m exists for each m. Next, write $s_n = s + \varepsilon_n$, where $\varepsilon_n \to 0$. Then

$$\sigma_m = \sum_n \alpha_{mn}(s + \varepsilon_n) = s\sum_n \alpha_{mn} + \sum_n \alpha_{mn}\varepsilon_n.$$

We have $s\sum_n \alpha_{mn} \to s$ by (ii), and it remains only to show that the expression

$$\rho_m = \sum_n \alpha_{mn}\varepsilon_n$$

tends to 0 as $m \to \infty$. Given $\delta > 0$, split ρ_m into two sums,

$$\rho_m = \sum_{n \le n_0} \alpha_{mn}\varepsilon_n + \sum_{n > n_0} \alpha_{mn}\varepsilon_n = \rho'_m + \rho''_m,$$

say, where n_0 is so large that $|\varepsilon_n| \le \delta$ for $n > n_0$. By (i),

$$|\rho''_m| \le \sum_{n>n_0} |\alpha_{mn}||\varepsilon_n| \le \sum_{n>n_0} |\alpha_{mn}|\delta \le A\delta.$$

On the other hand, ρ'_m consists of a fixed number of terms each of which tends to 0 as $m \to \infty$. Hence, $|\rho'_m| < A\delta$ for m large enough. Combining estimates, we see that $\rho_m \to 0$, which completes the proof.

It is useful to note that if $s = 0$, then condition (ii) is not required in the proof (and so in the statement of the theorem) above. It is also immediate from the proof that if $\{s_n\}$ depends on a parameter, and if s_n tends uniformly to s, then σ_m tends uniformly to s too.

If $\sigma_m \to s$, we shall say that the sequence $\{s_n\}$—or the series whose partial sums are the s_n—is *summable to limit (sum) s by means of the matrix* $(\mathscr{M})$, or simply is *summable* $(\mathscr{M})$ to s.

The matrix $(\mathscr{M})$ is called *positive* if $\alpha_{mn} \geq 0$ for all m,n. Condition (i) is then a corollary of (ii). For positive $(\mathscr{M})$, theorem (12.38) holds if $s = \pm\infty$; we leave the proof to the reader.

Two methods of summability are of special significance for Fourier series.

(a) *The method of the arithmetic mean.* Given $s_0, s_1, \ldots, s_n, \ldots$, consider the *arithmetic means* $\sigma_0, \sigma_1, \ldots, \sigma_m, \ldots$ defined by

$$\sigma_m = \frac{s_0 + s_1 + \cdots + s_m}{m+1} \qquad (m = 0, 1, 2, \ldots).$$

If $s_n \to s$ $(-\infty \leq s \leq +\infty)$, then $\sigma_m \to s$. This is clearly a special case of (12.38); the matrix is positive.

It is useful to note that if the s_n are the partial sums of a series $\sum_{k=0}^{\infty} u_k$, then

$$\sigma_m = \frac{s_0 + s_1 + \cdots + s_m}{m+1} = \frac{u_0 + (u_0 + u_1) + \cdots + (u_0 + u_1 + \cdots + u_m)}{m+1}$$

$$= \frac{1}{m+1}\sum_{k=0}^{m} (m + 1 - k)u_k.$$

Thus,

$$\textbf{(12.39)} \qquad \sigma_m = \sum_{k=0}^{m}\left(1 - \frac{k}{m+1}\right)u_k, \qquad s_m - \sigma_m = \frac{1}{m+1}\sum_{k=0}^{m} k u_k.$$

(b) *The method of Abel.* Given a series $u_0 + u_1 + \cdots + u_n + \cdots$, consider the power series

$$f(r) = \sum_{n=0}^{\infty} u_n r^n, \qquad 0 \leq r < 1,$$

assuming that it converges for $0 \leq r < 1$. If $f(r) \to s$ as $r \to 1$, we say that $\sum u_n$ is *Abel summable* (or *A-summable*) to sum s. The method can also be applied to sequences since any sequence $\{s_n\}$ can be written as a series $s_0 + (s_1 - s_0) + (s_2 - s_1) + \cdots$.

Let us now see the relation of Abel summability to the general scheme. We claim that for $0 \leq r < 1$, the formula

$$\textbf{(12.40)} \qquad \sum_{n=0}^{\infty} u_n r^n = (1 - r)\sum_{n=0}^{\infty} s_n r^n \qquad (s_n = u_0 + \cdots + u_n),$$

is valid assuming only that one of the two series which appear is convergent. If the right side converges, it equals

$$\sum_{n=0}^{\infty} s_n r^n - \sum_{n=0}^{\infty} s_n r^{n+1} = \sum_{n=0}^{\infty} s_n r^n - \sum_{n=1}^{\infty} s_{n-1} r^n$$

$$= s_0 + \sum_{n=1}^{\infty} (s_n - s_{n-1}) r^n = \sum_{n=0}^{\infty} u_n r^n.$$

Conversely, if $\sum_{n=0}^{\infty} u_n r^n$ converges for some r, $0 < r < 1$, its Cauchy product with the absolutely convergent series $\sum_{n=0}^{\infty} r^n = (1 - r)^{-1}$ converges to sum

$$\sum_{n=0}^{\infty} \left(\sum_{k=0}^{n} u_k r^k \cdot r^{n-k} \right) = \sum_{n=0}^{\infty} (u_0 + u_1 + \cdots + u_n) r^n = \sum_{n=0}^{\infty} s_n r^n.$$

This proves (12.40). Now, if $\{r_m\}$ is any sequence tending to 1, $0 < r_m < 1$, then the positive numbers

$$\alpha_{mn} = (1 - r_m) r_m^n$$

satisfy conditions (i), (ii), (iii) of (12.38). We leave the verification to the reader.

(12.41) Theorem (*Abel*) *If* $\sum_{n=0}^{\infty} u_n$ *converges to sum* s, $-\infty \le s \le +\infty$, *then it is A-summable to* s.

Proof. Suppose first that s is finite. Applying (12.40), we have to show that $(1 - r) \sum_{n=0}^{\infty} s_n r^n \to s$ as $r \to 1$. It is enough to prove that this relation holds for any sequence $r = r_m$, $m = 0, 1, \ldots$, where $0 < r_m^n < 1$, $r_m \to 1$. This is a corollary of (12.38) since the numbers $\alpha_{mn} = (1 - r_m) r_m^n$ satisfy (i), (ii), (iii). The matrix α_{mn} is positive, and so the proof holds for $s = \pm\infty$, the only prerequisite being that the series $\sum u_n r^n$ converges for $0 \le r < 1$.

We may also consider the power series

$$f(z) = \sum_{n=0}^{\infty} u_n z^n,$$

where z is a *complex* variable lying in the unit disc: $z = re^{ix}$, $0 \le r < 1$. If $f(z)$ tends to a limit s as z tends *nontangentially* to 1, i.e., as $z \to 1$ in such a way that

$$\frac{|1 - z|}{1 - |z|} \le C < +\infty \qquad (|z| < 1),$$

then $\sum_{n=0}^{\infty} u_n$ is said to be *nontangentially Abel summable* to sum s. The last inequality means that, in approaching 1, z remains between two chords of the unit circle through $z = 1$.

(12.42) Theorem (*Abel-Stolz*) *If* $\sum_{n=0}^{\infty} u_n$ *converges to a finite sum* s, *then it is nontangentially Abel summable to* s.

Proof. The proof is identical to that of Abel's theorem, except that now we use the formula $\sum u_n z^n = (1 - z) \sum s_n z^n$ and consider any sequence $\{z_m\}$ tending to 1 from the interior of the unit disc. The matrix α_{mn} is now $(1 - z_m)z_m^n$, conditions (ii) and (iii) are satisfied as before, and (i) takes the form

$$\frac{|1 - z_m|}{1 - |z_m|} \leq C.$$

(12.43) Theorem *If $\sum_{n=0}^{\infty} u_n$ is summable by the method of the arithmetic mean to sum s, then it is A-summable to s. If, in addition, s is finite, then $\sum_{n=0}^{\infty} u_n$ is nontangentially A-summable to s.*

Proof. Suppose that s is finite. By hypothesis,

$$\sigma_n = \frac{s_0 + s_1 + \cdots + s_n}{n + 1} \to s.$$

Write $s_0 + s_1 + \cdots + s_n = t_n$. Applying formula (12.40) twice, we have

$$\sum_{n=0}^{\infty} u_n r^n = (1 - r) \sum_{n=0}^{\infty} s_n r^n = (1 - r)^2 \sum_{n=0}^{\infty} t_n r^n = (1 - r)^2 \sum_{n=0}^{\infty} (n + 1)\sigma_n r^n.$$

Again, it is enough to consider any sequence $r_m \to 1$, $0 < r_m < 1$. We then have to apply (12.38) with matrix

$$\alpha_{mn} = (1 - r_m)^2(n + 1)\, r_m^n,$$

and we easily verify that (α_{mn}) satisfies conditions (i), (ii), (iii). The proof of the rest of the theorem is left to the reader.

While convergence of a series implies summability A, the converse is generally false: for example, $\sum_{n=0}^{\infty} (-1)^n$ diverges, but is A-summable to sum $\frac{1}{2}$ since $\sum_{n=0}^{\infty} (-r)^n = 1/(1 + r) \to \frac{1}{2}$ as $r \to 1-$. If one makes additional assumptions on the terms of the series, however, the converse will hold. The following result is both elementary and useful.

(12.44) Theorem (*Tauber*) *If $\sum u_n$ is A-summable to sum s, $-\infty \leq s \leq +\infty$, and if $u_n = o(1/n)$ as $n \to \infty$, then $\sum u_n$ converges to sum s.*

Proof. Write $u_n = \varepsilon_n/n$, $n \geq 1$, where $\varepsilon_n \to 0$. Let r_m be a sequence tending to 1 which we shall determine in a moment. Then $s_m - f(r_m)$ is a transformation of the sequence ε_n:

$$s_m - f(r_m) = \sum_{n=1}^{m} \frac{\varepsilon_n}{n} - \sum_{n=1}^{\infty} \frac{\varepsilon_n}{n} r_m^n = \sum_{n=1}^{\infty} \alpha_{mn}\varepsilon_n,$$

where

$$\alpha_{mn} = \frac{1}{n}(1 - r_m^n) \quad \text{if} \quad n \leq m, \qquad \alpha_{mn} = -\frac{1}{n} r_m^n \quad \text{if} \quad n > m.$$

If we verify conditions (i) and (iii) of (12.38), then the fact that $\varepsilon_n \to 0$ will give $s_m - f(r_m) \to 0$, and so also $s_m \to s$. Condition (iii) is obvious for any $\{r_m\} \to 1$. As for (i), observing that

$$1 - r^n = (1 - r)(1 + \cdots + r^{n-1}) \le (1 - r)n,$$

we have

$$\begin{aligned}\sum_n |\alpha_{mn}| &\le \sum_{n=1}^{m} \frac{1}{n}(1 - r_m)n + \sum_{n=m+1}^{\infty} \frac{1}{n} r_m^n \\ &\le m(1 - r_m) + \frac{1}{m+1}\sum_{n=0}^{\infty} r_m^n \\ &= m(1 - r_m) + \frac{1}{m+1}\frac{1}{1 - r_m}.\end{aligned}$$

Hence, if we choose $r_m = 1 - (1/m)$, then $\sum_n |\alpha_{mn}| \le 2$. Thus, condition (i) holds, and the theorem follows.

If $\sum u_n$ *is summable by the method of the arithmetic mean and* $nu_n \to 0$, *then* $\sum u_n$ *converges.* Of course, this is a corollary of (12.43) and (12.44), but a direct proof is on the surface: By (12.39),

$$s_m - \sigma_m = \frac{1}{m+1}\sum_{k=0}^{m} ku_k,$$

and the assumption $ku_k \to 0$ clearly implies $s_m - \sigma_m \to 0$. Thus, if $\sigma_m \to s$, then also $s_m \to s$. Actually, this argument shows that if $ku_k \to 0$, then whether $\{\sigma_m\}$ converges or not, the difference $s_m - \sigma_m$ tends to 0, i.e., the behavior of $\{s_m\}$ imitates that of $\{\sigma_m\}$. The same argument shows that *if* $\{\sigma_m\}$ *is a bounded sequence, and* $|u_n| \le A/n$ *for* $n = 1, 2, \ldots$, *then the sequence* $\{s_m\}$ *is bounded.* The result which follows lies deeper.

(12.45) Theorem (*Hardy*) *If* $\sum u_n$ *is summable by the method of the arithmetic mean to a finite sum s and if*

$$|u_n| \le \frac{A}{n} \qquad (n = 1, 2, \ldots),$$

then $\sum u_n$ *converges to s.*

Proof. Consider the expressions (which we shall call the *delayed arithmetic means*)

$$\sigma_{n,k} = \frac{s_{n+1} + s_{n+2} + \cdots + s_{n+k}}{k}.$$

They are easily expressible in terms of the σ_n:

$$\sigma_{n,k} = \frac{(s_0 + \cdots + s_{n+k}) - (s_0 + \cdots + s_n)}{k} = \frac{n+k+1}{k}\sigma_{n+k} - \frac{n}{k}\sigma_n$$
$$= \frac{n}{k}(\sigma_{n+k} - \sigma_n) + \frac{k+1}{k}\sigma_{n+k}.$$

It is clear that if k_n is any sequence of integers such that n/k_n is bounded as $n \to \infty$, then $\sigma_n \to s$ implies that $\sigma_{n,k_n} \to s$. Using the definition of $\sigma_{n,k}$, we also deduce that

$$\sigma_{n,k} = s_n + \frac{(s_{n+1} - s_n) + \cdots + (s_{n+k} - s_n)}{k}$$
$$= s_n + \frac{1}{k}\sum_{j=1}^{k}(k - j + 1)u_{n+j}.$$

Hence, assuming as we may that $A = 1$,

$$|\sigma_{n,k} - s_n| \le \sum_{j=1}^{k}|u_{n+j}| \le \frac{k}{n+1}.$$

Let $k = k_n = [\varepsilon n]$, where $\varepsilon > 0$ is arbitrarily small and fixed, and $[x]$ designates the integral part of x. Then n/k_n is bounded, and so $\sigma_{n,k_n} \to s$. But by taking $k_n = [\varepsilon n]$ in the last estimate, we obtain

$$\limsup_{n\to\infty} |\sigma_{n,k_n} - s_n| \le \varepsilon.$$

Hence, $s_n \to s$, and the proof is complete.

We remark in passing that the conclusion of Hardy's theorem is true if the assumption of summability by the method of the arithmetic mean is replaced by A-summability (theorem of Littlewood; see A. Zygmund, *Trigonometric Series*, vol. 1, 2nd ed., Cambridge University Press, Cambridge, 1968, p. 81).

6. Summability of $S[f]$ and $\tilde{S}[f]$ by the Method of the Arithmetic Mean

Given a periodic f, we denote by $s_n(x) = s_n(x,f)$ the partial sums of $S[f]$ and by $\sigma_n(x) = \sigma_n(x,f)$ their arithmetic means. Thus (see p. 222),

$$s_n(x) = \frac{1}{\pi}\int_{-\pi}^{\pi} f(t)D_n(x-t)\,dt = \frac{1}{\pi}\int_{-\pi}^{\pi} f(x+t)D_n(t)\,dt,$$
$$\sigma_n(x) = \frac{1}{\pi}\int_{-\pi}^{\pi} f(t)K_n(x-t)\,dt = \frac{1}{\pi}\int_{-\pi}^{\pi} f(x+t)K_n(t)\,dt,$$

where

$$K_n(t) = \frac{1}{n+1}\sum_{j=0}^{n} D_j(t) = \frac{1}{(n+1)2\sin\frac{1}{2}t}\sum_{j=0}^{n}\sin(j+\tfrac{1}{2})t.$$

Multiplying the last sum termwise by $2 \sin \frac{1}{2}t$ and using the equation $2 \sin (j + \frac{1}{2})t \sin \frac{1}{2}t = \cos jt - \cos (j + 1)t$, we get

$$\text{(12.46)} \qquad K_n(t) = \frac{1 - \cos (n + 1)t}{(n + 1)(2 \sin \frac{1}{2}t)^2} = \frac{2}{n + 1}\left(\frac{\sin [(n + 1)t/2]}{2 \sin \frac{1}{2}t}\right)^2.$$

The trigonometric polynomial $K_n(t)$ is called the nth *Fejér kernel*. An analogous kernel was considered in Chapter 9, (9.11), for nonperiodic functions. The formula

$$\sigma_n(x,f) = \frac{1}{\pi}\int_{-\pi}^{\pi} f(t)K_n(x - t)\,dt = \frac{1}{\pi}\int_{-\pi}^{\pi} f(x - t)K_n(t)\,dt$$

is a periodic version of the notion of convolving a function and a kernel. Some of the facts we will prove below are similar to ones we have already had in Chapter 9, but rather than connecting the present case with those results, we shall give brief direct proofs of the theorems we need.

Using the formula $D_j(t) = \frac{1}{2} + \sum_{m=1}^{j} \cos mt$, the Fejér kernel can be written [see (12.39)]

$$K_n(t) = \frac{1}{2} + \sum_{m=1}^{n}\left(1 - \frac{m}{n + 1}\right)\cos mt = \frac{1}{2}\sum_{m=-n}^{n}\left(1 - \frac{|m|}{n + 1}\right)e^{imt}.$$

It has the following properties:

(a) $K_n(t) \geq 0$; $K_n(-t) = K_n(t)$.
(b) $(1/\pi)\int_{-\pi}^{\pi} K_n(t)\,dt = 1$.
(c) $K_n(t) \leq (n + 1)/2$; $K_n(t) \leq A/(n + 1)t^2$ $(0 < t \leq \pi$; A an absolute constant).

Here, (a) and (b) are obvious from the various formulas above for K_n. The first part of (c) follows from the formula $K_n(t) = (n + 1)^{-1}\sum_{j=0}^{n} D_j(t)$ and the obvious estimate $|D_j| \leq j + \frac{1}{2}$. The second part follows from (12.46) if we note that $\sin u \geq (2/\pi)u$ for $0 \leq u \leq \frac{1}{2}\pi$.

From the second inequality (c), we immediately deduce

(c′) $\int_{\delta}^{\pi} K_n(t)\,dt \to 0$ as $n \to \infty$ for any fixed δ, $0 < \delta \leq \pi$.

These properties of K_n lead to the following basic result, which is related to (9.9) in Chapter 9.

(12.47) Theorem (*Fejér*) *Let f be integrable and periodic. Then*

$$\sigma_n(x) \to f(x)$$

at each point of continuity of f, and the convergence is uniform over every closed interval of continuity. In particular, $\sigma_n(x)$ tends to $f(x)$ uniformly everywhere if f is continuous everywhere. If f has a jump *discontinuity at x_0, then*

$$\sigma_n(x_0) \to \tfrac{1}{2}[f(x_0+) + f(x_0-)].$$

Proof. Suppose f is continuous on a closed interval $I = [a,b]$ (which may reduce to a point). Given $\varepsilon > 0$, we can find δ so that $|f(x + t) - f(x)| < \varepsilon$ for $x \in I$, $|t| < \delta$. Using (b) above, we can write

$$\sigma_n(x) - f(x) = \frac{1}{\pi}\int_{-\pi}^{\pi} [f(x + t) - f(x)]K_n(t)\,dt$$
$$= \frac{1}{\pi}\int_{|t|<\delta} + \frac{1}{\pi}\int_{\delta\le|t|\le\pi} = \alpha_n + \beta_n,$$

say. Clearly, if $x \in I$, then

$$|\alpha_n| \le \frac{1}{\pi}\int_{|t|<\delta} \varepsilon K_n(t)\,dt \le \frac{\varepsilon}{\pi}\int_{-\pi}^{\pi} K_n(t)\,dt = \varepsilon.$$

Let $M = \max|f|$ in I. Then, for $x \in I$,

$$|\beta_n| \le \frac{1}{\pi}\int_{\delta\le|t|\le\pi} (|f(x + t)| + M)K_n(t)\,dt$$
$$\le \frac{1}{\pi}[\max_{\delta\le|t|\le\pi} K_n(t)]\int_{\delta\le|t|\le\pi} [|f(x + t)| + M]\,dt.$$

The last integral is majorized by $\int_{-\pi}^{\pi}|f(x + t)|\,dt + 2\pi M = \int_{-\pi}^{\pi}|f(t)|\,dt + 2\pi M$, and by (c), the factor preceding it tends to 0. Hence $|\beta_n| \to 0$ uniformly for $x \in I$, and $|\alpha_n| + |\beta_n| < 2\varepsilon$ for n large enough and $x \in I$. This proves the first part of (12.47).

The proof of the second part is similar. We may assume that $f(x_0) = \tfrac{1}{2}[f(x_0+) + f(x_0-)]$. Then

$$\sigma_n(x_0) - f(x_0) = \frac{1}{\pi}\int_0^{\pi} [f(x_0 + t) + f(x_0 - t) - f(x_0+) - f(x_0-)]K_n(t)\,dt,$$

$$|\sigma_n(x_0) - f(x_0)| \le \frac{1}{\pi}\int_0^{\pi} |f(x_0 + t) - f(x_0+)|K_n(t)\,dt$$
$$+ \frac{1}{\pi}\int_0^{\pi} |f(x_0 - t) - f(x_0-)|K_n(t)\,dt = a_n + b_n.$$

To show, for example, that $a_n \to 0$, write $\int_0^{\pi} = \int_0^{\delta} + \int_{\delta}^{\pi}$ and use the fact that in $(0,\delta)$ the difference $|f(x_0 + t) - f(x_0+)|$ is small, while in (δ,π) we have $\max K_n(t)$ tending to zero.

The following result, although it is simple, deserves a statement.

(12.48) Theorem

(a) *Let f be periodic and integrable. If $f(x) \le B$ for all x, then also*

$\sigma_n(x) \le B$. *If* $f(x) \ge A$, *then* $\sigma_n(x) \ge A$. *If* $|f(x)| \le M$, *then* $|\sigma_n(x)| \le M$.

(b) *If* $f(x) \to \pm\infty$ *as* $x \to x_0$, *then* $\sigma_n(x_0) \to \pm\infty$ *as* $n \to \infty$.

We leave the proofs to the reader.

The following two results are corollaries of Fejér's theorem.

(12.49) Theorem *Let* f *be periodic and integrable,* $f \sim \sum c_k e^{ikx}$, *and let* F *be the indefinite integral of* f. *Then the series in the formula*

$$F(x) - c_0 x \sim C_0 + \sum{}' \frac{c_k}{ik} e^{ikx}$$

[see (12.14)] *converges uniformly.*

The series on the right is the Fourier series of a continuous function, and since its terms are $o(1/|k|)$ uniformly in x, the difference between its partial sums and the arithmetic means of its partial sums tends uniformly to 0 [see the discussion following the proof of (12.44)].

(12.50) Theorem (*Dirichlet-Jordan*) *If* f *is periodic and of bounded variation, then*

(a) $S[f]$ *converges to* $f(x)$ *at each point of continuity of* f, *and to* $\frac{1}{2}[f(x+) + f(x-)]$ *at each point of discontinuity.*

(b) *The convergence of* $S[f]$ *is uniform over every closed interval of continuity of* f.

(c) *The partial sums of* $S[f]$ *are uniformly bounded.*

Parts (a) and (b) follow immediately from Fejér's theorem if one uses (12.45) and the fact that the Fourier coefficients of a function of bounded variation are $O(1/|k|)$ [see (12.24)]. For (c), use (12.48) and the remark before (12.45).

Perhaps it is of interest to observe here that the classical theorem of Weierstrass about the uniform approximability of functions which are continuous in finite closed intervals by *power* polynomials can be easily deduced from Fejér's theorem. For suppose that $f(x)$ is continuous for $a \le x \le b$. The formula $x = \frac{1}{2}(a + b) + \frac{1}{2}(b - a)t$ establishes a mapping between the intervals $a \le x \le b$ and $-1 \le t \le +1$, and every $f(x)$ continuous in $[a,b]$ becomes a $g(t)$ continuous in $[-1,+1]$. If we approximate $g(t)$ by polynomials in t, we at the same time approximate $f(x)$ by polynomials in x. Hence, we may assume from the start that $f(x)$ is defined and continuous in $[-1,+1]$. Write $x = \cos\theta$. The function $h(\theta) = f(\cos\theta)$ is then defined and continuous in $[0,\pi]$, and if we extend it to $[-\pi,\pi]$ by the condition of evenness, and after that to $(-\infty,+\infty)$ by periodicity, then $h(\theta)$ can be approximated arbitrarily closely and uniformly on $[0,\pi]$ by *cosine* polynomials

$$T(\theta) = \sum_{k=0}^{n} \alpha_k \cos k\theta,$$

e.g., the arithmetic means of $S[h]$, these being cosine polynomials since h is even. It is easy to see that $\cos k\theta$ is a power polynomial of degree k in $x = \cos\theta$: for $k = 0, 1$, this is obvious, and for general k it follows by induction from the formula $\cos k\theta + \cos(k-2)\theta = 2\cos\theta\cos(k-1)\theta$. Thus, the polynomials $T(\theta)$ above are power polynomials $P(\cos\theta)$ in $\cos\theta$, and the approximability of $h(\theta)$ by $T(\theta)$ is the same thing as the approximability of $f(x)$ by $P(x)$.

We shall now consider the arithmetic means of $S[f]$ when f is merely integrable. In Chapter 7, we introduced the notion of a Lebesgue point of a function in $\mathbf{R}^n$, but here we are only interested in the case $n = 1$. We recall the definition. A point x_0 is a *Lebesgue point* for a locally integrable f if

$$\frac{1}{2h}\int_{-h}^{h} |f(x_0 + t) - f(x_0)|\,dt \to 0 \qquad (h \to 0),$$

and we proved that almost all points have this property.

Simultaneously with $S[f]$, we shall also consider $\tilde{S}[f]$, for f merely integrable. For $\varepsilon > 0$, we write

$$\begin{aligned}\tilde{f}_\varepsilon(x) &= -\frac{1}{\pi}\int_{\varepsilon\le|t|\le\pi} f(x+t)\frac{dt}{2\tan\frac{1}{2}t}\\ &= -\frac{1}{\pi}\int_{\varepsilon}^{\pi}\frac{f(x+t)-f(x-t)}{2\tan\frac{1}{2}t}\,dt.\end{aligned}$$

If $\lim_{\varepsilon\to 0}\tilde{f}_\varepsilon(x)$ exists, we will denote it $\tilde{f}(x)$ and call it the *function conjugate to* f:

$$\begin{aligned}\tilde{f}(x) &= \lim_{\varepsilon\to 0}\left(-\frac{1}{\pi}\int_{\varepsilon}^{\pi}\frac{f(x+t)-f(x-t)}{2\tan\frac{1}{2}t}\,dt\right)\\ &= \lim_{\varepsilon\to 0}\left(-\frac{1}{\pi}\int_{\varepsilon\le|t|\le\pi}\frac{f(x+t)}{2\tan\frac{1}{2}t}\,dt\right).\end{aligned}$$

We came across this function on p. 225 in connection with Dini's criterion. Occasionally, one also uses the notation

$$\tilde{f}(x) = -\frac{1}{\pi}\,\text{P.V.}\int_{-\pi}^{\pi}\frac{f(x+t)}{2\tan\frac{1}{2}t}\,dt,$$

where P.V. stands for *principal value*, indicating that the integral, which as a Lebesgue integral is generally divergent at $t = 0$, is given a new meaning by first removing a *symmetric* neighborhood around $t = 0$ and then making that neighborhood shrink to 0. Formally, $\tilde{f}$ is the convolution of f and $\frac{1}{2}\cot\frac{1}{2}t$, although the latter is not an integrable function. We will study the existence of $\tilde{f}$ later.

The arithmetic means of $\tilde{S}[f]$ will be denoted by $\tilde{\sigma}_n(x) = \tilde{\sigma}_n(x,f)$. From the formula on p. 222 for $\tilde{s}_n$, we obtain

$$\tilde{\sigma}_n(x) = -\frac{1}{\pi}\int_{-\pi}^{\pi} f(x+t)\tilde{K}_n(t)\,dt,$$

where

$$\tilde{K}_n(t) = \frac{1}{n+1}\sum_{j=0}^{n} \tilde{D}_j(t).$$

Of course, $\tilde{D}_0 = 0$.

(12.51) Theorem (*Lebesgue*) *Suppose that f is periodic and integrable. Then at every Lebesgue point x_0 of f (in particular, for almost every x_0),*

(i) $\sigma_n(x_0) \to f(x_0)$.
(ii) $\tilde{\sigma}_n(x_0) - \tilde{f}_{1/n}(x_0) \to 0$. $\quad (n \to \infty)$

Proof. We will use the estimates

$$K_n(t) \le n, \qquad K_n(t) \le \frac{A}{nt^2} \qquad (n \ge 1,\ 0 < t \le \pi), \tag{12.52}$$

which are just variants of (c), p. 236. If we have to use both estimates, then clearly the first is preferable for $t \le 1/n$ and the second for $t \ge 1/n$. The proof that follows is basically a repetition of the argument for (9.13), Chapter 9.

Let x_0 be a Lebesgue point of f (see p. 108). Assuming as we may that $f(x_0) = 0$, and letting

$$\phi(t) = |f(x_0+t)| + |f(x_0-t)|, \qquad \psi(t) = \int_0^t \phi(u)\,du,$$

the condition that x_0 is a Lebesgue point takes the form $\psi(h)/h \to 0$ as $h \to 0$. The formula for $\sigma_n(x_0)$ gives

$$|\sigma_n(x_0)| \le \frac{1}{\pi}\int_0^{\pi} \phi(t)K_n(t)\,dt = \frac{1}{\pi}\int_0^{1/n} + \frac{1}{\pi}\int_{1/n}^{\pi} = \alpha_n + \beta_n,$$

say. Clearly,

$$\alpha_n \le \int_0^{1/n} \phi(t)n\,dt = \frac{\psi(1/n)}{1/n} \to 0.$$

Next, using the second estimate for K_n and integrating by parts, we have

$$\beta_n \le \frac{A}{n}\int_{1/n}^{\pi} \frac{\phi(t)}{t^2}\,dt = \frac{A}{n}\left[\frac{\psi(t)}{t^2}\right]_{1/n}^{\pi} + \frac{2A}{n}\int_{1/n}^{\pi} \frac{\psi(t)}{t^3}\,dt.$$

The integrated term tends to 0 as $n \to \infty$. As for the last term, we will show that it also tends to 0. Given any $\varepsilon > 0$, take δ so small that $\psi(t)/t < \varepsilon$ if $0 < t < \delta$. Then

$$\frac{1}{n}\int_{1/n}^{\pi} \frac{\psi(t)}{t^3}\,dt \le \frac{1}{n}\int_{1/n}^{\delta} \frac{\varepsilon t}{t^3}\,dt + \frac{1}{n}\int_{\delta}^{\pi} \frac{\psi(t)}{t^3}\,dt.$$

The first term on the right is majorized by $(\varepsilon/n)\int_{1/n}^{\infty} t^{-2}\,dt = \varepsilon$, while the last term clearly tends to 0. Collecting results, we conclude that $\sigma_n(x_0) \to 0$. This proves (i).

To prove (ii), we need estimates for $\tilde{K}_n = [1/(n+1)]\sum_{j=0}^{n} \tilde{D}_j$. The obvious inequality $|\tilde{D}_j| \le j$ shows that

(12.53) $$|\tilde{K}_n(t)| \le n.$$

On the other hand, from the formula

$$\tilde{D}_j(t) = \frac{1}{2}\cot\frac{1}{2}t - \frac{\cos(j+\frac{1}{2})t}{2\sin\frac{1}{2}t}$$

(see (12.26)), we find that

$$\tilde{K}_n(t) - \frac{1}{2}\cot\frac{1}{2}t = \frac{\sin(n+\frac{3}{2})t}{(n+1)(2\sin\frac{1}{2}t)^2},$$

which shows that

(12.54) $$\left|\tilde{K}_n(t) - \frac{1}{2}\cot\frac{1}{2}t\right| \le \frac{A}{nt^2} \qquad (|t| \le \pi,\ n = 1, 2, \ldots).$$

The estimates (12.53) and (12.54) are analogues of (12.52), and they easily lead to (ii). We write

$$\begin{aligned}
&\tilde{\sigma}_n(x_0) - \tilde{f}_{1/n}(x_0)\\
&= -\frac{1}{\pi}\int_{-\pi}^{\pi} f(x_0+t)\tilde{K}_n(t)\,dt + \frac{1}{\pi}\int_{1/n \le |t| \le \pi} f(x_0+t)\frac{1}{2}\cot\frac{1}{2}t\,dt\\
&= -\frac{1}{\pi}\int_{|t|<1/n} f(x_0+t)\tilde{K}_n(t)\,dt + \frac{1}{\pi}\int_{1/n \le |t| \le \pi} f(x_0+t)\left[\frac{1}{2}\cot\frac{1}{2}t - \tilde{K}_n(t)\right]dt,
\end{aligned}$$

and use the estimates (12.53) and (12.54) in the last two integrals, respectively. An argument identical to that for α_n and β_n in the preceding proof shows that these integrals tend to 0. This completes the proof.

We remark that part (i) of Lebesgue's theorem leads to a new proof of the completeness of the trigonometric system [see (12.16)]. For if all the Fourier coefficients of f are 0, then $\sigma_n(x,f)$ vanishes identically.

Part (ii) of Lebesgue's theorem shows that $\lim \tilde{\sigma}_n(x_0)$ exists at every Lebesgue point of f at which the conjugate function

$$\tilde{f}(x_0) = \lim_{\varepsilon \to 0} \tilde{f}_\varepsilon(x_0)$$

exists. The converse is also true, though it requires an additional argument. Let $1/(n+1) \le \varepsilon \le 1/n$. Then

$$
\begin{aligned}
|\tilde{f}_\varepsilon(x_0) - \tilde{f}_{1/n}(x_0)| &\le \frac{1}{\pi}\int_{1/(n+1)}^{1/n} |f(x_0+t) - f(x_0-t)| \frac{1}{2}\cot\frac{1}{2}t\,dt \\
&\le \frac{1}{\pi}\int_{1/(n+1)}^{1/n} |f(x_0+t) - f(x_0-t)| \frac{dt}{t} \qquad (12.55)\\
&\le \frac{n+1}{\pi}\int_0^{1/n} |f(x_0+t) - f(x_0-t)|\,dt \to 0
\end{aligned}
$$

in view of the Lebesgue point condition. Hence, we obtain

(12.56) Theorem *At every Lebesgue point x_0 of an integrable f, the existence of $\tilde{f}(x_0)$ is equivalent to the summability of $\tilde{S}[f]$ by the method of the arithmetic mean, and $\tilde{f}(x_0) = \lim \tilde{\sigma}_n(x_0,f)$.*

Suppose now that f is not only integrable but also in L^2. If $f \sim \sum c_k e^{ikx}$, this means that $\sum |c_k|^2 < +\infty$. Observing that

$$\tilde{S}[f] = \sum c_k \varepsilon_k e^{ikx}, \qquad \varepsilon_k = -i \operatorname{sign} k$$

[see (12.12)], we see by the Riesz-Fisher theorem (8.30) that there is a function $g \in L^2$ such that $\tilde{S}[f] = S[g]$ and

$$\frac{1}{2\pi}\int_0^{2\pi} |g|^2 = \sum |c_k \varepsilon_k|^2.$$

Therefore,

$$\frac{1}{2\pi}\int_0^{2\pi} |g|^2 \le \sum |c_k|^2 = \frac{1}{2\pi}\int_0^{2\pi} |f|^2,$$

i.e., $\|g\|_2 \le \|f\|_2$. Since $\tilde{\sigma}_n(x,f) = \sigma_n(x,g)$ and $\lim \sigma_n(x,g)$ exists and equals g a.e. by (12.51), $\tilde{f}$ exists and equals g a.e. by (12.56), and we have proved the following result.

(12.57) Theorem *If f is periodic and in L^2, then the conjugate function*

$$\tilde{f}(x) = -\frac{1}{\pi}\int_0^{\pi} \frac{f(x+t) - f(x-t)}{2\tan\frac{1}{2}t}\,dt = -\frac{1}{\pi}\lim_{\varepsilon\to 0}\int_\varepsilon^{\pi}$$

exists almost everywhere and is in L^2. Moreover, $\|\tilde{f}\|_2 \le \|f\|_2$ (more precisely, $\|\tilde{f}\|_2 = \|f\|_2$ if $c_0 = 0$) and $\tilde{S}[f] = S[\tilde{f}]$.

The existence a.e. of $\tilde{f}$ is a remarkable result which shows that the odd part of f,

$$\psi_x(t) = \tfrac{1}{2}[f(x+t) - f(x-t)],$$

has special properties which are not immediate consequences of the theory of integration. Observing that $\frac{1}{2}\cot\frac{1}{2}t - (1/t)$ is bounded for $0 < t \le \pi$, we deduce from (12.57) that if $f \in L^2$, then the integral

$$\int_0^\pi \frac{f(x+t) - f(x-t)}{t}\,dt = \lim_{\varepsilon \to 0} \int_\varepsilon^\pi$$

exists almost everywhere, a result which is not obvious even for continuous f.

That $\tilde{f}$ exists almost everywhere for f merely integrable will be proved later [see (12.67)].

In the theorems which follow, we will use the notation

$$\|f\|_p = \left(\int_{-\pi}^{\pi} |f(x)|^p\,dx\right)^{1/p}, \ 1 \le p < \infty; \qquad \|f\|_\infty = \operatorname*{ess\,sup}_{|x| \le \pi} |f(x)|$$

(although sometimes it may be convenient to modify the definition of $\|f\|_p$ by inserting a numerical factor; e.g., by writing $\|f\|_p = [(1/2\pi)\int_{-\pi}^{\pi} |f|^p\,dx]^{1/p}$).

(12.58) Theorem *If $f \in L^p$, then*

(i) $\|\sigma_n\|_p \le \|f\|_p$, $1 \le p \le \infty$.
(ii) $\|f - \sigma_n\|_p \to 0$, $1 \le p < \infty$.

Proof. The theorem and its proof are repetitions of (9.1) and (9.6) of Chapter 9. If $p = \infty$, (i) is just (12.48)(a). If $1 < p < \infty$ and p' is the exponent conjugate to p, we have

$$\begin{aligned}|\sigma_n(x)| &\le \frac{1}{\pi}\int_{-\pi}^{\pi} |f(t)|K_n^{1/p}(x-t)\cdot K_n^{1/p'}(x-t)\,dt \\ &\le \left[\frac{1}{\pi}\int_{-\pi}^{\pi} |f(t)|^p K_n(x-t)\,dt\right]^{1/p}\left[\frac{1}{\pi}\int_{-\pi}^{\pi} K_n(x-t)\,dt\right]^{1/p'},\end{aligned}$$

by Hölder's inequality, and

$$|\sigma_n(x)|^p \le \frac{1}{\pi}\int_{-\pi}^{\pi} |f(t)|^p K_n(x-t)\,dt,$$

an inequality which clearly also holds for $p = 1$. Integrating both sides over $-\pi \le x \le \pi$, and interchanging the order of integration on the right, we obtain (i).

Part (ii) is proved similarly. We write

$$\begin{aligned}|\sigma_n(x) - f(x)| &\le \frac{1}{\pi}\int_{-\pi}^{\pi} |f(x+t) - f(x)|K_n(t)\,dt \\ &\le \left[\frac{1}{\pi}\int_{-\pi}^{\pi} |f(x+t) - f(x)|^p K_n(t)\,dt\right]^{1/p}\left[\frac{1}{\pi}\int_{-\pi}^{\pi} K_n(t)\,dt\right]^{1/p'},\end{aligned}$$

$$|\sigma_n(x) - f(x)|^p \le \frac{1}{\pi}\int_{-\pi}^{\pi} |f(x+t) - f(x)|^p K_n(t)\,dt.$$

Integrating both sides over $-\pi \le x \le \pi$ and interchanging the order of integration on the right, we obtain

$$\|\sigma_n - f\|_p^p \le \frac{1}{\pi}\int_{-\pi}^{\pi} \phi(t)K_n(t)\,dt$$

where

$$\phi(t) = \int_{-\pi}^{\pi} |f(x+t) - f(x)|^p\,dx.$$

Clearly, ϕ is a bounded function, and we know that it tends to 0 with t. Hence, by Fejér's theorem (12.47),

$$\frac{1}{\pi}\int_{-\pi}^{\pi} \phi(t)K_n(t)\,dt = \sigma_n(0,\phi) \to 0,$$

and (ii) follows.

We conclude this section by considering the maximal function

$$\sigma^*(x) = \sigma^*(x,f) = \sup_{n\ge 0} |\sigma_n(x,f)|.$$

In view of (12.51), $\sigma^*(x)$ is finite almost everywhere. It has properties not unlike those of the Hardy-Littlewood maximal function f^* considered in Chapters 7 and 9 and which are easily deducible from those of f^*. (Using similar symbols, σ^* and f^*, for different notions should not cause confusion.) First, we consider an adaptation of the definition of f^* to the case of periodic functions. For periodic f, it is natural to set

(12.59) $$f^*(x) = \sup_{0<h\le\pi} \frac{1}{2h}\int_{-h}^{h} |f(x+t)|\,dt.$$

Clearly, f^* is also periodic.

(12.60) Theorem *Let f be periodic and integrable. Then*

$$\|f^*\|_p \le c_p\|f\|_p, \qquad 1 < p \le \infty,$$

$$|\{x : 0 \le x \le 2\pi; f^*(x) > \alpha\}| \le \frac{c}{\alpha}\|f\|_1, \qquad \alpha > 0.$$

These inequalities are analogues (actually, corollaries) of (9.16) and (7.9). For let $g(x)$ be defined as equal to $f(x)$ in $(-\pi,3\pi)$ and to 0 elsewhere. Then, in $(0,2\pi)$, the maximal function f^* just defined is majorized by the Hardy-Littlewood maximal function of g, and the norms of g in $(-\infty,+\infty)$ are majorized by multiples of the corresponding norms of f in $(0,2\pi)$.

The first part of the following result is an analogue of (9.17) of Chapter 9.

(12.61) Theorem *Let f be periodic and integrable. Then there is an absolute constant c such that*

(i) $\sigma^*(x,f) \le cf^*(x)$.
(ii) $\sup_n |\tilde{\sigma}_n(x,f) - \tilde{f}_{1/n}(x)| \le cf^*(x)$.

Proof. The proof can be based on either Fubini's theorem [see the proof of (9.17)] or on the formula for integration by parts. We choose the second approach since it follows the same line as the proof of (12.51), but is actually easier since we do not have to consider Lebesgue points. Using the notation and proof of (12.51), we have

$$|\sigma_n(x,f)| \le \frac{1}{\pi}\int_0^\pi (|f(x+t)| + |f(x-t)|)K_n(t)\,dt$$
$$= \frac{1}{\pi}\int_0^{1/n} + \frac{1}{\pi}\int_{1/n}^{\pi} = \alpha_n + \beta_n,$$

where

$$\alpha_n \le \frac{1}{\pi} n \int_0^{1/n} \phi(t)\,dt, \qquad \beta_n \le \frac{A}{\pi n}\int_{1/n}^{\pi} \frac{\phi(t)}{t^2}\,dt$$

and $\phi(t) = |f(x+t)| + |f(x-t)|$. Let $\psi(t) = \int_0^t \phi(u)\,du$. The inequality $(\psi(t)/2t) \le f^*(x)$ shows that $\alpha_n \le f^*(x)$. If we integrate the integral majorizing β_n by parts so as to introduce $\psi(t)$ and again use the inequality $(\psi(t)/2t) \le f^*(x)$, we obtain $\beta_n \le Af^*(x)$, and (i) follows. The proof of (ii) is left to the reader.

The following result is a corollary of (12.61) and completes (12.57). It will be useful later.

(12.62) Theorem *If f is periodic and in L^2, then the maximal conjugate function*

$$\tilde{f}_*(x) = \sup_{0<\varepsilon\le\pi} |\tilde{f}_\varepsilon(x)|$$

is also in L^2 and

$$\|\tilde{f}_*\| \le A\|f\|_2 \qquad (A \text{ independent of } f).$$

We put the asterisk as a subscript here to avoid confusion with $(\tilde{f})^*$, the Hardy-Littlewood maximal function of $\tilde{f}$, which also appears in the proof below.

Proof. Let $1/(n+1) \le \varepsilon \le 1/n$. The inequalities in (12.55) give

$$|\tilde{f}_\varepsilon(x) - \tilde{f}_{1/n}(x)| \le \frac{n+1}{\pi}\int_0^{1/n} \{|f(x+t)| + |f(x-t)|\}\,dt \le f^*(x).$$

Combining this with (12.61)(ii), we find successively (with different A's at different places) that

$$|\tilde{f}_\varepsilon(x)| \le \sup_n |\tilde{\sigma}_n(x)| + Af^*(x) = \sup_n |\sigma_n(x,\tilde{f})| + Af^*(x),$$
$$|\tilde{f}_*(x)| \le A((\tilde{f})^*(x) + f^*(x)) \qquad \text{(by (12.61)(i))},$$
$$\|\tilde{f}_*\|_2 \le A(\|\tilde{f}\|_2 + \|f\|_2),$$
$$\|\tilde{f}_*\|_2 \le A\|f\|_2.$$

7. Summability of $S[f]$ by Abel Means

Given a periodic and integrable f,

$$f \sim \frac{1}{2}a_0 + \sum_{n=1}^{\infty} (a_n \cos nx + b_n \sin nx),$$

let

$$f(r,x) = \frac{1}{2}a_0 + \sum_{n=1}^{\infty} (a_n \cos nx + b_n \sin nx)r^n, \qquad 0 \le r < 1,$$

denote its Abel means. Since summability by arithmetic means implies Abel summability, the results of the preceding section immediately lead to results about A-summability of $S[f]$. For example, we have the relation

(12.63) $$f(r,x_0) \to f(x_0) \qquad (r \to 1)$$

at every Lebesgue point of f, and so almost everywhere. In particular, the last relation holds at each point of continuity of f, and uniformly over every closed interval of continuity.

However, an independent discussion of Abel summability has some merits, if only for the following two reasons: (a) the relation (12.63) holds at points which need not be Lebesgue points; (b) instead of (12.63), we may consider the more general relation

$$f(r,x) \to f(x_0)$$

as (r,x) tends to $(1,x_0)$ (i.e., as re^{ix} tends to e^{ix_0}) not only radially but also along more general curves, e.g., nontangentially (see p. 232).

We compute $f(r,x)$. Using the formulas for a_n and b_n, we have

$$f(r,x) = \frac{1}{2\pi}\int_{-\pi}^{\pi} f(t)\,dt + \sum_{n=1}^{\infty} r^n\left[\cos nx \frac{1}{\pi}\int_{-\pi}^{\pi} f(t)\cos nt\,dt + \sin nx \frac{1}{\pi}\int_{-\pi}^{\pi} f(t)\sin nt\,dt\right]$$

$$= \frac{1}{\pi}\int_{-\pi}^{\pi} f(t)\left[\frac{1}{2} + \sum_{1}^{\infty} r^n \cos n(x - t)\right] dt$$

$$= \frac{1}{\pi}\int_{-\pi}^{\pi} f(t)P(r, x - t)\, dt = \frac{1}{\pi}\int_{-\pi}^{\pi} f(x + t)P(r,t)\, dt,$$

where

$$P(r,t) = \frac{1}{2} + \sum_{n=1}^{\infty} r^n \cos nt$$

is called the periodic *Poisson kernel*. The function $f(r,x)$ is called the *Poisson integral of* f. All the formal operations above (like the interchange of the order of summation and integration) are easily justifiable.

We can write $P(r,t)$ in a finite form by observing that $\frac{1}{2} + \sum_{n=1}^{\infty} r^n \cos nt$ is the real part of the series

$$\frac{1}{2} + z + z^2 + \cdots = \frac{1}{2}\frac{1 + z}{1 - z} \qquad (z = re^{it}, 0 \le r < 1),$$

and a simple computation shows that

$$P(r,t) = \frac{1}{2}\frac{1 - r^2}{1 - 2r\cos t + r^2} = \frac{1}{2}\frac{1 - r^2}{(1 - r)^2 + 4r\sin^2 \frac{1}{2}t}.$$

This may be compared to the nonperiodic version of the Poisson kernel discussed on p. 151. The Poisson kernel has all the properties of the Fejér kernel, but is also much smoother. We list the properties:

(a) $P(r,t) \ge 0$; $P(r,-t) = P(r,t)$.
(b) $(1/\pi)\int_{-\pi}^{\pi} P(r,t)\, dt = 1$.
(c) $P(r,t) \le 1/(1 - r)$; $P(r,t) \le A(1 - r)/t^2$ ($\frac{1}{2} \le r < 1$, $|t| \le \pi$, A an absolute constant).

Properties (a) and (b) are obvious. The first part of (c) is a corollary of

$$P(r,t) \le \frac{1}{2} + r + r^2 + \cdots = \frac{1}{2}\frac{1 + r}{1 - r} \le \frac{1}{1 - r},$$

and the second part follows from

$$P(r,t) = \frac{1}{2}\frac{(1 - r)(1 + r)}{(1 - r)^2 + 4r\sin^2 \frac{1}{2}t} \le \frac{1 - r}{4r\sin^2 \frac{1}{2}t}.$$

The estimates (c) are analogues of the corresponding estimates for the Fejér kernel $K_n(t)$ if we identify $(1 - r)$ and $1/n$. Thus, results for the Fejér means have analogues for Abel means, and the proofs are basically the same. We shall, however, not dwell on this point and shall limit ourselves to several results of a somewhat different nature.

Given a periodic and integrable f, we shall systematically denote by F its indefinite integral; F need not be periodic. Besides the ordinary derivative of F at x,

$$F'(x) = \lim_{h \to 0} \frac{F(x+h) - F(x)}{h},$$

we shall also consider the *symmetric derivative*

$$F_s'(x) = \lim_{h \to 0+} \frac{F(x+h) - F(x-h)}{2h}.$$

Clearly, the existence of $F'(x)$ implies that of $F_s'(x)$ and $F_s'(x) = F'(x)$. The converse is not true, however, as shown by the simple example $F(x) = |x|$ at $x = 0$. Using the notions of the even and odd parts of a function introduced on p. 223, we see that F_s' is the ordinary derivative at 0 of the odd part of F at the point x (x is fixed, differentiation is with respect to h, at $h = 0$). Also, since

$$F_s'(x) = \lim_{h \to 0+} \frac{1}{h} \int_0^h \frac{f(x+t) + f(x-t)}{2}\, dt,$$

$F_s'(x)$ is the ordinary derivative at 0 of the integral of the even part of f at the point x.

(12.64) Theorem *Let f be periodic and integrable, and let F be the integral of f. Then*

(i) *at each point x_0 where $F_s'(x_0)$ exists, finite or infinite, $S[f]$ is Abel summable to the value $F_s'(x_0)$;*

(ii) *at each point x_0 where F has an ordinary and finite derivative $F'(x_0)$, the Poisson integral $f(r,x)$ tends to $F'(x_0)$ as (r,x) tends to $(1,x_0)$ nontangentially.*

Proof. (i) Suppose, as we may, that $x_0 = 0$. Write

$$\phi(t) = \tfrac{1}{2}[f(t) + f(-t)], \qquad \psi(t) = \int_0^t \phi(u)\, du.$$

We have

$$f(r,0) = \frac{2}{\pi} \int_0^\pi \phi(t) P(r,t)\, dt = \frac{2}{\pi} \int_0^\delta \phi(t) P(r,t)\, dt + o(1)$$

for any fixed δ, $0 < \delta \le \pi$, in view of the second part of property (c) for $P(r,t)$. Integration by parts shows that the last integral equals

$$-\frac{2}{\pi}\int_0^\delta \psi(t)P'(r,t)\,dt + o(1),$$

where

$$P'(r,t) = \frac{d}{dt}P(r,t) = -\frac{(1-r^2)r\sin t}{(1-2r\cos t + r^2)^2}.$$

Since $-P' \geq 0$ in $(0,\pi)$, if $\psi(t)/t$ is contained between m and M in $(0,\delta)$, then the last integral is contained between m and M multiplied by

$$\begin{aligned}-\frac{2}{\pi}\int_0^\delta tP'(r,t)\,dt &= \frac{2}{\pi}\int_0^\delta P(r,t)\,dt + o(1)\\ &= \frac{2}{\pi}\int_0^\pi P(r,t)\,dt + o(1) = 1 + o(1).\end{aligned}$$

Collecting results, we see that the limsup and liminf of $f(r,0)$ as $r \to 1$ are contained between m and M. This gives (i) when $F_s'(0)$ is either finite or infinite.

(ii) Assume again that $x_0 = 0$. Since the result is obvious when f is constant, we may also assume that $F'(0) = 0$. Suppose that $(r,x) \to (1,0)$ nontangentially, that is, $r \to 1$ and $x \to 0$ in such a way that

$$\frac{|x|}{1-r} \leq C. \tag{12.65}$$

Given $\varepsilon > 0$, take δ so small that $|F(u)/u| \leq \varepsilon$ for $|u| \leq 2\delta$. Write

$$\begin{aligned}f(r,x) &= \frac{1}{\pi}\int_{-\pi}^{\pi} f(x+t)P(r,t)\,dt = \frac{1}{\pi}\int_{-\delta}^{\delta} f(x+t)P(r,t)\,dt + o(1)\\ &= -\frac{1}{\pi}\int_{-\delta}^{\delta} F(x+t)P'(r,t)\,dt + o(1),\end{aligned}$$

using integration by parts and property (c) of P. If $|x| \leq \delta$ then $|x+t| \leq 2\delta$ in the last integral, and the integral itself is majorized in absolute value by

$$\begin{aligned}\varepsilon\int_{-\delta}^{\delta}(|x| + |t|)|P'(r,t)|\,dt &= 2\varepsilon|x|\int_0^\delta |P'(r,t)|\,dt\\ &\quad + 2\varepsilon\int_0^\delta t|P'(r,t)|\,dt.\end{aligned} \tag{12.66}$$

Since

$$\int_0^\delta |P'(r,t)|\,dt = P(r,0) - P(r,\delta) < P(r,0) \leq \frac{1}{1-r}$$

and

$$\int_0^\delta t|P'(r,t)|\,dt = -\int_0^\delta tP'(r,t)\,dt = -[tP(r,t)]_0^\delta + \int_0^\delta P(r,t)\,dt$$
$$\leq \int_0^\pi P(r,t)\,dt = \tfrac{1}{2}\pi,$$

condition (12.65) implies that the right side of (12.66) is less than a fixed multiple of ε. Hence, $f(r,x)$ tends to 0 under the hypothesis (12.65). This completes the proof of (ii).

8. Existence of $\tilde{f}$

In this section, we prove the following basic result.

(12.67) Theorem *If f is periodic and integrable, then the conjugate function*

$$\tilde{f}(x) = \lim_{\varepsilon\to 0} \tilde{f}_\varepsilon(x) = \lim_{\varepsilon\to 0}\left\{-\frac{1}{\pi}\int_{\varepsilon\leq|t|\leq\pi} \frac{f(x+t)}{2\tan\frac{1}{2}t}\,dt\right\}$$

exists almost everywhere, and is in weak L^1: for $\alpha > 0$

$$|\{x : |x| \leq \pi, |\tilde{f}(x)| > \alpha\}| \leq \frac{c}{\alpha}\|f\|_1,$$

where c is independent of f and α.

Remark: If f is integrable, $\tilde{f}$ need not be, as the following example shows. Let f be any periodic integrable function, nonnegative in $(0,\frac{1}{2}\pi)$ and zero elsewhere in $(-\pi,\pi)$. Then for $-\frac{1}{2}\pi < x < 0$,

$$\tilde{f}(x) = \frac{1}{\pi}\int_{-\pi}^{\pi} f(t)\frac{dt}{2\tan\frac{1}{2}(x-t)} = \frac{1}{\pi}\int_0^{\frac{1}{2}\pi} f(t)\frac{dt}{2\tan\frac{1}{2}(x-t)}$$
$$\leq \frac{1}{\pi}\int_0^{|x|} f(t)\frac{dt}{2\tan\frac{1}{2}(x-t)} = -\frac{1}{\pi}\int_0^{|x|} f(t)\frac{dt}{2\tan\frac{1}{2}(t-x)},$$
$$|\tilde{f}(x)| \geq \frac{1}{\pi}\int_0^{|x|} f(t)\frac{dt}{2\tan\frac{1}{2}(t+|x|)} \geq \frac{1}{\pi 2\tan|x|}\int_0^{|x|} f(t)\,dt.$$

Now choosing $f(t) = (t\log^2 t)^{-1}$ $(= (d/dt)[\log(1/t)]^{-1})$ for $0 < t < \frac{1}{2}$ and $f(t) = 0$ for $\frac{1}{2} \leq t < \pi$, we obtain $\tilde{f}(x) \geq c[|x|\log(1/|x|)]^{-1}$ near $x = 0$. Clearly, $f \in L$ but $\tilde{f} \notin L$.

The lemma that follows is essential for the proof of (12.67).

(12.68) Decomposition Lemma *Let Q be a finite interval in $\mathbf{R}^1$ and suppose that $f \in L(Q)$, $f \geq 0$. Then for any α satisfying*

(12.69)
$$\alpha \geq \frac{1}{|Q|}\int_Q f,$$

there is a sequence of nonoverlapping intervals $Q_1, Q_2, \ldots$ contained in Q such that

(i) $\alpha < \frac{1}{|Q_k|}\int_{Q_k} f \le 2\alpha \qquad (k = 1, 2, \ldots)$;

(ii) *almost everywhere in $P = Q - \bigcup Q_k$, we have $f(x) \le \alpha$;*

(iii) $|\bigcup Q_k| \le \frac{1}{\alpha}\int_{\bigcup Q_k} f \le \frac{1}{\alpha}\int_Q f.$

Proof. We split Q in half, obtaining 2 subintervals Q' of equal length. For each Q', there are only two possibilities: either $|Q'|^{-1}\int_{Q'} f \le \alpha$ or $|Q'|^{-1}\int_{Q'} f > \alpha$. Since $|Q'| = \frac{1}{2}|Q|$, the hypothesis (12.69) implies that $|Q'|^{-1}\int_{Q'} f \le 2\alpha$. Thus, for each Q', we have either

$$|Q'|^{-1}\int_{Q'} f \le \alpha \quad \text{or} \quad \alpha < |Q'|^{-1}\int_{Q'} f \le 2\alpha.$$

If Q' satisfies the first condition, we call it an interval of the first kind—otherwise, of the second kind.

We save any Q' of the second kind. If Q' is of the first kind, we may repeat the previous argument by splitting Q' into 2 equal parts Q''. For each Q'', we again have either

$$|Q''|^{-1}\int_{Q''} f \le \alpha \quad \text{or} \quad \alpha < |Q''|^{-1}\int_{Q''} f \le 2\alpha.$$

Saving those of the second kind, we repeat the procedure for each Q'' of the first kind, and so on.

Let $Q_1, Q_2, \ldots, Q_k, \ldots$ be the sequence of all the intervals of the second kind in the procedure above. Clearly, the Q_k are nonoverlapping and satisfy condition (i). Also, each $x \in P = Q - \bigcup Q_k$ belongs to a sequence of intervals $\{\bar{Q}\}$ with $|\bar{Q}|$ tending to 0 such that $|\bar{Q}|^{-1}\int_{\bar{Q}} f \le \alpha$. Since the ratio $|\bar{Q}|^{-1}\int_{\bar{Q}} f$ tends to $f(x)$ almost everywhere in P, (ii) follows. Finally, writing the first inequality (i) in the form $\alpha|Q_k| \le \int_{Q_k} f$, and summing over k, we deduce (iii).

Remarks:

(1) Lemma (12.68) holds for periodic functions of period 2π considered on the circumference of the unit circle, and $\alpha \ge (2\pi)^{-1}\int_{-\pi}^{\pi} f$. The proof is identical with that above.

(2) Lemma (12.68) is valid for $Q = \mathbf{R}^1$, $f \in L^1(\mathbf{R}^1)$, and *any* $\alpha > 0$. Moreover, it has an analogue in $\mathbf{R}^n$, $n > 1$, where Q and the Q_k are taken to be n-dimensional cubes with edges parallel to the coordinate axes. The proofs are left to the reader.

Proof of (12.67). Assume first that the periodic function $f \in L$ is ≥ 0. Fix any $\alpha \geq (2\pi)^{-1} \int_{-\pi}^{\pi} f$ and apply remark (1) above. We then obtain a sequence of nonoverlapping arcs $Q_1, Q_2, \ldots$ on the circumference Q of the unit circle such that

$$\text{(12.70)} \qquad \alpha < \frac{1}{|Q_k|}\int_{Q_k} f \leq 2\alpha, \qquad f \leq \alpha \text{ a.e. in } P = Q - \bigcup Q_k.$$

Make a decomposition

$$f = g + h,$$

where g is defined as equal to f in P and as $|Q_k|^{-1} \int_{Q_k} f$ on each arc Q_k. Hence, h equals 0 in P and $f - |Q_k|^{-1} \int_{Q_k} f$ on each Q_k. Using (12.70), we have

$$\text{(12.71)} \qquad \begin{aligned} &\text{(a) } 0 \leq g \leq \alpha \text{ a.e. in } P, \qquad 0 \leq g \leq 2\alpha \text{ in each } Q_k; \\ &\text{(b) } h = 0 \text{ in } P, \qquad \int_{Q_k} h = 0; \\ &\text{(c) } |h| \leq f + |Q_k|^{-1}\int_{Q_k} f \leq f + 2\alpha \text{ in each } Q_k. \end{aligned}$$

In particular, since g is bounded a.e., and so is in L^2, $\tilde{g}(x)$ exists a.e. and $\|\tilde{g}\|_2 \leq \|g\|_2$ by (12.57).

The study of $\tilde{h}$ requires the theorem (6.17) of Marcinkiewicz. That theorem was stated for functions defined on intervals of $\mathbf{R}^1$, while here we need it for functions defined on the circumference of the unit circle. Clearly, these two situations are identical. We restate the theorem.

(12.72) Theorem *Let F be a closed subset of the unit circumference Q, and let $G = Q - F$. Let $\delta(x)$ be the (circular) distance of the point $x \in Q$ from F. Then, for each $\lambda > 0$, the integral*

$$M_\lambda(x) = \int_Q \frac{\delta^\lambda(y)}{|x - y|^{1+\lambda}}\, dy = \int_G \frac{\delta^\lambda(y)}{|x - y|^{1+\lambda}}\, dy$$

is finite almost everywhere in F. Moreover,

$$\int_F M_\lambda(x)\, dx \leq 2\lambda^{-1}|G|.$$

We shall need the result only in case $\lambda = 1$.

We shall now study the existence of $\tilde{h} = \lim \tilde{h}_\varepsilon$, using properties (12.71)(b), (c) of h. Let Q_k^* denote the interior of the arc Q_k expanded concentrically twice, and let t_k denote the center of Q_k and d_k its length. Let $Q^* = \bigcup Q_k^*$ and let P^* be the complement of Q^* in Q; P^* is closed. We will consider only points $x \in P^*$.

Fix $x \in P^*$, and consider the equation

$$\tilde{h}_\varepsilon(x) = \frac{1}{\pi}\int_{|x-t|>\varepsilon} h(t)\,\frac{1}{2}\cot\frac{1}{2}(x - t)\, dt.$$

Since $h = 0$ outside $\bigcup Q_k$, the last integral is a sum

(i) of integrals extended over those Q_k which are totally outside $(x - \varepsilon, x + \varepsilon)$,
(ii) of at most two integrals extended over portions of those Q_k which contain the points $x \pm \varepsilon$.

We shall investigate these two cases separately, beginning with (ii). The interval Q_k which, say, contains $x + \varepsilon$ is distant from x by $\le \varepsilon$ and at the same time by $\ge \frac{1}{2}d_k$ (since $x \in P^*$). Hence, $\frac{1}{2}d_k \le \varepsilon$, and the integral under consideration is majorized in absolute value by

$$\frac{1}{\pi}\int_{x+\varepsilon}^{x+\varepsilon+d_k} \frac{|h(t)|}{|x - t|}\,dt \le \frac{1}{\pi\varepsilon}\int_{x+\varepsilon}^{x+3\varepsilon} |h(t)|\,dt \le \frac{1}{3\varepsilon}\int_0^{3\varepsilon} |h(x + t)|\,dt.$$

A similar estimate is valid for the interval Q_k containing $x - \varepsilon$, and using definition (12.59), we see that the contribution of (ii) is majorized by $2h^*(x)$. (We have just presupposed implicitly that $\varepsilon \le \frac{1}{3}\pi$, but since we are primarily interested in small ε, this is an unimportant restriction.)

We also notice that $h = 0$ in P^* so that by Lebesgue's theorem on the differentiability of integrals, the contribution of (ii) tends to 0 with ε at almost every $x \in P^*$.

Consider now any integral from (i). Since $\int_{Q_k} h = 0$ [see (12.71)(b)], it can be written in the form

$$\frac{1}{\pi}\int_{Q_k} h(t)\frac{1}{2}\left[\cot\frac{1}{2}(x - t) - \cot\frac{1}{2}(x - t_k)\right]dt$$

(t_k is the midpoint of Q_k). Let $\mathscr{I}_k(x)$ denote the last integral with the integrand replaced by its absolute value. Denoting absolute constants by A, we have

$$\begin{aligned}\mathscr{I}_k(x) &= \frac{1}{\pi}\int_{Q_k} |h(t)|\,\frac{1}{2}\,\frac{|\sin\frac{1}{2}(t - t_k)|}{|\sin\frac{1}{2}(x - t)\sin\frac{1}{2}(x - t_k)|}\,dt \\ &\le Ad_k\int_{Q_k} |h(t)|\,\frac{dt}{|x - t|\,|x - t_k|} \qquad (12.73) \\ &\le \frac{Ad_k}{(x - t_k)^2}\int_{Q_k} |h(t)|\,dt,\end{aligned}$$

where to arrive at the last term, we have used the fact that $|t - t_k| \le \frac{1}{2}d_k$ and $|x - t_k| \ge d_k$ (recall that $x \notin Q^*$ since $x \in P^*$), so that

(12.74) $$\frac{1}{2} \le \frac{|x - t|}{|x - t_k|} \le \frac{3}{2} \qquad (x \in P^*).$$

Using the first inequality in (12.71)(c) and (12.70), we also have

(12.75) $$\int_{Q_k} |h| \le 2\int_{Q_k} f \le 4\alpha|Q_k|.$$

If $\delta(t)$ denotes the distance from t to P^*, then $\delta(t) \geq \frac{1}{2}d_k$ for $t \in Q_k$. Collecting results and using (12.73)–(12.75), we get the final estimates

$$\mathscr{I}_k(x) \leq \frac{Ad_k}{(x - t_k)^2} 4\alpha|Q_k| \leq A\alpha \int_{Q_k} \frac{\delta(t)}{(x - t)^2} dt,$$

$$\sum_k \mathscr{I}_k(x) \leq A\alpha \sum_k \int_{Q_k} \frac{\delta(t)}{(x - t)^2} dt$$

for $x \in P^*$ and the intervals Q_k which are entirely contained in the complement of $(x - \varepsilon, x + \varepsilon)$.

The last sum is majorized by

$$A\alpha \int_Q \frac{\delta(t)}{(x - t)^2} dt,$$

a quantity which is finite almost everywhere in P^* by (12.72) with $\lambda = 1$. Hence, in view of our observation that the integrals in (ii) above tend to 0 almost everywhere in P^*, $\tilde{h} = \lim \tilde{h}_\varepsilon$ exists almost everywhere in P^* and

(12.76)
$$|\tilde{h}(x)| \leq A\alpha \int_Q \frac{\delta(t)}{(x - t)^2} dt \qquad \text{(a.e. in } P^*).$$

Since $\tilde{f} = \tilde{g} + \tilde{h}$, and $\tilde{g}$ exists almost everywhere, $\tilde{f}$ exists almost everywhere in P^*, i.e., everywhere in Q with the exception of a set of measure at most

(12.77)
$$|Q^*| \leq 2 \sum |Q_k| \leq \frac{2}{\alpha} \int_Q f.$$

Taking α arbitrarily large, we see that $\tilde{f}$ exists almost everywhere.

The assumption $f \geq 0$ can be dropped by considering the decomposition $f = f^+ - f^-$.

We still have to prove the weak integrability of $\tilde{f}$. This can be deduced from the estimates above, and we shall be brief. We may again assume that $f \geq 0$. Fix any $\alpha \geq (2\pi)^{-1} \int_Q f$, and consider the decomposition $f = g + h$ corresponding to this α. Then

$$\{x : |\tilde{f}| > \alpha\} \subset \{x : |\tilde{g}| > \tfrac{1}{2}\alpha\} \cup \{x : |\tilde{h}| > \tfrac{1}{2}\alpha\} = S \cup T,$$

say. Recall that $0 \leq g \leq 2\alpha$; we also have $\int_Q g = \int_Q f$ since $\int_Q h = 0$ by (12.71)(b). Hence, by Tchebyshev's inequality,

$$|S| \leq \left(\frac{1}{2}\alpha\right)^{-2} \int_Q \tilde{g}^2$$
$$\leq \frac{4}{\alpha^2} \int_Q g^2 \leq \frac{8}{\alpha} \int_Q g = \frac{8}{\alpha} \int_Q f.$$

To estimate $|T|$, let $T_1 = T \cap P^*$, $T_2 = T \cap Q^*$. Thus, $|T| = |T_1| + |T_2|$, and by (12.77),

$$|T_2| \le |Q^*| \le \frac{2}{\alpha}\int_Q f.$$

On T_1, $|\tilde{h}(x)|$ is majorized a.e. by [see (12.76)]

$$A\alpha \int_Q \frac{\delta(t)}{(x-t)^2}\,dt = A\alpha \int_{Q^*} \frac{\delta(t)}{(x-t)^2}\,dt = A\alpha M(x),$$

say, and if $|\tilde{h}(x)|$ is to be $\ge \frac{1}{2}\alpha$ there, then necessarily $M(x) \ge 1/(2A)$. But $M(x)$ is the integral $M_\lambda(x)$ of (12.72) corresponding to $\lambda = 1$, $G = Q^*$, $F = P^*$. Therefore, by the estimate given in (12.72) and (12.77),

$$\int_{P^*} M \le 2|Q^*| \le \frac{4}{\alpha}\int_Q f,$$

and by Tchebyshev's inequality, the subset of P^* where $M(x) \ge 1/(2A)$ has measure at most $2A\int_{P^*} M \le 8A\alpha^{-1}\int_Q f$. Thus, $|T_1| \le 8A\alpha^{-1}\int_Q f$, and collecting estimates we have

$$|\{x : |\tilde{f}(x)| > \alpha\}| \le \frac{c}{\alpha}\|f\|_1 \qquad (c = 10 + 8A).$$

This was proved for $\alpha \ge (2\pi)^{-1}\int_Q f$ but is trivially true for smaller α since our set certainly has measure $\le 2\pi$, while $2\pi \le \alpha^{-1}\int_Q f$ for such α.

This completes the proof of Theorem (12.67). The theorem is strengthened by the following result.

(12.78) Theorem *If $f \in L$, then the maximal conjugate function*

$$\tilde{f}_*(x) = \sup_{0<\varepsilon\le\pi} |\tilde{f}_\varepsilon(x)|$$

is in weak L^1, i.e.,

$$|\{x : \tilde{f}_*(x) > \alpha\}| \le \frac{c}{\alpha}\|f\|_1 \qquad (\alpha > 0),$$

where c is independent of f and α.

Proof. The proof is practically the same as that of (12.67). Leaving the details for the reader to fill, we argue as follows. We fix $\alpha \ge (2\pi)^{-1}\int_Q f$, and make the previous decomposition $f = g + h$, so that $\tilde{f}_* \le \tilde{g}_* + \tilde{h}_*$. By (12.62), $\|\tilde{g}_*\|_2 \le A\|g\|_2$, and the estimates we had for $\tilde{h}$ also hold for $\tilde{h}_*$. The restriction $\varepsilon \le \frac{1}{3}\pi$ that was imposed in the argument is unimportant since if $\varepsilon \ge \frac{1}{3}\pi$, then $|\tilde{h}_\varepsilon(x)| \le c\int_Q f$ for all x, so that $\sup_{\varepsilon\ge\pi/3} |\tilde{h}_\varepsilon(x)| \le c\int_Q f$.

Therefore, the set where $\sup_{\varepsilon\geq\pi/3} |\tilde{h}_\varepsilon(x)| > \alpha$ is empty if $\alpha \geq c \int_Q f$, while if $\alpha < c \int_Q f$, its measure is $\leq 2\pi \leq 2\pi c\alpha^{-1} \int_Q f$.

9. Properties of $\tilde{f}$ for $f \in L^p$, $1 < p < \infty$

(12.79) Theorem *If $f \in L^p$, $1 < p < \infty$, then $\tilde{f} \in L^p$ and $\tilde{S}[f] = S[\tilde{f}]$. Moreover,*

(12.80) (a) $\|\tilde{f}\|_p \leq A_p\|f\|_p$ and (b) $\|\tilde{f}_*\|_p \leq A_p\|f\|_p$.

The constant A_p depends only on p and is bounded for p away from 1 and ∞.

This is the central theorem of the section. Its proof is long and has to be split into several parts. Of course, (b) implies (a). The theorem is false for $p = 1$ [see the remark after the statement of (12.67)], and (12.67) is a substitute for this case. See also Exercise 21. The theorem is also false for $p = \infty$: see Exercises 19 and 20.

(12.81) Lemma *If $f \in L^p$, $1 < p \leq 2$, then $\tilde{f}_* \in L^p$ and* (12.80) *holds.*

Proof. For $p = 2$, this is Theorem (12.62). For $p = 1$, $\tilde{f}_*$ is in weak L^1 by theorem (12.67). The lemma will be obtained from these extreme cases by an interpolation argument similar to the one we used in the proof of the Hardy-Littlewood maximal theorem (9.16).

Let $f \in L^p$, $1 < p < 2$. For each fixed $\alpha > 0$, we make the decomposition $f = g + h$ with

$$g(x) = f(x) \text{ wherever } |f| \leq \alpha, \qquad g(x) = 0 \text{ otherwise,}$$
$$h(x) = f(x) \text{ wherever } |f| > \alpha, \qquad h(x) = 0 \text{ otherwise.}$$

Clearly, $g \in L^2$, $h \in L^1$, $\tilde{f}_* \leq \tilde{g}_* + \tilde{h}_*$. Let $\omega(\alpha) = |\{\tilde{f}_* > \alpha\}|$ be the distribution function of $\tilde{f}_*$. We have

$$\omega(\alpha) \leq |\{\tilde{g}_* > \tfrac{1}{2}\alpha\}| + |\{\tilde{h}_* > \tfrac{1}{2}\alpha\}|.$$

By (12.67),

$$|\{\tilde{h}_* > \tfrac{1}{2}\alpha\}| \leq (\tfrac{1}{2}\alpha)^{-1} \int_Q |h| = A\alpha^{-1} \int_{\{|f|>\alpha\}} |f|,$$

and by Tchebyshev's inequality and (12.62),

$$|\{\tilde{g}_* > \tfrac{1}{2}\alpha\}| \leq (\tfrac{1}{2}\alpha)^{-2} \int_Q \tilde{g}_*^2$$
$$\leq A\alpha^{-2} \int_Q g^2 = A\alpha^{-2} \int_{\{|f|\leq\alpha\}} f^2.$$

We have [see (5.51) and Exercise 16 of Chapter 5]

$$\|\tilde{f}_*\|_p^p = -\int_0^\infty \alpha^p \, d\omega(\alpha) = p \int_0^\infty \alpha^{p-1} \omega(\alpha)\, d\alpha.$$

Using the estimates above, the last integral is majorized by

$$Ap \int_0^\infty \alpha^{p-1}\left(\alpha^{-1} \int_{\{|f|>\alpha\}} |f|\, dx\right) d\alpha + Ap \int_0^\infty \alpha^{p-1}\left(\alpha^{-2} \int_{\{|f|\le\alpha\}} f^2\, dx\right) d\alpha.$$

Interchanging the order of integration, we can write this as

$$Ap \int_Q |f(x)| \left(\int_0^{|f(x)|} \alpha^{p-2}\, d\alpha\right) dx + Ap \int_Q f^2(x)\left(\int_{|f(x)|}^\infty \alpha^{p-3}\, d\alpha\right) dx$$
$$= Ap\left\{\frac{1}{p-1}\int_Q |f(x)|^p\, dx + \frac{1}{2-p}\int_Q |f(x)|^p\, dx\right\},$$

due to the fact that $1 < p < 2$. It follows that (12.80)(b), and so also (12.80(a), holds with

(12.82) $$A_p^p = Ap\left\{\frac{1}{p-1} + \frac{1}{2-p}\right\} \qquad (1 < p < 2).$$

It is not surprising that A_p becomes infinite as $p \to 1$ since as we have observed, the integrability of f does not imply that of $\tilde{f}$. On the other hand, in view of the validity of (12.80)(b) for $p = 2$, it is natural to expect that there is a better estimate for A_p near $p = 2$ than (12.82) (which becomes infinite as $p \to 2$). This we shall see below.

Meanwhile, note that the exponent 2 plays a rather accidental role in Lemma (12.81). In fact, if we knew (12.80)(b) for any $p_0 > 1$, then the argument above would give it for $1 < p \le p_0$, with A_p bounded for p away from $p = 1$ and $p = p_0$. The only change necessary in the proof is replacing the exponent 2 in the argument for g by p_0. Moreover, if instead of (12.80)(b) we only knew (12.80)(a) for some p_0, then by applying the same argument to $\tilde{f}$ instead of $\tilde{f}_*$, we would obtain (12.80)(a) for $1 < p \le p_0$, with A_p bounded for p away from 1 and p_0. We will use this idea below.

Inequality (12.80)(b) for any p implies

(12.83) $$\|\tilde{f}_\varepsilon\|_p \le A_p \|f\|_p$$

for the same p and all $\varepsilon > 0$. From the fact that $\|\tilde{f}_\varepsilon\|_p$ equals $\sup_g |\int_Q \tilde{f}_\varepsilon g|$ for all g with $\|g\|_{p'} \le 1$ $((1/p) + (1/p') = 1)$, and the easily verifiable formula

$$\int_Q \tilde{f}_\varepsilon g = -\int_Q f \tilde{g}_\varepsilon$$

(apply Fubini's theorem), it follows that if (12.83) holds for any p, $1 < p < \infty$, then it also holds for the conjugate exponent p', and $A_{p'} = A_p$ (see also Exercise 16, Chapter 10). But we proved (12.83) for $1 < p \le 2$. Hence, it also holds for $2 \le p < \infty$, and an observation analogous to the one made above (this time for $\tilde{f}_\varepsilon$ rather than $\tilde{f}$) shows that the A_p in (12.83) remains bounded for p away from 1 and ∞.

Using (12.82), we thus see that the A_p in (12.83) satisfies the inequality

(12.84) $$A_p \le \frac{A}{p-1} \qquad (1 < p \le 2),$$

and so also (since $A_p = A_{p'}$)

(12.85) $$A_p \le Ap \qquad \text{for } p \ge 2.$$

Since $\tilde{f}_\varepsilon(x) \to \tilde{f}(x)$ as $\varepsilon \to 0$ for any integrable f, (12.83) leads to the basic inequality $\|\tilde{f}\|_p \le A_p\|f\|_p$, $1 < p < \infty$, where A_p satisfies the two estimates above. This proves (12.80)(a); we still have to prove (12.80)(b) for $2 < p < \infty$ and the formula $\tilde{S}[f] = S[\tilde{f}]$ for $f \in L^p$, $p > 1$.

Let us show that if $f \in L^p$, $p > 1$, then $\tilde{S}[f] = S[\tilde{f}]$. We may assume that $p < 2$. In view of (12.56), $\lim \tilde{\sigma}_n(x)$ exists and equals $\tilde{f}(x)$ almost everywhere for f merely integrable. Next, by (12.61)(ii), $|\tilde{\sigma}_n(x)|$ is majorized by $\tilde{f}_*(x) + cf^*(x)$, an integrable function in our case since $f \in L^p$, $1 < p < 2$. Hence, by the dominated convergence theorem, each Fourier coefficient of $\tilde{\sigma}_n$ tends to the corresponding Fourier coefficient of $\tilde{f}$. But it also tends to the corresponding coefficient of $\tilde{S}[f]$. Thus, $\tilde{S}[f] = S[\tilde{f}]$.

To complete the proof of (12.79), it remains to show that (12.80)(b) holds for $1 < p < \infty$ (we have shown it only for $1 < p \le 2$), with A_p bounded for p away from 1 and ∞. It is immediate [see the proof of (12.62)] that if $1/(n+1) \le \varepsilon \le 1/n$, then $|\tilde{f}_\varepsilon(x) - \tilde{f}_{1/n}(x)|$ is majorized by $f^*(x)$. Hence, by (12.61)(ii),

$$\tilde{f}_*(x) \le cf^*(x) + \sup_n |\tilde{\sigma}_n(x)|.$$

By (12.61)(i) applied to $\tilde{S}[f] = S[\tilde{f}]$,

$$\sup_n |\tilde{\sigma}_n(x,f)| = \sigma^*(x,\tilde{f}) \le c(\tilde{f})^*.$$

Hence,

$$\begin{aligned} \tilde{f}_* &\le c(f^* + (\tilde{f})^*), \\ \|\tilde{f}_*\|_p &\le c(\|f^*\|_p + \|(\tilde{f})^*\|_p) \\ &\le cc_p(\|f\|_p + \|\tilde{f}\|_p), \end{aligned}$$

where c_p is the constant of the Hardy-Littlewood maximal theorem. In view of (12.80)(a), the inequality $\|\tilde{f}_*\|_p \le B_p\|f\|_p$ is thus established for all p, $1 < p < \infty$, with $B_p = cc_p[1 + A_p]$, A_p being the constant in (12.80)(a). Combining the estimates (12.84) and (12.85) for A_p with the estimate on p. 156 for c_p, it follows that B_p is bounded for p away from 1 and ∞. In particular, since c_p remains bounded as $p \to \infty$, it follows that B_p is $O(p)$ as $p \to \infty$. To estimate B_p as $p \to 1$, it is best to use the estimate derived in the proof of (12.81) [see (12.82)]; thus, $B_p = O(1/(p-1))$ as $p \to 1$, so that B_p satisfies the same sorts of estimates as A_p. This completes the proof of (12.79).

We have the following important corollary.

(12.86) Corollary *Let $f \in L^p$, $1 < p < \infty$. Then $\tilde{f}_\varepsilon$ converges in L^p norm to $\tilde{f}$.*

This is an immediate consequence of the dominated convergence theorem since $\tilde{f}_\varepsilon$ converges pointwise almost everywhere to $\tilde{f}$ and $|\tilde{f}_\varepsilon| \le \tilde{f}_* \in L^p$.

10. Application of Conjugate Functions to Partial Sums of $S[f]$

The behavior of the partial sums $s_n(x) = s_n(x,f)$ is a much more delicate topic than the behavior of the arithmetic means. We will consider only the question of the convergence of s_n to f in the metric L^p. The main tool here is a connection between the partial sums and conjugate functions.

Instead of s_n, it will be convenient to consider the expressions [see (12.27)]

$$\textbf{(12.87)}\quad s_n^{\#}(x) = \frac{s_n(x) + s_{n-1}(x)}{2} = s_n(x) - \frac{1}{2}(a_n \cos nx + b_n \sin nx).$$

We have

$$\begin{aligned} s_n^{\#}(x) &= \frac{1}{\pi}\int_{-\pi}^{\pi} f(t)\frac{D_n(x-t) + D_{n-1}(x-t)}{2}\,dt \\ &= \frac{1}{\pi}\int_{-\pi}^{\pi} f(t)\frac{\sin n(x-t)}{2\tan\frac{1}{2}(x-t)}\,dt \\ &= \sin nx \cdot \frac{1}{\pi}\int_{-\pi}^{\pi}\frac{f(t)\cos nt}{2\tan\frac{1}{2}(x-t)}\,dt - \cos nx\cdot\frac{1}{\pi}\int_{-\pi}^{\pi}\frac{f(t)\sin nt}{2\tan\frac{1}{2}(x-t)}\,dt. \end{aligned}$$

The last decomposition was purely formal, but from Section 8, we know that the cofactors of $\sin nx$ and $\cos nx$ exist almost everywhere in the principal value sense (and represent the conjugate functions of $f\cos nt$ and $f\sin nt$, respectively).

(12.88) Theorem (*M. Riesz*) *If $f \in L^p$, $1 < p < \infty$, then*

(i) $\|s_n\|_p \le c\|f\|_p$, $\quad \|f - s_n\|_p \to 0$,

(ii) $\|\tilde{s}_n\|_p \le c\|f\|_p$, $\quad \|\tilde{f} - \tilde{s}_n\|_p \to 0$,

where c depends only on p.

Proof. It is enough to prove (i), which implies (ii) since $\tilde{s}_n(x,f) = s_n(x,\tilde{f})$ and $\|\tilde{f}\|_p \le A_p\|f\|_p$, $1 < p < \infty$. The first part of (i) with $s_n^{\#}$ instead of s_n is an immediate corollary of the last formula for $s_n^{\#}$ and the inequality $\|\tilde{f}\|_p \le A_p\|f\|_p$: letting $g_n = f\cos nt$, $h_n = f\sin nt$, we have

$$s_n^{\#}(x) = \sin nx\,\tilde{g}_n(x) - \cos nx\,\tilde{h}_n(x) \qquad \text{(a.e.)},$$

$$\|s_n^{\#}\|_p \le \|\tilde{g}_n\|_p + \|\tilde{h}_n\|_p \le A_p(\|g_n\|_p + \|h_n\|_p) \le A_p\|f\|_p$$

for $1 < p < \infty$. Since, by (12.87), $\|s_n^{\#} - s_n\|_p \le A\|f\|_1 \le A\|f\|_p$, we obtain $\|s_n\|_p \le A_p\|f\|_p$.

The relation $\|f - s_n\|_p \to 0$ is obvious for functions which have a continuous derivative [see (12.20)], and since such functions are dense in L^p, also follows in the general case.

Exercises

1. Prove the following versions of theorems (12.13) and (12.14).
   (a) If $f \sim \frac{1}{2}a_0 + \sum_{k=1}^{\infty} (a_k \cos kx + b_k \sin kx)$, and f is the indefinite integral of f', then $f' \sim \sum_{k=1}^{\infty} k(b_k \cos kx - a_k \sin kx)$.
   (b) If $f \sim \frac{1}{2}a_0 + \sum_{k=1}^{\infty} (a_k \cos kx + b_k \sin kx)$, and F is the indefinite integral of f, then
$$F(x) - \frac{1}{2}a_0 x \sim \frac{1}{2}A_0 + \sum_{k=1}^{\infty} \frac{1}{k}(a_k \sin kx - b_k \cos kx).$$
2. Prove that if $f(x) \sim \sum c_k e^{ikx}$, then $f(x + \alpha) \sim \sum c_k e^{ik\alpha} e^{ikx}$.
3. If f is real or complex-valued and $f \sim \frac{1}{2}a_0 + \sum_{k=1}^{\infty} (a_k \cos kx + b_k \sin kx)$, show that
$$\frac{1}{\pi}\int_0^{2\pi} |f|^2 = \frac{1}{2}|a_0|^2 + \sum_{k=1}^{\infty} (|a_k|^2 + |b_k|^2).$$
4. Deduce from (12.18) that if $f,g \in L^2$, $f \sim \sum c_k e^{ikx}$, $g \sim \sum d_k e^{ikx}$, then
$$\frac{1}{2\pi}\int_0^{2\pi} f(t)g(x - t)\, dt = \sum c_k d_k e^{ikx}.$$
   Thus, a Fourier coefficient of the convolution of f and g equals the product of the corresponding coefficients of f and g.
5. Prove the following.
   (a) If f is periodic and equal to sign x in $(-\pi,\pi)$, then
$$f(x) \sim \frac{4}{\pi}\left\{\sin x + \frac{\sin 3x}{3} + \frac{\sin 5x}{5} + \cdots\right\}.$$
   (b) Let $0 < h < \frac{1}{2}\pi$, and let f be the "triangular" function defined as follows: f is periodic, even, continuous, $f(0) = 1$, $f(x) = 0$ for $2h \le x \le \pi$, f is linear in $(0,2h)$. Then
$$f \sim \frac{2h}{\pi}\left[\frac{1}{2} + \sum_{k=1}^{\infty}\left(\frac{\sin kh}{kh}\right)^2 \cos kx\right] = \frac{h}{\pi}\left[1 + \sum_{-\infty}^{+\infty}{}'\left(\frac{\sin kh}{kh}\right)^2 e^{ikx}\right].$$
   (c) Let g be periodic and equal to $\frac{1}{2}\log[1/|2\sin\frac{1}{2}x|]$ in $(-\pi,\pi)$. Then
$$g \sim \sum_{k=1}^{\infty} \frac{\cos kx}{k}.$$
   [For (b), the coefficients can be computed directly, or by using Exercise 4 and example (b), p. 215. Observe that the convolution of two characteristic functions of intervals is a "triangular" function. For (c), one may either

integrate by parts in the formula for the cosine coefficients of g, or consider the real part of the series

$$\sum_{k=1}^{\infty} \frac{z^k}{k} = \log \frac{1}{1-z}, \qquad z = re^{ix},$$

for $r < 1$, and then let $r \to 1$.]

6. Using the formula for the Fourier series of $\frac{1}{2}(\pi - x)$ given in Example (a), p. 215, prove the formulas

$$\sum_{n=1}^{\infty} \frac{1}{n^2} = \frac{\pi^2}{6}, \qquad \sum_{n=1}^{\infty} \frac{1}{n^4} = \frac{\pi^4}{90},$$

$$\sum_{n=1}^{\infty} \frac{1}{n^{2k}} = \pi^{2k} B_k \ (k = 1, 2, \ldots, B_k \text{ rational}).$$

7. Prove that each of the systems
   (a) $\frac{1}{2}, \cos x, \cos 2x, \ldots, \cos kx, \ldots$
   (b) $\sin x, \sin 2x, \ldots, \sin kx, \ldots$
   is orthogonal and complete over $(0,\pi)$.

8. Let $\{\phi_j(\mathbf{x})\}$ and $\{\psi_k(\mathbf{y})\}$ be two orthogonal systems, the first over a set $A \subset \mathbf{R}^m$, the second over a set $B \subset \mathbf{R}^n$. Then the (double) system

$$\omega_{j,k}(\mathbf{x},\mathbf{y}) = \phi_j(\mathbf{x})\psi_k(\mathbf{y})$$

is orthogonal over the Cartesian product $C = A \times B \subset \mathbf{R}^{m+n}$. If both $\{\phi_j\}$ and $\{\psi_k\}$ are complete, so is $\{\omega_{j,k}\}$.

Generalize this to the case of more orthogonal systems.

9. Let $(k_1,k_1,\ldots,k_n)$ be all lattice points in the space $\mathbf{R}^n$ (i.e., all distinct points with integral coordinates). Then the system

$$\exp i(k_1x_1 + k_2x_2 + \cdots + k_nx_n)$$

is orthogonal and complete over any cube in $\mathbf{R}^n$ with edge 2π.

10. If $f \sim \sum c_k e^{ikx} \in L^2$ and $g \sim \sum d_k e^{ikx} \in L^2$, then $h = fg$ is integrable, and if $h \sim \sum C_n e^{inx}$, then

$$C_n = \sum_{k=-\infty}^{+\infty} c_k d_{n-k},$$

where the series on the right converges absolutely. [For $n = 0$, this means $(1/2\pi)\int_0^{2\pi} fg = \sum c_k d_{-k}$, which is a variant of (12.18).] See also Exercises 17 and 18.

11. Let $f \sim \sum c_k e^{ikx} \in L^2$. For each n, let

$$\gamma_n = \sum_{k \neq n} c_k \frac{1}{n-k} = \sum_{k \neq 0} c_{n-k} \frac{1}{k}.$$

Show that $\{\gamma_n\} \in l^2$ and, more precisely, that $\sum |\gamma_n|^2 \leq \pi^2 \sum |c_k|^2$. The numbers γ_n are discrete (and formal) analogues of the integral $\int [f(t)/(x-t)]\, dt$. (The numbers γ_n are the Fourier coefficients of the function fg, where $g \sim \sum_{k \neq 0} e^{ikx}/k = i(\pi - x)$ for $0 < x < 2\pi$ [see Example (a), p. 215], and we have

$$\int_0^{2\pi} |fg|^2 \leq \pi^2 \int_0^{2\pi} |f|^2.)$$

12. Let f and g be periodic, $f \in L^p$, $g \in L^{p'}$, $1 \le p \le \infty$, $1/p + 1/p' = 1$. Consider the product $h_x(t) = f(x+t)g(t)$ as a function of t. Show that the nth Fourier coefficient of $h_x(t)$ tends to 0 as $n \to \infty$, uniformly in x. [Show that the L^1-modulus of continuity of h tends to 0 uniformly in x; apply (12.22).]

13. Let f be periodic and integrable. Show that for the partial sum s_n we have the formula
   (a) $s_n(x) = (1/\pi)\int_{-\pi}^{\pi} f(x+t)[(\sin nt)/t]\,dt + \varepsilon_n(x)$, where $\varepsilon_n(x)$ tends to 0 uniformly in x as $n \to \infty$. Likewise,
   (b) $\tilde{s}_n(x) = (1/\pi)\int_{-\pi}^{\pi} f(x+t)[(1-\cos nt)/t]\,dt + \eta_n(x)$, where $\eta_n(x)$ tends to 0 uniformly in x. [For (a), except for an error which is $o(1)$ uniformly in x, we have

$$s_n(x) = \frac{1}{\pi}\int_{-\pi}^{\pi} f(x+t)\frac{\sin nt}{2\tan\frac{1}{2}t}\,dt,$$

   and since $\frac{1}{2}\cot\frac{1}{2}t - 1/t$ is bounded in $(-\pi,\pi)$, the result follows easily from Exercise 12 with $p = 1$, $p' = \infty$.)

14. Let $\tilde{L}_n = 2\pi^{-1}\int_0^{\pi}|\tilde{D}_n(t)|\,dt$, $\tilde{D}_n$ denoting the conjugate Dirichlet kernel. Prove the following analogue of (12.36):

$$\tilde{L}_n \simeq \frac{2}{\pi}\log n.$$

15. Show that if x is a point of continuity of an integrable f, then $\tilde{s}_n(x) = o(\log n)$. If f has a jump discontinuity at x and the value of the jump is $d = f(x+) - f(x-)$, show that $\tilde{s}_n(x) \simeq -(d/\pi)\log n$. [The second statement follows from the first by considering (if, for example, $x = 0$) the function $h(t) = \frac{1}{2}(\pi - t)$ of Example (a), p. 215, whose conjugate function $\tilde{h}$ satisfies $\tilde{h} = -g$, where g is the function of Exercise 5(c).]

16. Prove the following generalization of (12.38). Suppose that $\liminf s_n = \underline{s}$ and $\limsup s_n = \bar{s}$ are finite. Then both $\liminf \sigma_n$ and $\limsup \sigma_n$ are contained between the numbers

$$\tfrac{1}{2}(\underline{s} + \bar{s}) \pm A\,\tfrac{1}{2}(\bar{s} - \underline{s}),$$

where A is the same as in condition (i) of (12.38).

17. Prove the following extension of Exercise 10. If $f \sim \sum c_k e^{ikx} \in L^p$, $g \sim \sum d_k e^{ikx} \in L^{p'}$, $1 < p < \infty$, $1/p + 1/p' = 1$ (thus also $1 < p' < \infty$), then the Fourier coefficients C_n of the (integrable) function $h = fg$ are given by

$$C_n = \sum_{k=-\infty}^{\infty} c_k d_{n-k} = \lim_{M\to+\infty}\sum_{k=-M}^{M}.$$

Thus, the C_n are the same as in Exercise 10, but the series representing the C_n are no longer claimed to be absolutely convergent, and we must consider the limits of their *symmetric* partial sums. [The proof is parallel to that of Exercise 10. We write

$$C_n = \frac{1}{2\pi}\int_0^{2\pi} fg\,e^{-inx} = \frac{1}{2\pi}\int_0^{2\pi}(f - s_M)ge^{-inx} + \frac{1}{2\pi}\int_0^{2\pi} s_M g e^{-inx},$$

where $s_M = s_M(x,f)$, observe that the first integral on the right is majorized by $\int_0^{2\pi}|f - s_M||g|$, apply Hölder's inequality, and use the fact that $\|f - s_M\|_p \to 0$ by (12.88).]

18. The result of Exercise 17 is valid when $p = 1, p' = \infty$ (or when $p = \infty, p' = 1$), but the series defining the C_n must be taken in the sense of the (symmetric) first arithmetic means:

$$C_n = \lim_{M \to +\infty} \sum_{k=-M}^{M} c_k\left(1 - \frac{|k|}{M+1}\right) d_{n-k}.$$

19. We know that theorem (12.79) is false for $p = 1$. Show that it is also false for $p = \infty$, i.e., that the conjugate function of a bounded function need not be bounded. [Consider, for example, the two series $\sum_{k=1}^{\infty} (\sin kx)/k \sim \frac{1}{2}(\pi - x)$, $0 < x < 2\pi$, and $\sum_{k=1}^{\infty} (\cos kx)/k \sim \frac{1}{2} \log (1/|2 \sin \frac{1}{2}x|)$, $-\pi < x < \pi$ (see Exercise 5).]

20. There is a substitute result for theorem (12.79) in case $p = \infty$. Let f be a periodic function with $|f| \le 1$. Then there are absolute constants $\lambda, \mu > 0$ such that

$$\int_{-\pi}^{\pi} e^{\lambda|\tilde{f}|} \le \mu.$$

[Write

$$e^{\lambda|\tilde{f}|} - \lambda|\tilde{f}| - 1 = \sum_{n=2}^{\infty} \frac{\lambda^n|\tilde{f}|^n}{n!}$$

and integrate termwise. Use (12.80)(a), (12.85) and the fact that $n^n \le c^n n!$]

21. Suppose that f is a periodic function for which

$$\int_{-\pi}^{\pi} |f| \log^+ |f| < +\infty,$$

where $\log^+$ stands for the positive part of log. This clearly implies that $f \in L^1$. Show that $\tilde{f} \in L^1$ and that there are absolute constants A and B such that

$$\int_{-\pi}^{\pi} |\tilde{f}| \le A \int_{-\pi}^{\pi} |f| \log^+ |f| \, dx + B.$$

[Assume as we may that $|\tilde{f}| \ge 2$, say, and write $\omega(\alpha) = |\{x : |x| < \pi, |f(x)| > \alpha\}|$. Then

$$\int_{-\pi}^{\pi} |\tilde{f}| = \int_0^{\infty} \omega(\alpha)\, d\alpha = \int_0^2 + \int_2^{\infty}.$$

For the first integral of the right, use the fact that $\omega(\alpha) \le 2\pi$, and for the second, use an argument like that in the proof of (12.81.)

Notation

$x + y$	addition
$\mathbf{x} \cdot \mathbf{y}$	dot product
$\lvert x \rvert$	absolute value
x^+	positive part
x^-	negative part
$f(x+)$	limit from the right
$f(x-)$	limit from the left
$\rightarrow$	converges to
$\xrightarrow{m}$	converges in measure to
$\nearrow$	increases to
$\searrow$	decreases to
$\{x_k\}$	sequence
$\{x\colon \ldots\}$	set of x satisfying . . .
$x \in E$	x an element of E
$x \notin E$	x not an element of E
$E_1 \cup E_2$, $\bigcup_k E_k$	union
$E_1 \cap E_2$, $\bigcap_k E_k$	intersection
$E_1 - E_2$	difference, relative complement
$E_1 \subset E_2$	subset
CE	complement
$\bar{E}$	closure
$\mathring{E}$	interior
diam E	diameter
$\delta(E)$	diameter
$\delta(x)$	distance function
$\lvert E \rvert$	measure
$\lvert E \rvert_e$	outermeasure

$[a,b]$	closed interval
(a,b)	open interval
$(a,b]$, $[a,b)$	partly open intervals
G_δ	countable intersection of open sets
F_σ	countable union of closed sets
$\varnothing$	empty set
sup	supremum, least upper bound
inf	infimum, greatest lower bound
limsup	limit superior
liminf	limit inferior
ess sup	essential supremum
ess inf	essential infimum
L, L^1, L^p, L^∞	classes of functions
l^p	classes of sequences
R^1	real number system
R^n	n-dimensional Euclidean space
C	Complex number system
$\int_E f$	Lebesgue integral in R^n
$\int_E f\, d\mu$	Lebesgue integral in a measure space
$\int_a^b f\, d\phi$	Riemann-Stieltjes integral
$\lVert \cdot \rVert$	norm

$\lVert\cdot\rVert_p$	L^p norm
$f * g$	convolution
χ_E	characteristic function
a.e.	almost everywhere
(S,Σ,μ)	measure space
$o(\phi(x))$, $O(\phi(x))$	order of magnitude relations
$V(E)$, $\underline{V}(E)$, $\bar{V}(E)$, $V[f; a,b]$	variations
$S[f]$	Fourier series
$S[f]$	conjugate Fourier series
usc	upper semicontinuous
lsc	lower semicontinuous
R_Γ	Riemann-Stieltjes sum
S_Γ	sum of increments
$\omega(f; \delta)$, $\omega_p(f; \delta)$	moduli of continuity
$\omega_{f,E}(\alpha)$	distribution function
f^*	Hardy-Littlewood maximal function
$H_\alpha(A)$	Hausdorff measure
$\Lambda(A)$, $\Lambda^*(A)$	Lebesgue-Stieltjes measure, outer measure
$v(I)$	volume

Index

D

I

J

K

L

M

N

O

P

R

S

T

U

V